Springer Series in Physical Environment 4

T0414149

Wolfgang Dreybrodt

Processes in Karst Systems
Physics, Chemistry, and Geology

With 184 Figures

Springer-Verlag
Berlin Heidelberg New York
London Paris Tokyo

Professor Dr. WOLFGANG DREYBRODT
Universität Bremen
FB 1 Physik/Elektrotechnik
Postfach 33 04 40
2800 Bremen 33, FRG

ISBN-13: 978-3-642-83354-0   e-ISBN-13: 978-3-642-83352-6
DOI: 10.1007/978-3-642-83352-6

Library of Congress Cataloging-in-Publication Data. Dreybrodt, Wolfgang. Processes in
Karst system. (Springer series in physical environment ; 4) 1. Karst. I. Title. II. Series.
GB600.D75   1988   551.4′47   88-15982

The use of registered names, trademarks, etc. in this publication does not imply, even in
the absence of a specific statement, that such names are exempt from the relevant pro-
tective laws and regulations and therefore free for general use.

Typesetting: ASCO Trade Typesetting Ltd., North Point, Hong Kong

2132/3130-543210 – Printed on acid-free paper

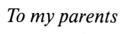

To my parents

Preface

Karst is a complex system, the evolution of which is governed by a complicated network of feedback loops. A two-way interaction exists, such that processes change properties of the system, and vice versa, changed properties of the system may activate new processes, and so on. To fully understand karst systems one must know as much as possible about the processes which determine karst evolution. Many of these have their basis in general laws of physics and chemistry, and can often be expressed in terms of mathematical equations.

When I started research on karst processes, looking at regular stalagmites, I wondered whether, in all the variety of shapes existing in speleothems, there might be a general law which could explain principle properties such as shape and rate of growth.

To a physicist it is challenging to try out his rather abstract and simplifying scientific thinking on real objects in nature. This starting point led me to deposition and then to dissolution of calcite and to its application in limestone rocks, and a new picture of karstification emerged, which I believe leads to a deeper understanding of what happens in karst. Many terms, which have often been used as descriptions, such as, for instance, fractures easily penetrable by water, or solutional widening of fractures by calcite aggressive solution, now gain a well-defined meaning, derived from the physics and chemistry underlying these descriptive terms. They can thus be used more concisely when discussing karst processes. I have therefore attempted the adventure of writing a book dealing with basic principles of physics, physical chemistry, and also geology, and combining them to understand dissolution and precipitation of limestone and the consequences of this in the process of karstification. The result is that the book covers a wide range of knowledge from several disciplines and may possibly, because of its heterogeneity, not be so easy to read.

This, however, is the price to be paid if one starts to look at complex natural systems from an interdisciplinary point of view. The book is organized in three distinct parts, which may be read separately for a first understanding. This might then give motivation to look into the other parts with increased interest.

The idea of writing this book came to us, when we met in 1986 at a colloquium of the "Schwerpunktprogramm: Hydrogeochemische Vorgänge im Wasserkreislauf in der gesättigten und ungesättigten Zone", supported by the Deutsche Forschungsgemeinschaft. This program brought many experts from different disciplines together and provided a most inspiring atmosphere, for which I thank the DFG and Prof. G. Mattheß, the coordinator of the program.

I want to express my thanks to Prof. H. W. Franke, Prof. M. A. Geyh, Prof. A. Bögli, Dr. S. Kempe, and W. Uffrecht for their encouragement when I started to work in this field. Many people have devoted their time to showing me their caves and I have learned a lot from them. I express may gratitute to K. H. Pielsticker, D. Stoffels, K. Stübs, U. Tauchert, S. Gamsjäger, and especially to B. Schillat.

During a visit to karst areas in the United States, when I was starting to write this book, I profited greatly from most inspiring discussions and field trips. I wish to express my thanks for this and for their generous hospitality to Dr. R. O. Ewers, R. A. Jameson, R. Kerbo, Dr. A. N. Palmer, Dr. J. F. Quinlan, Dr. J. Thrailkill, and Dr. W. B. White. I also thank Dr. P. T. Milanovic for showing to me the karst and its land use projects in the region of Dubrovnik, Yugoslavia.

Many people have helped to complete this work. I want to thank all of them. Dr. D. Buhmann, who worked with me for several years on problems of calcite dissolution, has contributed much in this field and has also designed the many programs for the computational work. His careful reading of the manuscript has led to many valuable suggestions, which have much improved the text. I also thank Prof. E. Usdowski and Prof. H. D. Schulz for their cooperation and fruitful discussions. They also have contributed much to the improvement of the text. G. Ankele has drawn many of the diagrams, and B. Bödeker has produced the typescript.

Finally, I thank my wife and my children for suffering the frame of mind into which one sinks when writing a book, which one hopes is worthwile.

Bremen, June 1988   WOLFGANG DREYBRODT

Contents

1   **Introduction**   1

1.1 The Process of Karstification   2
1.2 Approaches to an Explanation of Karst
     Development   5
1.3 Organization of the Book   8

Part I   Basic Principles from Physics and Chemistry

2   **Chemistry of the System $H_2O - CO_2 - CaCO_3$** . .   13

2.1 Reactions and Equilibria   13
2.2 Boundary Conditions for Achieving Equilibrium   21
     2.2.1 The General Case   21
     2.2.2 Change of Boundary Conditions   25
     2.2.3 Saturated $CaCO_3$ Solutions   26
     2.2.4 Mixing of Saturated Waters in the Closed
            System   29
     2.2.5 Influence of Foreign Ions on Calcite
            Dissolution   30

3   **Mass Transport**   43

3.1 The Random Walk Problem as Principle
     for Diffusional Mass Transport   43
3.2 The Laws of Diffusion   46
     3.2.1 Some Basic Solutions   48
3.3 Diffusive and Advective Mass Transport   53
     3.3.1 Diffusion in Turbulent Flow   54
     3.3.2 Hydrodynamic Dispersion   54
3.4 Diffusion Coefficients and Their Magnitudes . .   56
     3.4.1 Molecular Diffusion   56
     3.4.2 Eddy Diffusion   58
     3.4.3 Hydrodynamic Dispersion   58

4   Chemical Kinetics 59

4.1   Rate Laws of Elementary and Overall Reactions .   59
      4.1.1   Elementary Reactions   59
      4.1.2   Overall Reactions   61
      4.1.3   Temperature Dependence of Rate
              Constants   64
4.2   Approaching Equilibrium   65
      4.2.1   Irreversible Reactions   66
      4.2.2   Reversible Reactions   69
4.3   The Kinetics of the Reaction
      $H_2O + CO_2 \rightleftharpoons H^+ + HCO_3^-$ 70
      4.3.1   Examples   73
4.4   Mixed Kinetics   75

5   Hydrodynamics of Flow   80

5.1   Laminar and Turbulent Flow   82
      5.1.1   The Law of Bernoulli   83
      5.1.2   Laminar Flow   85
      5.1.3   Turbulent Flow   86
      5.1.4   An Example of Hydraulic Characteristics
              of Karst Aquifers   92
      5.1.5   Flow Through Porous Media   93

Part II   Principles of Dissolution and Precipitation of CaCO$_3$

**6   Dissolution and Precipitation of Calcite:
      The Chemistry of the Heterogeneous Surface** . . 103

6.1   Experimental Methods of Calcite Dissolution . . 104
      6.1.1   Rotating Disc System 105
      6.1.2   Batch Experiments 108
      6.1.3   Measurement of Dissolution Rates   . . . 110
6.2   The Kinetics of Calcite Dissolution 112
      6.2.1   Three Regions of Calcite Dissolution . . 112
      6.2.2   The Mechanistic Model of Plummer,
              Wigley and Parkhurst (PWP Model) . . . 116
      6.2.3   Comparison of the Mechanistic Model
              with Other Experiments 127
6.3   Kinetics of Calcite Precipitation 130
6.4   Dissolution Close to Equilibrium 136

**7   Modelling the Kinetics of Calcite Dissolution
      and Precipitation in Natural Environments
      of Karst Areas** 140

7.1   Statement of the Problem 140
7.2   The Mass Transport Equation 142
7.3   Dissolution and Precipitation Kinetics
      in the Open System 144
      7.3.1   Calculation of Dissolution and Precipitation
              Rates 144
      7.3.2   Results for the Open System 149
7.4   Dissolution and Precipitation Kinetics
      in the Closed System 161
      7.4.1   Calculation of the Rates 161
      7.4.2   Results for the Closed System 162
7.5   The Influence of the Diffusion Boundary Layer
      to Dissolution and Precipitation Rates
      in Turbulent Flow 170
      7.5.1   Calculation of Dissolution and Precipitation
              Rates 173
      7.5.2   Experimental Verification 175
7.6   Dissolution in Natural Systems 176
      7.6.1   Influence of Foreign Ions on Dissolution
              Rates Far from Equilibrium 176
      7.6.2   Influence of Lithology on Dissolution
              Rates 178
      7.6.3   Dissolution in Porous Media 180

Part III   Conceptual Models of Karst Processes

8   Karst Systems 185

8.1   Fractures and Discontinuities in Limestone Rocks   187
      8.1.1   Joints 188
      8.1.2   Faults 189
      8.1.3   Bedding Planes 191
8.2   Structural Segments and Tectonic Control
      of Caves 191
8.3   Karst Aquifers 200
8.4   Caves 206
8.5   Development of Caves 209

**9   Models of Karst Development from the Initial State
        to Maturity** 218

9.1   Pressure Fields in Fractures and Their Influence
       on the Development of Early Channels 218
9.2   The Network Linking Models of Ewers 223
       9.2.1   The Low Dip Model 223
       9.2.2   The High Dip Model 225
       9.2.3   The Restricted Input Model 227
9.3   Dissolution Kinetics and the Concept
       of Penetration Length 229
       9.3.1   The Concept of Penetration Length . . . 230
       9.3.2   Penetration Lengths in Turbulent Flow . 234
9.4   Karst Development from the Initial State
       to Maturity 236
9.5   Geomorphological Theories from the
       Viewpoint of the Mathematical Model 240
9.6   Experimental Models 243
9.7   Karst Processes on the Surface 245
       9.7.1   The Formation of Dolines and Shafts . . 246
       9.7.2   Karst Denudation 248
       9.7.3   Denudation Rates on Bare Rock
               at the Surface and in Caves 251

10   Precipitation of Calcite in Natural Environments . 256

10.1   Speleothems: Growth and Morphology 256
10.2   Stalagmites 257
       10.2.1 Morphology of Stalagmites 257
       10.2.2 Growth Rates of Stalagmites and Related
               Diameters 264
10.3   Calcite Precipitation in Surface Streams 268

11   Conclusion and Future Perspective 273

References 275

Subject Index 283

1 Introduction

There is nothing in the world softer and more gentle than water.
But in subduing the firm and the hard nothing is equal to it.

That the soft defeats the strong and the firm is overcome by the
gentle, everybody knows, but nobody acts accordingly.

Laotse

Carbon dioxide is an ubiquitious gas present in the earth's atmosphere, in the soil air, and in dissolved form in rainwater as well as in groundwater, in rivers, lakes and in the ocean. In dissolved form it constitutes H_2CO_3, carbonic acid, which is the motor of erosion and weathering of rocks. Figure 1.1 shows the CO_2 pressure as its negative decadic logarithm for various environments. For CO_2 dissolved in water this is the pressure which would be in equilibrium with the aqueous $CO_2^{(aq)}$ dissolved.

The major part of CO_2 responsible for weathering and erosion of carbonate rocks is derived from CO_2 in the soil air, which enters precipitation water percolating through the soil. In soils, CO_2 is produced by bacterial decomposition of organic matter, and also by root respiration.

Therefore biological activity is an important determinant of weathering and erosion. An extensive review of the role of CO_2 in the freshwater and in the rock cycle has been given by Kempe (1979a,b) and Degens et al. (1984).

This book deals with the solutional activity of carbon dioxide containing water to limestone rocks exposed at the earth's surface. The interaction of limestone and aggressive water penetrating into its primary fissures and voids creates an unique landform known as karst. Since limestone is an abundantly occurring sedimentary rock karst landforms are widely distributed and cover about 20% of the earth's surface. Depending on many factors such as the climatic regime of the area, amount of water infiltrating into the rock, geological setting of the rock, position of inflows and outflows of water and the type of primary fracture system in the rock, karst landforms can show a variety of differing features. Thus, alpine karst in Europe, where the strata are folded, is distinctly different from karst landforms in the United States, where the strata extend evenly without any folding over large distances.

There is one common characteristic, however, to all karst areas: the development of a subsurface drainage system as the process of karstification proceeds. Thus, in all areas of karst initially active surface drainage by fluvial systems is replaced by subsurface circulation of water. In the mature state of karst all meteoric water after very short travel distances sinks into the ground and occurs in karst springs at the base level of erosion, leaving the elevated parts of the landscapes entirely dry without any surface water. This book is concerned with the question as to how this complex process of a subsurface drainage system development can be described from its initial state to maturity by the physics and chemistry underlying the geological conditions with determine karstification.

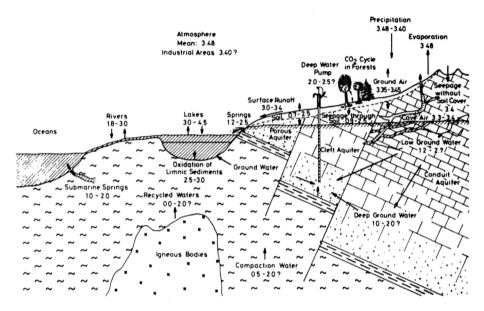

Fig. 1.1. $-\log P_{CO_2}$ of different freshwater compartments. P_{CO_2} is in atmospheres, i.e. $-\log P_{CO_2} = 3$ corresponds to $1 \cdot 10^{-3}$ atm (Kempe 1979)

1.1 The Process of Karstification

Figure 1.2 shows a sequence of limestone landform evolution from which the basic process of karstification can be envisaged. Initially (a) a limestone block is situated between two impermeable, insoluble layers of rock. Consequently, in the first stage a fluvial system drains the surface of the area. Onset of karstification begins when at some place in this area limestone is exposed to the surface, thus providing direct access of surface water into the limestone rock (b). This may happen by fluvial erosion of the caprock. Furthermore, some output of water has to be provided at a lower level, which delivers the hydraulic head by gravity driving the water through the primary narrow fissures in the limestone block. It is of no concern, how this first input-output configuration is established. It may as well be created in an area where limestone is exposed to the surface initially and a lower lying output is provided by valley incision. Therefore, Fig. 1.2 gives just one example.

Once a suitable input-output configuration exists, surface water penetrates into the fissures of the rock establishing a water table close to the surface. Since the width of the primary fissures is in the order of several $10~\mu$, only an insignificant fraction of the meteoric and surface water can be transported through the limestone. In contrast to insoluble, non-carbonate rocks, limestone is dissolved by carbonic acid-containing water and a gradual enlarging of the primary fissures increases the amount of water transported through the limestone. This in turn enhances the amount of limestone removed by dissolution. Thus, a positive feedback loop leads to a progressive enlargement of secondary porosity and a complex limestone aquifer

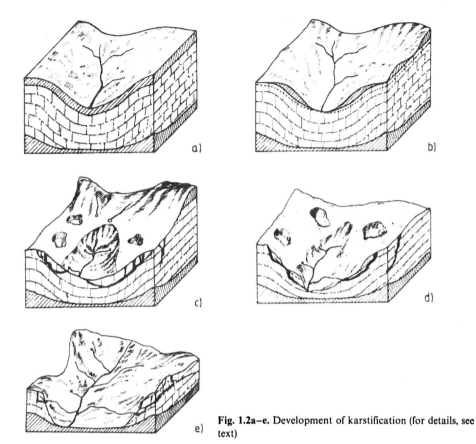

Fig. 1.2a–e. Development of karstification (for details, see text)

results. Large amounts of water can be stored in the many fissures and voids in the rock which have been enlarged to only small dimensions below 0.1 cm. On the other hand, complicated three-dimensional conduit systems, i.e. water-filled caves develop, which emerge as large karst springs at the output and drain the aquifer. This process changes the surface landforms. With the development of effective conduit drainage, sinkholes are formed on the surface, which guide the surface water undergound. Erosion of the covering caprock proceeds until the bare limestone rock is exposed to the surface (c) accepting meteoric water and circulating it underground by percolation, thus establishing new routes of groundwater flow.

During this process of increasing permeability of the limestone mass, lowering of the groundwater table occurs until base level is reached. At the same time, the complicated interplay of changing groundwater routes and changing input-output configurations dissects the limestone mass (d). When finally base level is reached again, a fluvial system is established at this level and residual limestone hills are left isolated on a fluvial plane (e). It is obvious that differing boundary conditions exert significant influence on the evolution of the landform. Where the limestone mass reaches much deeper into the ground than the lowest possible base level, quite different landforms will result, as is the case in the Dinaric karst in Yugoslavia.

Furthermore, the regional tectonics of joint systems will largely determine the first exploitable flow parts. Sediments such as flysch and dolomite may serve as hydrological barriers, thus enhancing the complexity of karst systems.

This variety of karst landforms has led to several attempts to classify karstlands. Cvijic (1924–1926) divided karst into three types. Holokarst is encountered in areas consisting entirely of soluble carbonate rocks. It is the most completely developed karst and is constituted of well-jointed massive and pure limestone with a karstification level reaching below existing base levels. Holokarst exhibits vast, bare and rocky land with poor vegetation and ample characteristics of karst, such as large caves, closed depressions and lack of river valleys. A typical example of holokarst is the Dinaric karst in Yugoslavia and Greece.

Merokarst is an incomplete karst with properties of non-karstic areas. The process of karstification has not yet proceeded very much and the land misses the typical karst characteristics. Commonly, the rock is covered by arable soil and rich vegetation. Although caves are present, drainage patterns are not very complex and underground river courses can be followed.

These two types of karst show an important difference with respect to the carbon dioxide contents of the water entering into the limestone. In merokarst meteoric water percolates through the vegetated soil zone and therefore attains high CO_2 concentrations and consequently a high capacity of dissolving limestone. In holokarst the CO_2 content of meteoric water is derived from the atmosphere and is several orders of magnitude lower.

Merokarst and holokarst are defined as states belonging to the early and final sequences of karstification respectively. It is suitable therefore to define transitional types of karst as a karst landform exhibiting features of both holo- and merokarst respectively.

Differences resulting from the geological setting of the rock lead to hydrogeological consequences. From this two important classifications can be given (Milanovic 1981).

Platform karst is characterized by horizontal or gently sloping strata which extend over large areas. Often the base level of karstification is determined by the underlying impervious rock. Typical examples are karst landforms in the USA (Kentucky).

In contrast to platform karst, geosyncline karst develops in distinctly folded structures, which also consequently exhibit faulting and ruptured carbonate rocks. This geological setting is ideally suited for the development of holokarst. A third classification scheme distinguishes deep and shallow karst (Katzer 1909). In deep karst impervious beds are located at great depth reaching often below sea level. Therefore, karstification is not limited by such strata as in the case of shallow karst, where carbonate layers are of limited thickness and eventually become exposed by karstification and concurrent surface erosion, as in the example of Fig. 1.1.

There are many other classifications relating to climate or to the morphological appearance of specific karst features, which have been summarized by Milanovic

(1981) or Sweeting (1972). In this context only a few cases are presented in order to show some of the structural conditions which determine karstification.

1.2　Approaches to an Explanation of Karst Development

Understanding the process of karstification requires as a first step a descriptive formulation of karst systems. Therefore, the first question to be answered is "what is it like"? From analogues, which are observed in various regions, explanations in terms of "why is it so", then may be inferred.

One important principle in the descriptive approach is to relate different properties of karst systems to each other. In the following a few such relations are discussed as typical examples.

1. The circulation of water in karst terranes is related to their geological characteristics. The relation of the directions of cave passages, which constitute early groundwater routes of the system, to the tectonics of the joint systems in the rock can give information as to how and under which conditions joint systems control development of karst. There are many other relations between properties of karst water systems and geological structure, such as anticlines, synclines and other structural features.

2. The response of karst springs to flood events gives information on the type of aquifers comprising the karst water system. There are karst springs with discharge responding almost immediately to flood events as one limiting case in contrast to those which show only a very slow and delayed increase of discharge. The first type can be related to a conduit drainage system, where large passages act like pipes. The other extreme is a diffused type of aquifer where water flows through the many small, interconnected fissures and voids. These have high flow resistance and large storage properties. Therefore, they show a sluggish reaction to pulse flood events like a heavy rainstorm. From detailed analysis of the hydrographic properties of those springs, one agrees nowadays that karst systems are constituted of two interconnected aquifers. The first is a conduit aquifer, which is most effective in draining the system, whereas the second, diffused aquifer represents a storage reservoir.

3. Surface karst development is closely related to the state of underground karstification. Development of collapse features is only possible in areas where large caverns exist underground. The connection of water drainage systems on the surface to those in the subsurface, e.g. in merokarst, reflects the state of karstification underground. Dye tracing of surface waters and the observation of reappearance gives valuable information on underground karstification and helps to identify groundwater basins.

4. The investigation of caves and the morphology of cave passages gives valuable information on cave development. Two phases of cave development can be recognized. In the early state the groundwater table is high and cave conduits can develop, completely filled with water in the phreatic zone. These passages show circular or lenticular shapes and their morphology indicates clearly the dominance

of solutional corrosion in their origin. In a later state cave passages will, in part, be abandoned by groundwater, and water flow with a free surface exists in the now vadose region. Cave passages originating under these conditions show vertical entrenchment and develop into canyons.

Relations of cave levels to the position of former river valleys give further valuable information on the evolution of cave systems.

From the tremendous amount of field observations, which has been reviewed to some extent in recent textbooks, e.g. Bögli (1980), Jennings (1985), Jakucz (1977), Milanovic (1981), Pfeffer (1978), and Trudgill (1985) as well as review articles by Hanshaw and Back (1979) and Stringfield et al. (1979) and a recent V.T. Stringfield Symposium edited by Back and La Moreaux (1983), a general picture of the process of karstification has emerged: Karstification is a process due to the dissolving action of groundwater as it penetrates into soluble carbonate rocks. Onset of this process occurs at the moment when a configuration of input and output of groundwater has been established in the karstifiable rock mass, which provides the hydraulic gradient to drive water through an interconnected system of primary fissures of microsize in the order of several 10 μ.

Little is known on the initial phase of karstification since direct observation is not possible. It is inferred, however, that a system of penetrable primary fissures and fractures must exist, which is enlarged by solutional aggressivity of groundwater, thus initiating the first secondary permeability for the future karst water aquifer. Flow in these fractures is certainly laminar as can be concluded from their presumable size and the existing hydraulic gradients. Under favourable conditions some of the solutionally enlarged, initial microchannels may evolve into a network of first conduit pipes of a few millimetre diameter and turbulent flow will set in. Under these flow conditions the amount of limestone rock removed by solution will increase for two reasons: (1) Solution rates will increase due to the turbulence which effects fast transport of dissolved species into the bulk solution. (2) The throughput of aggressive water increases and thus the capacity of dissolving limestone.

Therefore, once a certain diameter of first channels is exceeded, an effective drainage system develops. This eventually changes the input-output configuration and the relation between drainage at the surface and underground creates the well-known karst features. At the same time, the diffuse system is altered, increasing the permeability of the rock comprising it. The development of both the diffuse and conduit aquifer system proceeds with mutual influence between the two and will finally lead to the mature karst system.

In this picture many questions remain open. Thus, the term penetrable or favourable fissures and joints is not an explanation but merely a word stating the fact that at some time ago karst initiation has started. The question to be asked would be: What is a penetrable fracture in terms of its dimension and the field of hydraulic gradients in the initial "fracture aquifer"? Even if this question could be answered, the next point to be determined is, how far can water flow in a fracture without losing its power of solutional enlargement of this fracture; and related to this, how much limestone can be removed by a certain volume of water flowing through the fissures?

These questions can no longer be answered by the descriptive approach. Instead, an interdisciplinary approach has to be taken, employing the chemistry of the system limestone-water-carbon dioxide and also the physics of hydrodynamics of flow in fractured systems as well as in pipes constituting the conduit aquifer.

Thrailkill, in a classical paper (1968), discussed those "chemical and hydrological factors in the excavation of limestone caves".

He investigated flow patterns in pipe networks simulating laminar and turbulent flow in karst aquifers and concluded that the flow pattern is similar under laminar and turbulent flow, provided the lateral extension of the karst aquifer is wide relative to its depth and permeability is distributed evenly. He investigated the chemical evolution of vadose water percolating through the rock on its way to the water table and concluded that most of this water is saturated with respect to calcite when arriving at the water table. To explain enlargement of karst porosity in the shallow phreatic zone he looked for reasons of renewed undersaturation. By postulating a minimum cave of 500-m length and 1-m average diameter to be excavated in 100 000 years by receiving rain water of the infiltration of the 1-km^2 area above, he defined a standard minimum undersaturation of 0.0108 ppm Ca^{2+}. Undersaturations of this magnitude are possible by the temperature effect. When water cools down by only 1°C, undersaturations already 50 times as high are predicted from the mass action laws of CO_2–H_2O–$CaCO_3$ systems. A similar undersaturation results by the mixing effect (*Mischungskorrosion*) proposed by Bögli (1964). This effect states that the mixture of two saturated $CaCO_3$ solutions, with differing CO_2 concentrations and consequently differing Ca^{2+} concentrations regain renewed aggressivity upon mixing. Although these considerations show that the development of caverns and karst aquifers is not in contradiction to the hydrodynamics of flow and the laws of equilibrium chemistry, as one should expect, they are still lacking important principles which govern the evolution of karst systems.

Since karst evolution is a process in space and time two important questions have to be asked, which cannot be answered from equilibrium chemistry.

The first question pertains to the spatial extension of karst systems. This is related to the question, how far under given conditions calcite-aggressive water can flow until it becomes saturated and is no longer capable of dissolving limestone, thus enlarging the cross-section of its flow path. The second question deals with time which is needed to develop a mature karst system from its initial state.

The key to answering these questions is the kinetics of calcite dissolution. If the reaction of calcite dissolution is extremely fast, water once in contact with calcite (limestone) becomes saturated in a very short time. Therefore, dissolution of limestone by water penetrating into primary fractures will stop after very short travel distances and only surface denudation in limestone areas would result. One would expect an evenly distributed lowering of limestone surfaces. Thus, the characteristics of karst, i.e. underground circulation of water, could never develop. In other words, if dissolution were an infinitely fast process, karst landforms would not exist at all. On the other hand, if one assumes that the reactions proceed extremely slow, as a consequence water penetrating into primary fractures would keep its solutional power evenly over extremely long distances. Enlargement of flow routes is then expected to proceed with equal rates everywhere and a homogeneous distribution of secondary permeability would result, in contrast to that observed in nature. Furthermore, extremely slow kinetics of the reaction as a consequence would show extremely small amounts of limestone dissolved per area and time unit. Thus, the rates of enlargement would be extremely small and the time required to establish karst aquifers could be in principle infinitely long in case of infinitely slow reactions. To understand karst processes detailed knowledge of calcite dissolution kinetics is therefore necessary as was first realized by White and Longyear (1962).

It is one of the main purposes of this book to discuss in detail the complicated matter of the kinetics of dissolution and also the precipitation of calcite and to relate these results to the development of karst systems.

1.3   Organization of the Book

The book is divided into three parts. The first part comprises Chapters 2, 3, 4 and 5 and provides an introduction to the equilibrium chemistry of the system $H_2O-CO_2-CaCO_3$ (Chap. 2) and an introduction to the basic principles of mass transport (Chap. 3) and chemical kinetics (Chap. 4). Chapter 5 deals with the principles of laminar and turbulent flow in fractures and pipes and introduces the basic elements of Darcy flow.

The next two chapters, constituting part two, deal with the kinetics of calcite dissolution and precipitation. Chapter 6 gives a critical review of the current literature and introduces into empirical expressions from which dissolution rates can be calculated once the chemical composition of the solution at the dissolving calcite surface is known. Chapter 7 investigates calcite dissolution under boundary conditions which are similar to those existing in karst aquifers. By combining principles of mass transport by molecular or turbulent diffusion (Chap. 3), equilibrium chemistry (Chap. 2) and the chemical kinetics of conversion of CO_2 into aggressive carbonic acid H_2CO_3, with the results of Chapter 6, dissolution rates of calcite are derived, which later can be applied to specific problems in karstification. From this theory it is also possible to predict precipitation rates of calcite from supersaturated solutions.

The last part of the book deals with the consequences of the results of calcite dissolution kinetics on the development of karst systems. Chapter 8 introduces general principles constituting the elements of karst systems. It discusses the relation of fracture systems to the orientation and the type of cave passages. An introduction to the hydrologic properties of karst aquifers is given. Finally, descriptive models of cave development and karstification are reviewed from the recent literature. Chapter 9 combines the principles of calcite dissolution as discussed in Chapter 7 and those of hydrodynamics of Chapter 5 to give in the framework of the elements presented in Chapter 8 models of initial karst development. This is done by considering Darcy flow in two-dimensional porous media, modelling bedding plane partings or joints. Examples are given how flow fields change upon enlargement by dissolution. In a next step the concept of penetration length is introduced to derive numbers, giving the length water can flow until its dissolution rates have dropped to 10% of the initial value. The consequence of these numbers derived for different regimes of dissolution kinetics, close and far from equilibrium, to initial and mature karstification processes are discussed. Mathematical models are presented showing the evolution of first initial flow paths in space and time. These results are also discussed in view of recent geomorphological theories of karst evolution. The last section of Chapter 9 discusses rates of surface denudation in terms of equilibrium chemistry and dissolution kinetics.

Chapter 10 deals with the precipitation of calcite in karst environments. In the first part the morphology and growth rates of regular stalagmites are discussed in some detail and compared to field observations. In the second part results of calcite precipitation in surface streams are presented and interpreted in terms of the theory represented in Chapter 7.

Part I
Basic Principles from Physics and Chemistry

2 Chemistry of the System $H_2O-CO_2-CaCO_3$

2.1 Reactions and Equilibria

Weathering of limestone rocks, at the surface and underground, is due to the reaction:

$$CaCO_3 + CO_2 + H_2O \rightarrow Ca^{2+} + 2\,HCO_3^- \,.$$

This is a global reaction equation and, according to Plummer et al. (1978), it comprises three different attacks of the system H_2O-CO_2 to the calcite surface. They are:

$$CaCO_3 + H^+ \rightleftharpoons Ca^{2+} + HCO_3^- \,;$$

$$CaCO_3 + H_2CO_3 \rightleftharpoons Ca^{2+} + 2\,HCO_3^- \,;$$

$$CaCO_3 + H_2O \rightleftharpoons Ca^{2+} + CO_3^{2-} + H_2O \,.$$

From these equations it is evident that the reaction of the system H_2O-CO_2 with $CaCO_3$ is governed by the concentrations of the species $H^+, HCO_3^-, CO_3^{2-}, H_2CO_3$ and Ca^{2+} at the calcite surface.

To elucidate the chemical equilibria between these species, we first treat the pure carbonic acid system H_2O-CO_2 prior to any dissolution of $CaCO_3$.

The chemical reactions in this system are:

$$CO_2^g + H_2O \rightleftharpoons CO_2^{aq} + H_2O \,.$$

This process describes transfer of CO_2 from the gas phase into the solution, where the aqueous species CO_2^{aq} is formed. The equilibrium between CO_2^g and CO_2^{aq} is given by:

$$K_D = \frac{(CO_2^{aq})}{(CO_2^g)}, \tag{2.1}$$

or by Henry's law, which relates the partial pressure p_{CO_2} of the surrounding atmosphere to the activity of the dissolved CO_2

$$(CO_2^{aq}) = K_H \cdot p_{CO_2} \,. \tag{2.1a}$$

The parentheses indicate activities, which are related to concentrations by activity coefficients γ as will be discussed below. Concentrations are given by square brackets. K_H and K_D are related by:

$$K_D = K_H \cdot R \cdot T \,. \tag{2.1b}$$

T is the absolute temperature and R the gas constant. CO_2 reacts with water to form carbonic acid:

$$H_2O + CO_2 \rightleftharpoons H_2CO_3,$$

with the mass action equation:

$$(CO_2^{aq}) = K_0 \cdot (H_2CO_3). \tag{2.2}$$

Often one defines $[H_2CO_3^*] = [CO_2^{aq}] + [H_2CO_3]$. Then Eq. (2.2) reads:

$$\frac{(CO_2^{aq})}{(H_2CO_3^*)} = \left(1 + \frac{1}{K_0}\right)^{-1}. \tag{2.2a}$$

H_2CO_3 dissociates into H^+ and HCO_3^-:

$$H_2CO_3 \rightleftharpoons H^+ + HCO_3^-,$$

with the mass action law

$$\frac{(H^+)(HCO_3^-)}{(H_2CO_3^*)} = K_1$$

or   (2.3)

$$\frac{(H^+)(HCO_3^-)}{(H_2CO_3)} = K_1 \cdot (1 + K_0) = K_{H_2CO_3}.$$

The next dissociation step is:

$$HCO_3^- \rightleftharpoons H^+ + CO_3^{2-}$$

with   (2.4)

$$\frac{(H^+)(CO_3^{2-})}{(HCO_3^-)} = K_2.$$

Finally, we have the dissociation of water with:

$$(H^+)(OH^-) = K_w. \tag{2.5}$$

In Eqs. 2.1 to 2.5 the mass action equations have been written using activities. These are related to the concentrations by activity coefficients which depend on the ionic strength I of the solution, defined by:

$$I = \frac{1}{2}\sum_i Z_i^2 \cdot c_i. \tag{2.6}$$

Z_i is the charge of the i-th species in the solution and c_i its concentration in mol l^{-1}. For a solution of CO_2 in pure H_2O:

$$I = \frac{1}{2}([H^+] + [OH^-] + [HCO_3^-] + 4[CO_3^{2-}]). \tag{2.6a}$$

Activities and concentrations are related to each other (Garrels and Christ 1965) by:

$$(i) = \gamma_i \cdot [i]. \tag{2.7}$$

Table 2.1. Activity coefficient γ for various ions encountered in karst water at T = 15°C calculated by the extended Debye-Hückel expression (intermediate values can be estimated by linear interpolation)

Ion	Ionic strength I					Ionic radius
	0.001	0.005	0.01	0.05	0.1	a[Å]
H^+	0.967	0.935	0.915	0.856	0.828	9
Mg^{2+}	0.874	0.760	0.694	0.522	0.450	8
Ca^{2+}	0.872	0.751	0.680	0.489	0.407	6
CO_3^{2-}	0.870	0.745	0.669	0.461	0.370	4.5
SO_4^{2-}	0.870	0.743	0.667	0.456	0.363	4
HCO_3^-, Na^+	0.965	0.929	0.904	0.822	0.776	4
OH^-, Cl^-	0.965	0.927	0.902	0.815	0.765	3

The individual ion activity coefficients are usually calculated by the extended Debye-Hückel equation as:

$$\log \gamma_i = -Az_i^2 \cdot \frac{\sqrt{I}}{1 + Ba_i\sqrt{I}} . \tag{2.8}$$

The values A and B are dependent on temperature (A = 0.4883 + 8.074 × 10^{-4}t, B = 0.3241 + 1.6 × 10^{-4}t), t is the temperature in °C. The values a_i represent ionic radii. Table 2.1 lists ionic radii and gives activity coefficients calculated from Eq. (2.8) for various ionic strengths and ions encountered in karst water. As can be seen from Eq. (2.8) and Table 2.1 $\gamma_i \rightarrow 1$ with I approaching zero. In natural karst water the ionic strength is well below 0.1 and the extended Debye-Hückel theory can be used with confidence.

For uncharged species such as CO_2 and H_2CO_3 the activity coefficients are given (Plummer and Mackenzie 1974) by:

$$\gamma_i = 10^{0.11} \approx 1, \quad \text{if} \quad I < 0.1 . \tag{2.9}$$

The dissociation of carbonic acid into bicarbonate HCO_3^- and carbonate CO_3^{2-} is governed by pH as a master variable. This can be derived as follows.

The total amount of carbon in a CO_2–H_2O system is given by:

$$C_T = [H_2CO_3^*] + [HCO_3^-] + [CO_3^{2-}] . \tag{2.10}$$

We define the molar fractions:

$$\alpha_0 = \frac{[H_2CO_3^*]}{C_T}; \quad \alpha_1 = \frac{[HCO_3^-]}{C_T}; \quad \alpha_2 = \frac{[CO_3^{2-}]}{C_T} . \tag{2.11}$$

Combining these equations we find:

$$\frac{1}{\alpha_0} = 1 + \frac{[HCO_3^-]}{[H_2CO_3^*]} + \frac{[CO_3^{2-}]}{[H_2CO_3^*]} . \tag{2.12}$$

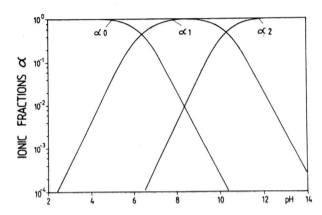

Using mass action Eqs. (2.3) and (2.4) and converting activities into concentrations by Eq. (2.7) yields:

$$\alpha_0 = \left(1 + \frac{K_1}{\gamma_{HCO_3}\gamma_H[H^+]} + \frac{K_1 \cdot K_2}{\gamma_H^2\gamma_{CO_3}[H^+]^2}\right)^{-1}. \tag{2.13a}$$

Similarly, one obtains:

$$\alpha_1 = \left(1 + \frac{[H^+]\gamma_{HCO_3}\gamma_H}{K_1} + \frac{K_2\gamma_{HCO_3}}{\gamma_H\gamma_{CO_3}[H^+]}\right)^{-1}; \tag{2.13b}$$

$$\alpha_2 = \left(1 + \frac{\gamma_H^2\gamma_{CO_3}[H^+]^2}{K_1K_2} + \frac{[H^+]\gamma_H\gamma_{CO_3}}{K_2\gamma_{HCO_3}}\right)^{-1}. \tag{2.13c}$$

Figure 2.1 visualizes these ionic fractions as a function of pH for the case of a very dilute solution, where all the γ_i are equal to 1.

In the region pH < 4 virtually no HCO_3^- and CO_3^{2-} is present and only $H_2CO_3^*$ exists. With increasing pH carbonic acid dissociates, forming HCO_3^-. Thus, below pH = 8.3 practically no CO_3^{2-} is present. Above pH = 8.3 HCO_3^- starts to dissociate until at pH 12 only CO_3^{2-} ions exist.

It should be noted here that this behaviour depends only very weakly on temperature and ionic strength (Loewenthal and Marais 1978). The data in Fig. 2.1 therefore are representative with sufficient accuracy for the realm of karst waters. One important conclusion from this diagram is that in most cases of karst waters the species CO_3^{2-} can be neglected, since karst waters very rarely exhibit pH values higher than 8.3.

If $CaCO_3$ is dissolved, the equilibrium between dissolved species and the solid is given by the solubility product K_c:

$$K_c = (Ca^{2+})_{eq}(CO_3^{2-})_{eq}, \tag{2.14}$$

where $(Ca^{2+})_{eq}$ and $(CO_3^{2-})_{eq}$ are the activities at equilibrium. To describe the saturation state of a solution, one uses the saturation state:

$$\Omega = \frac{(Ca^{2+})(CO_3^{2-})}{K_c}. \tag{2.15}$$

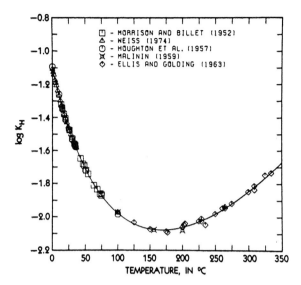

Fig. 2.2. Temperature dependence of K_H. The *full line* is calculated from the data in Table 2.2 (Plummer and Busenberg 1982)

The solution is undersaturated for $\Omega < 1$, saturated for $\Omega = 1$ and supersaturated for $\Omega > 1$.

Ca^{2+} reacts with HCO_3^- and CO_3^{2-} to form ion pairs. The corresponding ion pair activities are related to mass action constants according to the equations:

$$K_3 \cdot (Ca^{2+}) \cdot (HCO_3^-) = (CaHCO_3^+) \,;$$

$$K_4 \cdot (Ca^{2+}) \cdot (CO_3^{2-}) = (CaCO_3^0) \,. \tag{2.16}$$

In the case of natural karst water these species can be neglected safely in all calculations and will therefore not be included in our further considerations.

The most recent data of mass action constants and their temperature dependence for the system H_2O–CO_2–$CaCO_3$ are given in Figs. 2.2 to 2.7 (Plummer and Busenberg 1982). It should be noted that very often the values of mass action constants K are given as pK related to K by:

$$pK = -\log K, \qquad K = 10^{-pK} \,. \tag{2.17}$$

Table 2.2 summarizes the empirical temperature dependence of all mass action constants defined above.

To calculate equilibria in an electrolytic system one has to observe the neutrality of electrical charges in the solution. The equation of electroneutrality states that the sum of all positive ionic charges is to be balanced by that of all the negative ones. In the case of a pure H_2O–CO_2–$CaCO_3$ system and neglecting ion pairs, this is formulated as:

$$2[Ca^{2+}] + [H^+] = [HCO_3^-] + 2[CO_3^{2-}] + [OH^-] \,. \tag{2.18}$$

In the region between pH = 4 to pH = 8.4 with an accuracy of 2%, this equation can be relaxed to:

$$2 \cdot [Ca^{2+}] = [HCO_3^-] \,. \tag{2.18a}$$

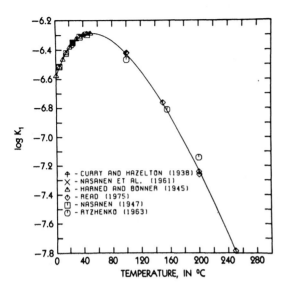

Fig. 2.3. Temperature dependence of K_1. The *full line* is calculated from the data in Table 2.2 (Plummer and Busenberg 1982)

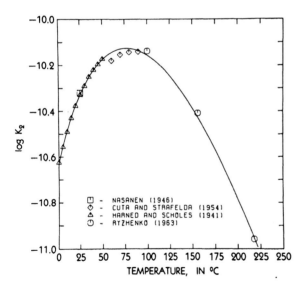

Fig. 2.4. Temperature dependence of K_2. The *full line* is calculated from the data in Table 2.2 (Plummer and Busenberg 1982)

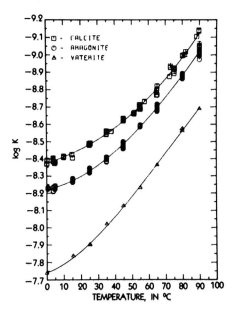

Fig. 2.5. Temperature dependence of the solubility product K for the three $CaCO_3$ modifications: calcite (K_C), aragonite (K_A) and vaterite (K_V). The *full lines* are calculated from the data in Table 2.2 (Plummer and Busenberg 1982)

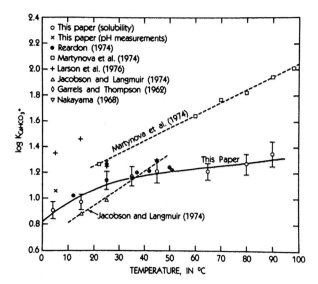

Fig. 2.6. Temperature dependence of $K_{CaHCO_3^+} = K_3$ (Plummer and Busenberg 1982)

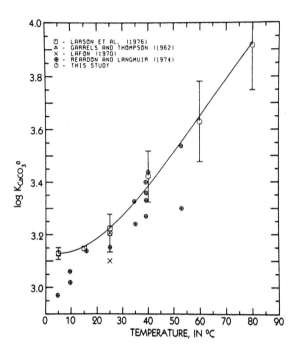

Fig. 2.7. Temperature dependence of $K_{CaCO_3^0} = K_4$ (Plummer and Busenberg 1982)

Table 2.2. Temperature (K) dependence of mass action constants

	References[a]
$\log K_H = 108.3865 + 0.01985076T - 6919.53/T - 40.45154 \log T + 669365/T^2$	(1)
$\log K_1 = -356.3094 - 0.06091964T + 21834.37/T - 126.8339 \log T - 1684915/T^2$	(1)
$\log K_2 = -107.8871 - 0.03252849T + 5151.79/T + 38.92561 \log T - 56371.9/T^2$	(1)
$K_0 = 1.7 \times 10^{-4}/K_1$	(2)
$\log K_W = 22.801 - 0.010365T - 4787.3/T - 7.1321 \log T$	(3)
$\log K_3 = 1209.12 + 0.31294T - 34765.05/T - 478.782 \log T$	(1)
$\log K_4 = -1228.732 - 0.299444T + 35512.75/T + 485.818 \log T$	(1)
$\log K_C = -171.9065 - 0.077993T + 2839.319/T + 71.595 \log T$	(1)

[a] (1) Plummer and Busenberg (1982); (2) Wissbrun et al. (1954); (3) Harned and Hamer (1933).

This can be seen from Fig. 2.1, which shows $[CO_3^{2-}] \ll [HCO_3^-]$ for pH < 8. From this an important conclusion on ionic strengths in karst waters can be drawn. In aqueous solutions of pure $CaCO_3$ the ionic strength is given by a sufficiently accurate approximation (Eq. 2.6a) as:

$$I = 3 \cdot [Ca^{2+}]. \tag{2.19}$$

In the presence of Mg^{2+} derived from dolomite or magnesian calcites, then:

$$I = 3 \cdot ([Ca^{2+}] + [Mg^{2+}]). \tag{2.20}$$

This is of high practical interest as it allows one to obtain the ionic strength I directly from the concentration of the metal ions. In equilibrium calculations, therefore, once the concentrations of the metal ions are known, the activity coefficients no longer depend on the concentration of the carbonic ion species and iterative procedures to calculate I and the related activity coefficients are not necessary.

2.2   Boundary Conditions for Achieving Equilibrium

Dissolution of calcite by CO_2-containing water in natural systems proceeds under a variety of different boundary conditions:

1. In the saturated zone of karst aquifers or porous calcareous soils there is no interface between an atmosphere and the solution and consequently no mass transport of CO_2 into the solution. Therefore, during dissolution of calcite, CO_2 is consumed from the solution.
2. In all situations where water flows in contact with limestone rocks and a CO_2-gas phase, e.g. in rivers, mass transport of CO_2 is effective between the liquid-gas interface. Thus, the CO_2 consumed by dissolution of calcite is replaced.

These two situations are generally defined as closed and open system conditions and most of the geological dissolution processes are described in terms of these two extreme situations.

There is a general intermediate condition, however, the extreme limits of which constitute the open and the closed system.

3. In the unsaturated zone of any aquifer a situation exists where a limited volume V_1 of solution is in contact with calcite and also forms a surface to a limited volume V_g of a CO_2-containing atmosphere. In this case, in contrast to the open system where V_g is assumed to be extremely large, any mass exchange between this atmosphere and the solution changes the CO_2 composition of both the atmosphere and the solution. Thus, open system conditions are the limiting case with $V_g \rightarrow \infty$ and the closed system is achieved with $V_g \rightarrow 0$.

2.2.1   The General Case

The purpose of this section is to formulate the chemical composition of the H_2O–CO_2–$CaCO_3$ system during the process of dissolution for the general intermediate condition. As stated above, this includes the two extreme cases of open and closed systems.

Figure 2.8 illustrates this general case in the unsaturated zone of a porous calcareous medium. The volume of the solution is V_1. This solution is in contact to a limited volume of gas, V_g. Dissolution of $CaCO_3$ proceeds sufficiently slow, so that all carbonic species are in equilibrium to each other. Furthermore, we assume that the aqueous CO_2 is in equilibrium with the atmospheric CO_2.

In an idealized pure system H_2O–$CaCO_3$–CO_2 the chemical composition of the solution can be calculated once the initial CO_2 pressure of the atmosphere and

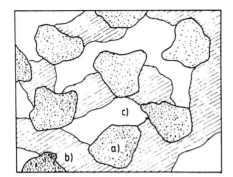

a) calcareous grain
b) Volume of solution V_l ⌷
c) Volume of gas V_g ⌷

Fig. 2.8. Illustration of the intermediate case in a porous medium. The total value of the gas phase is V_g, the total volume of the solution is V_l

the Ca^{2+} concentration of the solution are known. This is of utmost importance, since in many geologically relevant situations (cf. Chap. 6) the dissolution rates are determined by this composition.

Each mole of $CaCO_3$ dissolved releases 1 mol Ca^{2+} ions and 1 mol carbonate ions. The total amount of carbon in the system is therefore given by conservation of carbonate species, i.e. C-atoms:

$$M_T = ([Ca^{2+}] + [HCO_3^-]_i + [H_2CO_3^*]_i + [CO_3^{2-}]_i) \cdot V_l + [CO_2^g]_i V_g$$

$$= ([HCO_3^-] + [H_2CO_3^*] + [CO_3^{2-}]) \cdot V_l + [CO_2^g] \cdot V_g . \qquad (2.21)$$

The right hand side of Eq. (2.21) gives the amount of C as calculated from the concentrations of all carbon-containing species in the liquid and in the gas phase. This is equal to amount of carbon atoms present in the pure H_2O-CO_2 solution, prior to dissolution of calcite, augmented by the amount of Ca^{2+}-ions dissolved, since each $CaCO_3$ released from the solid adds also one carbon atom to the solution. According to Henry's law we have:

$$[CO_2^g] = \frac{[H_2CO_3^*]}{K_H RT} \cdot \left(1 + \frac{1}{K_0}\right)^{-1} . \qquad (2.22)$$

The initial values $[\ ]_i$ are those of the initial H_2O-CO_2 system prior to dissolution, i.e., with $[Ca^{2+}] = 0$. They are given by:

$$[H_2CO_3^*]_i = K_H p_{CO_2}^i(1 + 1/K_0);$$

$$[HCO_3^-]_i = ([H_2CO_3^*]_i \cdot K_1/\gamma_H^i \gamma_{HCO_3}^i)^{1/2};$$

$$[CO_3^{2-}]_i \approx 0 . \qquad (2.23)$$

As a second master equation for the purpose of our calculation we have to observe the equation of electroneutrality, which reads:

$$2[Ca^{2+}] + [H^+] = [HCO_3^-] + 2[CO_3^{2-}] + [OH^-] . \qquad (2.24)$$

Subtracting Eq. (2.21) from Eq. (2.24), and employing the mass action laws, Eq. (2.3) and Eq. (2.4), we obtain:

$$[Ca^{2+}] + [H^+] - v - \frac{K_w'}{[H^+]} + \left(w \cdot [H^+] - \frac{K_2'}{[H^+]}\right) \cdot [HCO_3^-] = 0 . \qquad (2.25)$$

and similarly by multiplying Eq. (2.21) by 2 and subtracting Eq. (2.24):

$$2v - [H^+] + \frac{K_w'}{[H^+]} = [HCO_3^-](1 + 2w \cdot [H^+]) \tag{2.26}$$

The abbreviations are:

$$v = f \cdot [H_2CO_3^*]_i + [HCO_3^-]_i; \qquad f = 1 + \frac{(1 + 1/K_0)^{-1}}{K_H RT} \varkappa;$$

$$\varkappa = V_g/V_l; \qquad K_w' = K_w/\gamma_H \gamma_{OH};$$

$$K_2' = K_2 \gamma_{HCO_3}/\gamma_H \gamma_{CO_3}; \qquad w = f \cdot \frac{\gamma_H \gamma_{HCO_3}}{K_1}. \tag{2.27}$$

Combining Eqs. (2.25) and (2.26) leads to a fourth-order polynomial in hydrogenion concentration which can be solved numerically:

$$[Ca^{2+}] + [H^+] - v - K_w'/[H^+] + (w[H^+] - K_2'/[H^+])$$
$$\cdot (2v - [H^+] + K_w'/[H^+]) \cdot (1 + 2w \cdot [H^+])^{-1} = 0. \tag{2.28}$$

The activities involved in Eqs. (2.26, 2.27, 2.28) can be calculated with sufficient accuracy by approximating the ionic strength:

$$I = 3 \cdot [Ca^{2+}]. \tag{2.29}$$

Therefore, no iteration procedure in computing $[H^+]$ is necessary. Once $[H^+]$ is known $[HCO_3^-]$ can be obtained by Eq. (2.25). $[H_2CO_3^*]$, $[CO_3^{2-}]$ and $[CO_2^g]$ then are easily derived from the mass balance equations.

Open and closed system conditions are included in this general formulation by using $f = 1$ in the closed system and $f = \infty$ (i.e. $f \approx 10^5$) in the open system.

In most cases of practical interest in a natural system pH is below 8 and the carbonate and OH^- species can be neglected in Eqs. (2.21) and (2.24). Then Eq. (2.28) is reduced to a quadratic which can be employed much more conveniently:

$$[H^+]^2 + [H^+] \cdot (2[Ca^{2+}] + 1/w) + \frac{1}{w}([Ca^{2+}] - v) = 0. \tag{2.30}$$

The solution of this equation can be used to calculate all the other species as described above. It is given by:

$$[H^+] = -\frac{1}{2}\left(2[Ca^{2+}] + \frac{1}{w}\right) + \frac{1}{2}\sqrt{4[Ca^{2+}]^2 + \frac{4v}{w} + \frac{1}{w^2}}. \tag{2.30a}$$

Figure 2.9 shows the hydrogen ion concentration $[H^+]$ as a function of dissolved $[Ca^{2+}]$. The results of the exact calculation employing Eq. (2.28) and the approximation of Eq. (2.30) are compared for the evolution of the closed system, i.e. $f = 1$. Deviations in the values of $[H^+]$ are only significant for pH > 8. Below this value the approximation can be used with confidence in *all* cases. In the closed system $[H^+]$ drops very rapidly with increasing Ca^{2+} concentration and is very sensitive to changes in $[Ca^{2+}]$. The evolution of $[CO_3^{2-}]$ for a variety of intermediate cases is illustrated in Fig. 2.10. The carbonate concentration is shown in a mirror-

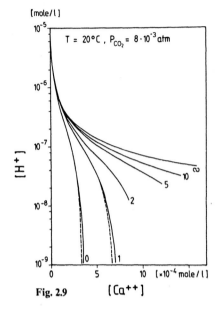

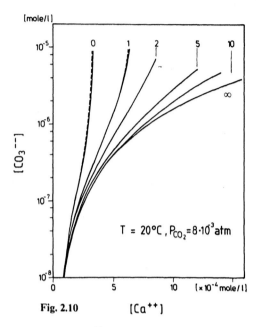

Fig. 2.9

Fig. 2.9. Hydrogen ion concentration [H^+] as a function of [Ca^{2+}] in the solution as the system approaches equilibrium. The *numbers* give the values of $V_g/V_1 = \varkappa$. The *dashed lines* show the approximation according to Eq. (2.30). The initial CO_2 pressure is 8×10^{-3} atm

Fig. 2.10. Carbonate ion concentration [CO_3^{2-}] as a function of [Ca^{2+}] in the solution as the system approaches equilibrium. The *numbers* give the values of $V_g/V_1 = \varkappa$. The *dashed lines* show the approximation according to Eq. (2.30). Initial conditions as in Fig. 2.9

like curve compared to [H^+]. A steep increase of [CO_3^{2-}] is observed as [H^+] decreases.

For the open system the decrease in [H^+] is much less steep as the system approaches equilibrium and so is the increase in [CO_3^{2-}]. The intermediate cases show a behaviour which is between the two extremes, tending more to be like a closed system for $\varkappa \leq 2$ and like an open system for $\varkappa \geq 10$. Figure 2.11 shows the saturation state Ω (cf. Eq. 2.15) for the same conditions as in Fig. 2.10.

Figure 2.12 gives the results for the $H_2CO_3^*$ concentration in the solution which is in equilibrium with the p_{CO_2} of the surrounding atmosphere. Only in the open system does this value remain constant. In all other cases there is a linear decrease of this value as [Ca^{2+}] increases. This behaviour results from the stoichiometry of calcite dissolution, which consumes one CO_2 molecule for each Ca^{2+} ion being released. It can also be derived as an approximation of Eq. (2.21). In all cases where $p_{CO_2} \geq 3 \times 10^{-4}$ atm, [HCO_3^-]$_i$ is small compared to [$H_2CO_3^*$]$_i$ and can be neglected. For pH ≤ 8.2 we also can neglect [CO_3^{2-}] and OH^- in Eqs. (2.21) and (2.24). Combining these relaxed equations, we obtain:

$$[H_2CO_3^*] = [H_2CO_3^*]_i - [Ca^{2+}] \cdot \frac{1}{f}. \tag{2.31}$$

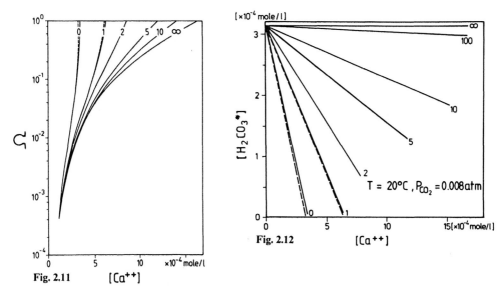

Fig. 2.11 $[Ca^{++}]$

Fig. 2.11. Fig. 2.12

Fig. 2.11. Saturation state Ω as a function of $[Ca^{2+}]$ in the solution as the system approaches equilibrium. The *numbers* give the values of $V_g/V_1 = \varkappa$. The *dashed lines* show the approximation according to Eq. (2.30). Initial conditions as in Fig. 2.9

Fig. 2.12. $[H_2CO_3^*]$ as a function of $[Ca^{2+}]$ in the solution as the system approaches equilibrium. The *numbers* give the values of $V_g/V_1 = \varkappa$. The *dashed lines* show the approximation according to Eq. (2.31). Initial conditions as in Fig. 2.9

Thus, the slope of the lines in Fig. 2.12 is given by $1/f$, where f is defined in Eq. (2.27). There are slight deviations between the exact solution and the approximation of Eq. (2.31) as is indicated by the dotted lines for $\varkappa = 0$ and $\varkappa = 1$. For $\varkappa > 2$ these deviations are negligible.

2.2.2 Change of Boundary Conditions

In karst systems one very often encounters the situation where water flows with a free surface dissolving limestone under open system conditions. Then the water sinks into the ground and the conditions change to the general case, i.e. water dissolves limestone in the presence of a limited volume of gas with initially the same partial pressure as in the open system. To deal with this problem, we assume that open system conditions with a partial pressure $p_{CO_2}^0$ in the atmosphere prevail until the Ca^{2+} concentration has reached the value $[Ca^{2+}]^0$. Then dissolution shall proceed in the general case with $V_g/V_1 = \varkappa$.

Conservation of carbon species, cf. Eq. (2.21), then reads:

$$([Ca^{2+}] - [Ca^{2+}]^0 + [HCO_3^-]^0 + [H_2CO_3^*]^0 + [CO_3^{2-}]^0) \cdot V_1 + V_g \cdot [CO_2^g]^0$$

$$= ([H_2CO_3^*] + [HCO_3^-] + [CO_3^{2-}]) \cdot V_1 + V_g \cdot [CO_2^g] . \qquad (2.32)$$

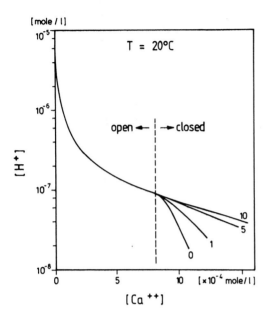

Fig. 2.13. $[H^+]$ as a function of $[Ca^{2+}]$ in the solution, when dissolution proceeds first in the open system and is then under conditions of the closed or intermediate case. *Numbers* on curves designate $\varkappa$. Initial conditions as in Fig. 2.9

The $[\ ]^0$ refer to the chemical composition attained when dissolution conditions change from the open system to the intermediate case. The $[\ ]$ refer to dissolution in the intermediate condition. In all cases $[Ca^{2+}] > [Ca^{2+}]^0$.

Proceeding as in section 2.2.1 we obtain similarly equations analogous to Eqs. (2.25), (2.26), (2.28) and (2.30) provided we replace v by:

$$v^s = f \cdot [H_2CO_3^*]^0 + [HCO_3^-]^0 + [CO_3^{2-}]^0 - [Ca^{2+}]^0 \ .$$

To calculate the chemical evolution under these circumstances, we first calculate the evolution for the open system as in section 2.2.1 up to the value $[Ca^{2+}]^0$. From this we obtain $[H_2CO_3^*]^0$, $[HCO_3^-]^0$ and $[CO_3^{2-}]^0$. Using these to find v^s we use the calculation for the general case with $\varkappa$ for approaching equilibrium. Figure 2.13 gives the evolution $[H^+]$ for such a case for various $\varkappa$.

2.2.3 Saturated $CaCO_3$ Solutions

In principle the solubility of calcite can be calculated from the above expressions by additionaly employing the solubility product of Eq. (2.14) and determining numerically the corresponding saturation concentration $[Ca^{2+}]_{eq}$ with $\Omega = 1$.

In all cases of practical interest, however, approximations can be used which are sufficiently accurate.

To find the pH value in equilibrium with calcite at a given p_{CO_2} we use the mass balance Eqs. (2.1) to (2.4) expressing activities by concentrations.

$$[H_2CO_3^*] = K_H \cdot p_{CO_2} \cdot (1 + 1/K_0) ; \qquad (2.1a, 2.2a)$$

$$\gamma_H[H^+]\gamma_{HCO_3}[HCO_3^-] = K_1[H_2CO_3^*] \ . \qquad (2.3)$$

Combining these yields:

$$[HCO_3^-] = \frac{K_1 K_H p_{CO_2}}{\gamma_{HCO_3} \gamma_H [H^+]} .$$

Expressing

$$[HCO_3^-] = \frac{\gamma_H \gamma_{CO_3} [H^+][CO_3^{2-}]}{\gamma_{HCO_3} K_2} , \tag{2.4}$$

one obtains $[CO_3^{2-}]$ as a function of p_{CO_2} and $[H^+]$. Substituting this into the equation describing the solubility product:

$$K_c = \gamma_{Ca} \gamma_{CO_3} \cdot [Ca^{2+}]_{eq} \cdot [CO_3^{2-}]_{eq} , \tag{2.14}$$

one has

$$K_c = \frac{\gamma_{Ca} [Ca^{2+}]_{eq} K_1 K_2 K_H p_{CO_2}}{\gamma_H^2 \cdot [H^+]_{eq}^2} . \tag{2.33}$$

With the relaxed charge balance equation:

$$2 \cdot [Ca^{2+}] = [HCO_3^-] \tag{2.18a}$$

substituted into Eq. (2.33) we obtain the final result for the hydrogen ion activity in equilibrium with a calcite-saturated solution, and with a CO_2-containing atmosphere of p_{CO_2}:

$$(H^+)_{eq}^3 = \frac{K_1^2 K_2 K_H^2 \gamma_{Ca}}{2 K_c \gamma_{HCO_3}} \cdot (p_{CO_2})^2 . \tag{2.34}$$

The saturation value of $[Ca^{2+}]_{eq}$ can be obtained similarly. Dividing Eq. (2.3) by (2.4) gives:

$$\frac{K_1}{K_2} = \frac{\gamma_{HCO_3}^2 \cdot [HCO_3^-]^2}{\gamma_{CO_3} [H_2 CO_3^*][CO_3^{2-}]} . \tag{2.35a}$$

Substituting $[H_2 CO_3^*]$ by Eq. (2.1) and $[CO_3^{2-}]$ by Eq. (2.14) leads to:

$$\frac{K_1}{K_2} = \frac{\gamma_{HCO_3}^2 \gamma_{Ca} \cdot [HCO_3^-]_{eq}^2 \cdot [Ca^{2+}]_{eq}}{K_c K_H p_{CO_2}} . \tag{2.35b}$$

With relaxed charge balance, Eq. (2.18a), we find

$$[Ca^{2+}]_{eq}^3 = p_{CO_2} \cdot \frac{K_1 K_c K_H}{4 K_2 \gamma_{Ca} \gamma_{HCO_3}^2} . \tag{2.35c}$$

Equations (2.34) and (2.35) can be used directly for the open system, since p_{CO_2} remains constant during the dissolution process. To obtain the saturation values for all the other cases as a function of the initial $p_{CO_2}^i$, one has to remember that p_{CO_2} drops during dissolution, since each Ca^{2+} released into solution consumes one CO_2 molecule.

Recalling that prior to dissolution pH is low, such that $[H_2 CO_3]_i \gg [HCO_3^-]_i \gg [CO_3^{2-}]_i$ and that for pH < 8, $[HCO_3^-] \gg [CO_3^{2-}]$ by combining Eqs.

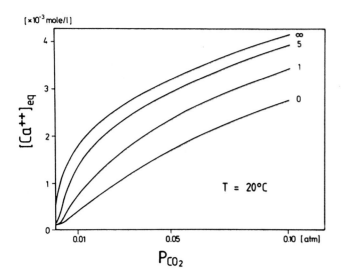

Fig. 2.14. Equilibrium concentrations of $[Ca^{2+}]_{eq}$ as a function of initial $p^i_{CO_2}$ prior to dissolution for various values of $V_g/V_l = \varkappa$, given by the *numbers* on the curves

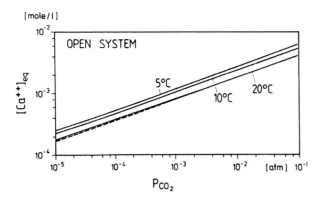

Fig. 2.15. Equilibrium values of $[Ca^{2+}]_{eq}$ as a function of p_{CO_2} in equilibrium with the solution at different temperatures. The *dashed line* gives the approximation according to Eq. (2.35c)

(2.21) to (2.24), we find:

$$p_{CO_2} = p^i_{CO_2} - \frac{[Ca^{2+}]}{fK_H(1 + 1/K_0)} . \tag{2.36}$$

Introduction this into Eq. (2.35) for $[Ca^{2+}]_{eq}$ leads to a cubic equation from which $[Ca^{2+}]_{eq}$ can be calculated.

Figure 2.14 plots the solubility of $CaCO_3$ in terms of $[Ca^{2+}]$ as a function of the initial $p^i_{CO_2}$ in the atmosphere of the system prior to any dissolution, for a variety of ratios V_g/V_l. With increasing volume of the gas the curves approach that of the open system. All the curves have been calculated by an exact equilibrium calculation.

For the open system the curve is redrawn on a logarithmic plot (Fig. 2.15). The dashed curve is calculated from Eq. (2.35), for T = 20°C. For $p_{CO_2} > 10^{-3}$ atm the agreement between the exact and approximate calculation is almost perfect. A slight deviation becomes noticeable for $p_{CO_2} < 10^{-3}$ atm.

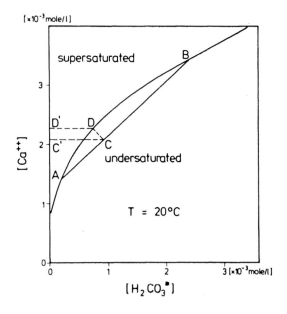

Fig. 2.16. The equilibrium curve divides the $[H_2CO_3^*] - [Ca^{2+}]$ diagram into two parts. *Above*: The curve solutions are supersaturated; *below*: undersaturation exists. Mixing of two saturated solutions (A) and (B) leads to an undersaturated solution (C). The additional amount of Ca^{2+}, which can be dissolved after mixing is given by $\overline{C'D'}$ (see text)

Figure 2.16 replots relation (2.35) for the open system as a function of $[H_2CO_3^*]$. The region below the curve denotes undersaturated solutions; the area above corresponds to supersaturated solutions. From the non-linearity of the saturation curve a very important conclusion can be drawn. Consider an equilibrated solution (point A) under closed system conditions. If this solution mixes with a saturated solution of a different composition (point B) under closed system conditions, then the composition of the resulting solution is represented by point C. Depending on the ratio of the volumes of the two mixed solutions, this point lies on the straight line, connecting A and B in the region of undersaturation. Therefore, mixing of two differently saturated waters leads to renewed calcite aggressivity of the solution in a pure $H_2O-CO_2-CaCO_3$ system. In general for natural waters containing foreign ions, mixing can also result in supersaturation, depending on the particular compositions of the waters involved (Wigley and Plummer 1976). The importance of undersaturation by mixing for processes of karstification was first pointed out by Bögli (1963) and was considered in more detail by Dreybrodt (1981a).

2.2.4 Mixing of Saturated Waters in the Closed System

To estimate the effect of renewed dissolution by mixing of saturated waters in a closed system, one has to know: (1) How much $CaCO_3$ can be dissolved? (2) What is the evolution of the chemical composition during the approach to equilibrium? The answer to the first question addresses the problem, how much limestone can be removed in a karst system, and has been discussed by Thrailkill (1968). Knowledge on the chemical composition of the solution is important, since from this dissolution rates determining the evolution of a karst system in time can be calculated.

In principle, both problems can be solved by using the procedure described in section 2.2.1. The conservation of carbonate species, Eq. (2.21), is to be replaced by:

$$C_T = [Ca^{2+}] + \sum_{n=1}^{N} (H_2CO_3^*]_{i,n} + [HCO_3^-]_{i,n}) \cdot \frac{V_n}{V}$$

$$= [HCO_3^-] + [H_2CO_3^*] + [CO_3^{2-}]. \tag{2.37}$$

This refers to N different saturated solutions with volumes V_n, Ca^2 concentrations $[Ca^{2+}]_{eq,n}$ and initial $[H_2CO_3^*]_{i,n}$. $[Ca^{2+}]$, $[HCO_3^-]$ and $[CO_3^{2-}]$ refer to the concentration of the mixed solution with total volume $V = \sum V_n$. Equations (2.10) to (2.24) remain unchanged. Consequently, in order to obtain the composition of the mixed solution, one has to replace v in Eq. (2.28) by:

$$v^m = \sum_{n=1}^{N} ([H_2CO_3^*]_{i,n} + [HCO_3^-]_{i,n}) \cdot \frac{V_n}{V} \tag{2.38}$$

and use all the corresponding equations in section 2.2.1 with this new value of v^m.

A much simpler way, however, is a graphic method represented in Fig. 2.16. Here, we plot the calcite equilibrium curve in dependence of $H_2CO_3^*$. For simplicity, we consider the case of two waters mixing, represented by points A and B. Before mixing, their concentrations are $[Ca^{2+}]_n$, $[HCO_3^-]_n$, $[H_2CO_3^*]_n$, n = 1, 2. Within the approximation of Eq. (2.13) we obtain for the concentrations immediately after mixing:

$$[Ca^{2+}]_m = \frac{V_1}{V} \cdot [Ca^{2+}]_1 + \frac{V_2}{V}[Ca^{2+}]_2$$

and   (2.39)

$$[H_2CO_3^*]_m = \frac{V_1}{V}[H_2CO_3^*]_1 + \frac{V_2}{V}[H_2CO_3^*]_2 .$$

This is represented by point C, where $AC/CB = V_1/V_2$.

Since further dissolution proceeds according to Eq. (2.31), we have:

$$[H_2CO_3^*] = [H_2CO_3^*]_m - [Ca^{2+}] + [Ca^{2+}]_m \tag{2.40}$$

as reaction path. This path is drawn as a straight line with slope -1 starting at point C. The intersection of this line with the equilibrium curve gives the new equilibrium at point D. Thus, the amount of $CaCO_3$ which can be dissolved by renewed aggressivity can be read from the point C' and D' at the ordinate.

2.2.5   Influence of Foreign Ions on Calcite Dissolution

Karst waters are rarely derived from the pure system H_2O-$CaCO_3$-CO_2. Mostly due to dolomite or magnesian calcites and due to gypsum and anhydrite, which usually are also encountered in karst areas, one finds Mg^{2+} and SO_4^{2-} in quite considerable amounts. In many cases one also observes Na^+, K^+ and Cl^-. All these ions change the ionic strength of the solution, thus affecting the ionic equilibria. Furthermore, if minerals are dissolved which have one ion in common with $CaCO_3$, such as $MgCO_3$ or $CaSO_4$, equilibria are changed by the common-ion effect.

Therefore, not only does the chemical evolution of the system change as dissolution proceeds, but also the solubility of calcite is altered in the presence of foreign ions.

To discuss this in more detail, we assume that in addition to the species resulting from the pure system the following substances might have been added:

1. strong acid $H_{n_A}A$, e.g. HCl or H_2SO_4;
2. strong base $B(OH)_{n_B}$, e.g. NaOH or $Ca(OH)_2$;
3. substance $Me_{n_M}CO_3$ with the CO_3^{2-} ion in common with $CaCO_3$, e.g. $MgCO_3$, Na_2CO_3;
4. substance $Ca(An)_{n_{An}}$ with the Ca^{2+} as the common ion;
5. substance, which dissociates into indifferent ions X^{+Z_x} and Y^{-Z_y} which do not react with either Ca^{2+} or CO_3^{2-}, e.g. resulting from NaCl, etc. ($+Z_x$ and $-Z_y$ denote the charges of these ions).

We furthermore assume that all substances added are fully dissociated. In a system with $\varkappa = V_g/V_l$ as described in section 2.2.1 the conservation of carbonic species now reads:

$$\left([Ca^{2+}] - \frac{1}{n_{An}}[An] + \frac{1}{n_M}[Me] + [H_2CO_3^*]_i + [HCO_3^-]_i + [CO_3^{2-}]_i\right)\cdot V_l$$
$$+ [CO_2^g]_i\cdot V_g = ([H_2CO_3^*] + [HCO_3^-] + [CO_3^{2-}])\cdot V_l + [CO_2^g]\cdot V_g .$$
(2.41)

If we assume that the solution evolves from meteoric water in equilibrium with a given initial $[CO_2^g]_i$, the initial conditions are those of Eq. (2.23).

The equation of charge balance is:

$$2[Ca^{2+}] + [H^+] + [F] = [HCO_3^-] + 2[CO_3^{2-}] + [OH^-]$$
(2.42)

with the total charge [F] of the foreign ions as

$$[F] = Z_B[B] + Z_{Me}[Me] + Z_x[X] - Z_A[A] - Z_{An}[An] - Z_y[Y]$$
(2.42a)

Z relates to the charge of the corresponding ion.

Using the same procedure as in section 2.2.1, we obtain:

$$[Ca^{2+}] + [F] + [H^+] - v^F - \frac{K'_w}{[H^+]}$$
$$+ \frac{(w\cdot[H^+] + K'_2/[H^+])\cdot(2v^F - [H^+] - [F] + K'_w/[H^+])}{(2w\cdot[H^+] + 1)} = 0$$
(2.43)

from which $[H^+]$ can be calculated. $[HCO_3^-]$ is found from the relation:

$$[HCO_3^-]\cdot(2w\cdot[H^+] + 1) = 2v^F - [H^+] - [F] + \frac{K'_w}{[H^+]} .$$
(2.44)

$[CO_3^{2-}]$, $[H_2CO_3^*]$, $[CO_2^g]$ can then by computed by observing the mass action laws, v^F is given by:

$$v^F = f\cdot[H_2CO_3^*]_i + [HCO_3^-]_i + \frac{1}{n_M}[Me] - \frac{1}{n_{An}}[An] .$$
(2.45)

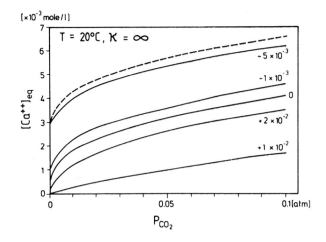

[$\times 10^{-3}$ mole/l]

Fig. 2.17. Equilibrium values of [Ca^{2+}] in the presence of common ions, i.e. F $\neq$ 0, as function of p_{CO_2}. *Numbers* on the curves designate F in mol l^{-1}. The *dashed curve* results by shifting the curve of the pure system (F = 0) vertically by 2.5 $\times$ 10^{-3} mol l^{-1} and designates the equilibrium curve upon addition of Ca^{2+} as a common ion with F = $-5 \times$ 10^{-3} mol l^{-1} without the common ion effect present. The corresponding curve with the common ion effect effective is below the dashed curve, F = -5×10^{-3} mol l^{-1} and shows that calcite solubility decreases

The ionic activity coefficients have to be calculated by using the ionic strength, including all the species dissolved.

For pH < 8.3, again by neglecting [CO_3^{2-}] and [OH^-], the fourth-order polynomial in [H^+], eq. (2.43) can be approximated quite accurately by the quadratic:

$$[H^+]^2 + [H^+](2[Ca^{2+}] + [F] + 1/w) + \frac{1}{w}([Ca^{2+}] + [F] - v^F) = 0 . \quad (2.46)$$

Thus, Eqs. (2.43) to (2.46) constitute the most general case of the chemical evolution of karst water.

The calcite saturation values can now be obtained by calculating the chemical composition for a given Ca concentration. From this the saturation state Ω is found. Using a half-step procedure, the value of [Ca^{2+}] corresponding to $\Omega = 1$ is found numerically.

The result for such a calculation for the open system is visualized in Fig. 2.17, where the equilibrium concentration [Ca^{2+}]$_{eq}$ is plotted as a function of p_{CO_2} in equilibrium with the solution, for various values of [F].

Note that [F] < 0 corresponds to the presence of acids or substances of the type $Ca(An)_{n_{An}}$, whereas [F] > 0 derives from bases or $Me_{n_M}CO_3$. If only indifferent substances are added, [F] = 0. An approximative expression for the equilibrium concentration can be easily derived as in section 2.2.3. In the presence of foreign ions the relaxed equation of charge balance reads:

$$2[Ca^{2+}] + [F] = [HCO_3^-] . \quad (2.47)$$

Substituting this into Eq. (2.35b) leads to:

$$[Ca^{2+}]_{eq} \cdot (2[Ca^{2+}]_{eq} + [F])^2 = \frac{K_1 K_c K_H}{K_2 \gamma_{Ca} \gamma_{HCO_3}^2} \cdot p_{CO_2} , \quad (2.48)$$

which can be also written by using Eq. (2.35c):

$$[Ca^{2+}]_{eq} \left([Ca^{2+}]_{eq} + \frac{[F]}{2}\right)^2 = ([Ca^{2+}]_{eq}^P)^3 \cdot \frac{\gamma_{Ca}^P (\gamma_{HCO_3}^P)^2}{\gamma_{Ca}(\gamma_{HCO_3})^2} . \quad (2.48a)$$

The index p refers to the corresponding equilibrium value of $[Ca^{2+}]$ and activity coefficients in the pure $H_2O-CO_2-CaCO_3$ system. If Ca^{2+} is added as a common ion, $[F] < 0$. The corresponding saturation curves are shown in Fig. 2.17. They are located above the saturation curve with $[F] = 0$. If the addition of Ca^{2+} as a common ion would not change the solubility of $CaCO_3$, then the new equilibrium curve would just be the sum of the two concentrations. This is shown for $[F] = -5 \times 10^{-3}$ as a dashed curve. The real saturation curve lies below this dashed line, thus showing that the calcite solubility is decreased.

The influence of foreign ions to calcite solubility can be classified into several independent mechanisms.

1. *Ionic strength effect*: If indifferent substances such as NaCl are added $[F] = 0$ because of charge balance. Ionic strength, however, is changed and therefore the activity coefficients become smaller. Thus, the change of ionic strength increases calcite solubility, an effect which is always operative, but is small in karst waters.

2. *Common-ion effect*: Addition of common ions into a pure calcite-saturated solution increases the ionic activity product and therefore the solution becomes supersaturated. Thus, the common-ion effect in any case decreases calcite solubility. This can be easily seen from Fig. 2.17. For $[F] > 0$, i.e. CO_3^{2-} added as a common ion, this is evident. For $[F] < 0$, i.e. when Ca^{2+} is added as common ion this can be seen by comparing the dashed upper curve with the curve resulting from $[F] = -5 \cdot 10^{-3}$. The dashed curve results by adding $2.5 \, mol\, l^{-1}$ of Ca^{2+} as common ion to the pure system $CaCO_3-H_2O-CO_2$ in equilibrium and assuming no common ion effect to exist. The curve with $F = -5 \times 10^{-3} \, mol \, l^{-1}$ results also from adding $2.5 \, mol\, l^{-1}$ of Ca^{2+} as a common ion but with the common ion effect present. The reduced calcite solubility is seen from the fact that it is situated below the dashed curve.

3. *Acid effect*: Adding an acid such as HCl to a saturated solution shifts the pH to a considerably lower value. This drastically decreases the activity of CO_3^{2-} (cf. Fig. 2.1) and therefore additional calcite is dissolved. As we can see from Eqs. (2.48a) and (2.42a) adding 2 mmol HCl to 1 litre solution is equivalent to adding 1 mmol $CaCl_2$ and leads to the same value of $[Ca^{2+}]_{eq}$. In both cases $[F] = -1 \, mmol\, l^{-1}$.

4. *Base effect*: If bases are added, the pH is shifted to higher values, thus increasing the activity of CO_3^{2-} (cf. Fig. 2.1). Therefore, the ion activity product is increased and calcite precipitates from the solution. Again, as in the case of the acid effect, addition of a base is equivalent to addition of CO_3^{2-} as a common ion provided the corresponding values of $[F]$ are equal.

5. *Ion-pair effect*: As we have discussed in section 2.1, Eq. (2.16), the presence of ion pairs, such as $CaHCO_3^+$, $CaHCO_3^0$, does not change the calculations in any significant amount in the case of natural karst water. In the presence of SO_4^{2-}, however, ion pairs $CaSO_4^0$ are generated to quite a considerable amount with the mass action law:

$$(Ca^{2+})(SO_4^{2-}) = K_7 \cdot (CaSO_4^0) . \tag{2.49}$$

At $20°C\, K_7 = 5 \times 10^{-4} \, mol\, l^{-1}$. Thus, at a concentration $[Ca^{2+}] = 2 \times 10^{-3} \, mol\, l^{-1}$ and $[SO_4^{2-}] = 5 \times 10^{-4} \, mol \, l^{-1}$, the amount of Ca^{2+} incorporated into ion pairs is about 10% of total Ca^{2+}. This reduces the ionic activity product and enhances calcite solubility. If we add $Me_{n_M}(SO_4^{2-})$ the concentration of Ca^{2+} in the solution is:

$$[Ca^{2+}] = [Ca^{2+}]^t - [CaSO_4^0] , \qquad (2.50)$$

where t refers to total Ca.

From the relaxed electroneutrality we have:

$$2[Ca^{2+}] + Z_M[Me] = [HCO_3^-] + 2[SO_4^{2-}] . \qquad (2.51)$$

Conservation of sulphur mass gives:

$$[SO_4^{2-}] + [CaSO_4^0] = [Me] \cdot \frac{1}{n_M} . \qquad (2.52)$$

Combining these equations yields:

$$[HCO_3^-] = 2[Ca^{2+}] + 2 \cdot \frac{[Ca^{2+}][SO_4^{2-}]\gamma_{Ca}\gamma_{SO_4}}{K_7} . \qquad (2.53)$$

Introduction into Eq. (2.35b) yields:

$$[Ca^{2+}]_{eq} = [Ca^{2+}]_{eq}^P \cdot \left(1 + \frac{[SO_4^{2-}] \cdot \gamma_{Ca}\gamma_{SO_4}}{K_7}\right)^{-2/3} \cdot \left(\frac{\gamma_{Ca}^P(\gamma_{HCO_3}^P)^2}{\gamma_{Ca}(\gamma_{HCO_3})^2}\right)^{1/3} . \qquad (2.54)$$

Since, however, we are interested in total dissolved Ca, expressing $[Ca^{2+}]$, using Eq. (2.50), (2.49) by $[Ca^{2+}]^t$ and by regarding the approximation we obtain:

$$[Ca]^t = [Ca^{2+}]_{eq}^P \cdot \left(1 + \frac{[SO_4^{2-}]\gamma_{Ca}\gamma_{SO_4}}{K_7}\right)^{1/3} \cdot \left(\frac{\gamma_{Ca}^P(\gamma_{HCO_3}^P)^2}{\gamma_{Ca}(\gamma_{HCO_3})^2}\right)^{1/3} . \qquad (2.55)$$

From this equation one immediately visualizes that calcite solubility is increased quite considerably in the presence of sulphates.

A similar effect exists in the presence of Mg^{2+} due to the reaction with HCO_3^- according to the mass action law:

$$(Mg^{2+})(HCO_3^-) = K_8 \cdot (MgHCO_3^+) . \qquad (2.56)$$

$K_8 \approx 0.1$ mol l^{-1}. This reaction reduces $[HCO_3^-]$, the result being a slight increase in calcite solubility in the order of 1%. The saturation value resulting from the presence of a substance containing Mg^{2+}, but not the common ion CO_3^{2-}, can be calculated as above by observing charge balance and Eq. (2.56):

$$[HCO_3^-] \cdot \left(1 + \frac{[Mg^{2+}] \cdot \gamma_{Mg}\gamma_{HCO_3}}{K_8\gamma_{MgHCO_3}}\right) = 2[Ca^{2+}] \qquad (2.57)$$

as

$$[Ca]_{eq}^t = [Ca^{2+}]_{eq}^P \cdot \left(1 + \frac{[Mg^{2+}]\gamma_{Mg}\gamma_{HCO_3}}{\gamma_{MgHCO_3}}\right)^{2/3} \cdot \left(\frac{\gamma_{Ca}^P(\gamma_{HCO_3}^P)^2}{\gamma_{Ca}(\gamma_{HCO_3})^2}\right)^{1/3} . \qquad (2.58)$$

All the mathematics of the evolution of chemical cmposition of a $CaCO_3$–H_2O–CO_2 system in the presence of foreign ions can be condensed into one computer program (TURBO-PASCAL). This program is given in Table 2.3.

Table 2.3. Computer program for the calculation of the chemistry of an $H_2O-CO_2-CaCO_3$ system including the influence of a foreign ion for the open, closed and intermediate system

```
PROGRAM equilibrium (Input,Output);

USES transcend;

CONST  K5=1.707E-4;

VAR acid,alpha,alphas,an,Ausdruck,B,base,Ca,Caend,cat,CO2I,CO3,Ctot,
    deltaCa,elneu,f,foreign,gammaCa,gammaCO3,gammaH,gammaHCO3,gammaOH,
    HCO3,HCO3I,Hplus,H2CO3S,H2CO3SI,ionicstrength,Izus,kappa1,kappa2,kappa3,
    kappa4,kappa4s,KH,Kriterium,KS,KW,KWs,K0,K1,K2,K6,Loga,OH,omega,PCO2,
    PCO2end,PHcenter,PHend,PHstart,PH1,R,rate,T,TC,Vgas,X,xcat,yan :REAL;
    Factor,K,Krit,nan,ncat,X1,X2,zacid,zan,zbase,zcat,zx,zy        :INTEGER;
    Q1,Q2,Q3,Q4,Q5,Q6,Q7,Q8                                       :CHAR;
    Drucker                                                       :INTERACTIVE;

PROCEDURE Input2;
begin
      write ('Common ion with CaCO3 ? <y/n> ');        readln (Q5);
      if  Q5  in  ['Y','y']  then begin
         write ('Cation  <y/n> ');                     readln (Q6);
         write ('Anion   <y/n> ');                     readln (Q7);
         if  Q6  in  ['Y','y']  then begin
            write ('Give concentration of cation :  '); readln (cat);
            write ('Give charge of cation (z) :  ');    readln (zcat);
            end               else cat:=0.0;
         if  Q7  in  ['Y','y']  then begin
            write ('Give concentration of anion :  '); readln (an);
            write ('Give charge of cation (z) : ');    readln (zan);
            end               else an:=0.0;
         end;
      write ('other foreign ions (no common ions with CaCO3) ? <y/n> ');
                                                       readln (Q8);
      if  Q8  in  ['Y','y']  then begin
         write ('Cations :    concentration = ');      readln (xcat);
         write (' ':16,'charge (z) = ');               readln (zx);
         write ('Anions :     concentration = ');      readln (yan);
         write (' ':16,'charge (z) = ');               readln (zy);
         end               else begin   zx:=0;
                                         zy:=0;
                                end;
end;
                    (*  ----------------  *)
```

Table 2.3. (*continued*)

```
PROCEDURE Input1;
begin
   write ('Give temperature (C)      ');
   readln (TC);
   write ('Give CO2 pressure (atm)  ');
   readln (PCO2);
   write ('Calculation of rates in steps of DeltaCa.  Give DeltaCa ',
         '(mol/l) :   ');
   readln (DeltaCa);
   writeln ('Give ratio of volume of gas to volume of solution.');
   write (' ':20,' V(gas) : V(solution) = ');
   readln (Vgas);
   writeln ('neglect OH- und CO3-- in the equation of electro neutrality ?');
   write ('(= approximation by a quadratic equation)  <y/n> : ');
   readln (Q1);
   writeln; write ('Foreign ions present ?  <y/n> ');
   readln (Q2);
   If Q2 in ['Y','y'] then begin
      write ('acid ? <y/n> ');        readln (Q3);
      if Q3 in ['Y','y'] then begin
         write ('concentration of acid (mol/l) :  '); readln (acid);
         write ('charge of acid (z) :  ');             readln (zacid);
         end                 else zacid:=0;
      write ('base ? <y/n> ');        readln (Q4);
      if Q4 in ['Y','y'] then begin
         write ('concentration of base (mol/l) :  '); readln (base);
         write ('charge of base (z) :  ');             readln (zbase);
         end                 else zbase:=0;
      input2;
   end;
end;
                     (* ----------------- *)
PROCEDURE Init;
begin
   rewrite (Drucker,'printer:');
   T    := TC + 273.16;
   R    := 0.082057;
   LOGA := Ln(10);
   KRIT := 0;
   Ca   := 1.0E-4;
   KS:= exp (-LOGA*(8.15087602 + TC*0.0136633623 - TC*TC*3.5812701E-5));
    X:= 108.3865 - 6919.53/T + 0.01985076*T - 40.45154*Log(T);
```

Table 2.3. (continued)

```
  KH:= exp (LOGA*(669365.0/(T*T)+X));
  KW:= exp (LOGA*(22.801 - 4787.3/T - 0.010365*T -7.1321*Log(T)));
  K6:= exp (LOGA*(-356.3094 + 21834.37/T -6.0919964E-2*T + 126.8339
           *Log(T) - 1684915.0/(T*T)));
  K2:= exp (LOGA*(-107.8871 + 5151.79/T - 0.03252849*T + 38.92561
           *Log(T) - 563713.9/(T*T)));
  K0:= K5 / K6;
  K1:= K5*K6 / (K5+K6);
   B:= 3.077 - 0.0146*TC;
   f:= 1 + Vgas / (( 1+1/K0) * KH * R * T );
  kappa1 := exp (LOGA*(0.198 - 444.0/T));
  kappa2 := exp (LOGA*(2.84 - 2177.0/T));
  kappa3 := exp (LOGA*(-5.86 - 317.0/T));
  kappa4s:= exp (LOGA*(-2.38 + 0.0252*TC));
  if Q1 in ['Y','y'] then factor:= 0
                      else factor:= 1;
  writeln (Drucker,Chr(14),' Open / Closed - mixed system');
  writeln (Drucker);
  writeln (Drucker,'   Calcium',' ':10,'Rate',' ':6,'satur.index',
                  ' ':6,'Hplus',' ':11,'CO3--');
  writeln (Drucker,' ':12,'HCO3-',' ':11,'H2CO3',' ':11,'C(total)');
  writeln (Drucker);
end;
                  (* --------------- *)
PROCEDURE activities;
VAR A1,B1,rootI:real;
begin
  A1 := 0.48809 + TC * 8.074E-4;
  B1 := 0.3241  + TC * 1.6E-4;
  if Krit = 0 then ionicstrength:= 3*Ca
           else ionicstrength:= 0.5 * (4*Ca + 4*CO3 + OH + Hplus + HCO3);
  If Q2 in ['Y','y'] then ionicstrength := ionicstrength + Izus;
  rootI    := sqrt (ionicstrength);

  gammaCa  := exp (-Loga*(A1*4*rootI / (1+B1*5*rootI) + 0.165*ionicstrength));
  gammaHCO3:= exp (-Loga* A1*  rootI / (1+B1*5.4*rootI));
  gammaOH  := exp (-Loga* A1*  rootI / (1+B1*3.5*rootI));
  gammaH   := exp (-Loga* A1*  rootI / (1 + B1*9*rootI));
  gammaCO3 := exp (-Loga* A1*4*rootI / (1+B1*5.4*rootI));

  KWs:= KW / (gammaH*gammaOH);
end;
                  (* --------------- *)
```

Table 2.3. (*continued*)

```
PROCEDURE Initialvalues;
begin
  CO2I    := KH * PCO2;
  H2CO3SI:= CO2I * (1 + 1/K0);
  HCO3I   := sqrt (H2CO3SI * K1);
  Alpha   := f * H2CO3SI + HCO3I;
  If Q2 in ['Y','y'] then begin  if zcat = 1  then  ncat:= 2
                                                else  ncat:= 1;
                                 if zan = 1   then  nan := 2
                                                else  nan := 1;
                                 AlphaS := cat/ncat - an/nan;
                                 Alpha:= Alpha + AlphaS;
                          end;
end;
                   (* --------------- *)
PROCEDURE PHvalue (PHvalue:real);
begin
  Hplus    := (exp (-Loga*PHvalue))/gammaH;
  Ausdruck:= Ca + Hplus + Foreign - Alpha - Factor*KWs/Hplus +
             (f * gammaH*gammaHCO3*Hplus/K1 - Factor*K2*gammaHCO3/
             (gammaH*gammaCO3*Hplus)) * (2*Alpha - Hplus - Foreign +
             Factor*KWs/Hplus) / (1 + 2*f*gammaH*gammaHCO3*Hplus/K1);
end;
                   (* --------------- *)
PROCEDURE dissolutionrate;
begin
  kappa4:= (exp (-B*Loga)) * exp((Ln(KH*(1+1/K0)/(H2CO3S)))*0.611);
  if (H2CO3S/KH) >= 0.05  then kappa4:= kappa4s;
  rate  := kappa1*gammaH*Hplus + kappa2*H2CO3S + kappa3 - kappa4*Ca*HCO3
           *gammaCa*gammaHCO3;
end;
                   (* --------------- *)
PROCEDURE iterations;
begin
  PHcenter:=0.0;
  repeat
   activities;
   PHstart := 0.0;
   PHend   :=12.0;
   PH1     := PHcenter;
   PHcenter:= 0.5 * (PHstart + PHend);
```

Table 2.3. (*continued*)

```
    repeat
       PHvalue (PHstart);
          if Ausdruck < 0 then    X1:=-1
                          else    X1:=1;
       PHvalue (PHcenter);
          if Ausdruck < 0 then    X2:=-1
                          else    X2:=1;
       if (X1+X2) = 0  then  begin    PHend    := PHcenter;
                                      PHcenter := 0.5 * (PHstart+PHend)
                            end
                       else  begin    PHstart  := PHcenter;
                                      PHcenter := 0.5 * (PHstart+PHend)
                            end
    until  (PHend - PHstart) <= 1E-3;

    Kriterium := abs ((PH1-PHcenter) / PHcenter);
    Krit      := Krit + 1;
    OH   := KWs / Hplus;
    HCO3 := (2*Alpha - Hplus - Foreign + Factor*OH ) / ( 1 +
            2*f*gammaH*gammaHCO3*Hplus/K1);
    CO3  := K2*HCO3*gammaHCO3 / (gammaCO3*gammaH*Hplus);
    H2CO3S:= Hplus*gammaH*HCO3*gammaHCO3/K1;
    CToT := HCO3 + CO3 + H2CO3S;

  until Kriterium <= 0.01;
end;
                (*  ---------------  *)
PROCEDURE print1;
begin
   writeln (Drucker);
   writeln (Drucker,'   concentrations at Ca saturation:');
   writeln (Drucker,' ':4,'pH-value: ',PHcenter:5:3,' ':10,'calcium: ',
                                                              Caend);
   writeln (Drucker,' ':9,'CO3: ',CO3,' ':8,'HCO3: ',HCO3);
   writeln (Drucker,'       PCO2 : ',PCO2end,' atm');
   writeln (Drucker);
   writeln (Drucker,'   constants:');
   writeln (Drucker,' ':7,'KH= ',KH,' ':7,'   KW= ',KW);
   writeln (Drucker,' ':7,'K0= ',K0,' ':8,'   K1= ',K1);
   writeln (Drucker,' ':7,'K2= ',K2,' ':6,'   K6= ',K6);
   writeln (Drucker,' ':7,'KS= ',KS,'    -LOG(KS)= ',-LOG(KS));
   writeln (Drucker,'   kappa1= ',kappa1,' ':7,'kappa2= ',kappa2);
```

Table 2.3. (*continued*)

```
    writeln (Drucker,'   kappa3= ',kappa3,' ':7,'kappa4= ',kappa4);
    writeln (Drucker);
    writeln (Drucker,'   activity coefficients:');
    writeln (Drucker,' ':7,'H+',' ':7,'OH-',' ':7,'HCO3-',' ':5,'CO3--',
            ' ':5,'Ca++',' ':5,'ionic strength');
end;
                     (* --------------- *)
PROCEDURE print4;
begin
    writeln (Drucker); writeln (Drucker);
    writeln (Drucker,' ':4,'CO2 pressure: ',PCO2:7:5,' atm',' ':5,
            'temperature: ',TC:5:2,' C');
    writeln (Drucker,' ':4,'ratio of volume of gas to volume of solution:');
    writeln (Drucker,' ':8,'Vg / Vs = ',Vgas:8:1);
    print1;
    writeln (Drucker,'   ',gammaH:8:5,'   ',gammaOH:8:5,'   ',gammaHCO3:8:5,
            '   ',gammaCO3:8:5,'   ',gammaCa:8:5,'    ',ionicstrength);
    writeln (Drucker);
    writeln (Drucker,'   dissolution rate at the end of the calculation: ',
                  rate);
    writeln (Drucker,'   electroneutrality: ',elneu);
    writeln (Drucker,'   saturation index: ',omega:7:4,'       ( IAP= ',
                  omega*KS,' )');
    If Q1 in ['Y','y'] then  writeln (Drucker,'approximation by quadratic',
                          ' equation = OH- and CO3-- omitted in electro',
                          'neutrality');
end;
                     (* --------------- *)
PROCEDURE print2;
begin
    print4;
    If  Q2  in  ['Y','y']  then begin
        writeln (Drucker);
        writeln (Drucker,'   concentration and charge of foreign ions:');
        if Q3 in ['Y','y'] then writeln (Drucker,' ':7,'acid: ',acid,' mol/l',
                              ' ':6,'charge: ',zacid);
        if Q4 in ['Y','y'] then writeln (Drucker,' ':7,'base: ',base,' mol/l',
                              ' ':6,'charge: ',zbase);
        if Q6 in ['Y','y'] then writeln (Drucker,' ':7,'common ion; cation: ',
                              cat,' mol/l',' ':4,'charge: ',zcat);
        if Q7 in ['Y','y'] then writeln (Drucker,' ':7,'common ion; anion: ',
                              an,' mol/l',' ':4,'charge: ',zan);
```

Table 2.3. (*continued*)

```
        if Q8 in ['Y','y'] then begin
                writeln (Drucker,' ':4,'no common ion with CaCO3 :');
                writeln (Drucker,' ':7,'cation: ',xcat,' mol/l',' ':5,
                        'anion: ',yan,' mol/l');
        end;
        writeln (Drucker,'  calculated parameters:',' ':8,'I (additional)= ',
                Izus);
        writeln (Drucker,'  Charge F = ',foreign,' ':4,'alpha (additional)= '
                ,alphaS);
    end;
end;
                    (* --------------- *)
PROCEDURE print3;
begin
  writeln (Drucker,' ',Ca:12,' ':3,Rate:12,' ':4,Omega:8:5,' ':5
            ,Hplus:12,' ':4,Co3:12);
  writeln (Drucker,' ':10,HCO3:12,' ':4,H2CO3S:12,' ':2,CToT:12);
end;
                    (* --- main  program  --- *)
begin
  Input1;      Init;        Initialvalues;
  if  Q2  in  ['Y','y']  then  begin
      Izus := 0.5 * (zacid*zacid*acid + zbase*zbase*base + zcat*zcat*cat +
                    zan*zan*an + zx*zx*xcat + zy*zy*yan);
      Foreign := zbase*base + zcat*cat + zx*xcat - zacid*acid - zan*an
                    - zy*yan;
                            end
                        else  Foreign:= 0.0;
    K:= 0;
    repeat
      Iterations;
      dissolutionrate;
      Omega := GammaCa * GammaCo3 * Ca * CO3 / KS;
      Caend := Ca;
      print3;
      if Omega  >= 1.0  then  begin    deltaCa:= 0.5*deltaCa;
                                       Ca     := Ca - deltaCa;
                                       K:=1;
                            end
                        else  begin    if K = 1 then  deltaCa:= 0.5*deltaCa;
                                       Ca:= Ca+deltaCa;
                            end;
```

Table 2.3. (*continued*)

```
      Krit:=0;
   until abs( Omega - 1.0 )   <= 1E-2;
      elneu    := Hplus + 2*Caend - OH - HCO3 - 2*CO3 + Foreign;
      PCO2end := H2CO3S / (KH * (1 + 1 / KO));
      print2;
      close (Drucker);
 end.
```

It asks in an interactive way for all the relevant input parameters such as $\varkappa$, p_{CO_2}, temperature, presence of common ions, their concentrations, other indifferent ions, their corresponding charges and concentrations, presence of bases and acids and the corresponding parameters. The chemical composition of the system can then be computed as a function of Ca^{2+} concentration from 0 to saturation in steps of ΔCa^{2+}. The output of the program lists dissolution rates due to the PWP equation (cf. Chap. 6), saturation state Ω, $[H_2CO_3]$, $[HCO_3^-]$, $[CO_3^{2-}]$, $[C_T]$ and $[H^+]$.

Note that the ion-pair effect is not included in the computer program listed above.

Attention: Due to the expression Ausdruck in the procedure PH-value numerical problems may result for very large $f \to \infty$. Therefore the ratio of volume of gas to volume of solution $V_g/V_e = \varkappa$ (cf. Eq. (2.27)) should be limited to 10^5, which is an upper value to calculate the limiting case of the open system with high precision.

3 Mass Transport

Mass exchange between a liquid phase and a solid, such as in dissolution and precipitation of calcite, requires some kind of transport mechanism, which in the case of dissolution removes the ionic species released from the solid surface into the bulk of the solution, and vice versa in the case of precipitation transports the species from the bulk to the surface, where they are built into the crystal lattice.

If such transport mechanisms did not exist, dissolution would not be possible, since close to the surface there would be an increasing concentration of the ionic species dissolved, until the ionic activity product reaches the value of the solubility product and dissolution would stop. Correspondingly in the case of precipitation, a depletion layer close to the solid surface would stop further deposition.

There are two ways to establish mass transport: The first way is by convection or more generally by flow of the liquid, which displaces a parcel of the solution due to the influence of external forces, such as gravity, away from the solid surface, thus removing dissolved species. The other is diffusion, which results from a statistical movement of the dissolved species. It may be driven by Brownian motion, or by other random motions such as eddies in turbulent flow or mechanical motions as they occur statistically when a solution is driven through porous media.

In this chapter we will discuss the principles of diffusion and will develop the equations of mass transport including both diffusion and convection.

3.1 The Random Walk Problem as Principle for Diffusional Mass Transport

We consider a particle, which has been removed from the solid, and is positioned at its surface. Due to the thermal motion, the particle is kicked irregularly by the surrounding molecules. Each kick displaces the particle by an average distance l into a direction selected by chance with equal probability. If the particle hits the surface of the solid, it is reflected. We further assume that the average time between two kicks is τ. The question is now, how far in average does the particle move away from the surface after a time t? From Fig. 3.1 it is quite clear that from this, mass transport of particles away from the surface results, since the surface cannot be crossed by the particles, and therefore one direction of the motion is preferred. To discuss this problem from a more general view, we consider the situation of Fig. 3.2, where we assume that a particle in the bulk of the solution can move freely in any direction. Let us assume that after $N - 1$ kicks the particle has moved to the position

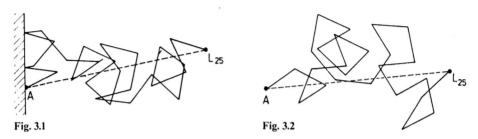

Fig. 3.1   **Fig. 3.2**

Fig. 3.1. Transport of a particle released from a solid surface by random walk. L_{25} is the distance after 25 steps

Fig. 3.2. Random walk of a particle in the bulk of the solution

$(N-1)$ with coordinates $(x_{N-1}, y_{N-1}, z_{N-1})$. After one more kick, which displaces the particle by the distance l, it has arrived at a new position (N) with coordinates $(x_{N-1} + l_x, y_{N-1} + l_y, z_{N-1} + l_z)$. The new distance L is then:

$$L_N^2 = (x_{N-1}^2 + y_{N-1}^2 + z_{N-1}^2) + (l_x^2 + l_y^2 + l_z^2) + 2(l_x x_{N-1} + l_y y_{N-1} + l_z z_{N-1})$$

$$= L_{N-1}^2 + l^2 + 2(l_x x_{N-1} + l_y y_{N-1} + l_z z_{N-1}). \tag{3.1}$$

Now we put the particle back to its position $(N-1)$ and repeat the process, considering that each direction of displacement by the kick is equally probable. Doing this may times and taking the average of all the results of the distances L obtained, we get:

$$\overline{L_N^2} = \overline{L_{N-1}^2} + l^2. \tag{3.2}$$

The last term in Eq. (3.1) vanishes since for each value of l_x there is an equally probable event with the negative value $(-l_x)$. This is also true for l_y and l_z, since all directions occur with equal probability.

It is obvious that for $N = 1$, i.e. after the first kick, the average distance $\overline{L_1^2}$ is equal to l^2. Thus, after two steps $\overline{L_2^2} = l^2 + \overline{L_1^2} = 2 \cdot l^2$. Repeating this up to N, we obtain:

$$\overline{L_N^2} = Nl^2, \tag{3.3}$$

a particularly simple result.

If the average time between two kicks is τ, then after N kicks the time, $t = N\tau$ has elapsed. Therefore, the average distance after that time is:

$$\overline{L_t^2} = \frac{l^2}{\tau} \cdot t. \tag{3.4}$$

The average velocity v of the particle is:

$$v = l/\tau \tag{3.5}$$

and we finally obtain

$$\overline{L_t^2} = v \cdot l \cdot t. \tag{3.6}$$

Thus, the square of the average distance is proportional to the time. The constant $v \cdot l$ contains the statistical parameters of average velocity and the average distance between two kicks.

So far we have observed only one particle, but we have repeated the process many times. We might further assume that there are many particles concentrated in a small volume around the origin of the coordinate system at time zero, and that there are no particles elsewhere. The question now to be answered is: How many particles will be found at exactly the distance L after time t? The answer to this question is: There will be none, since there is no chance that a series of steps sums up to a particular number out of an infinite amount. Therefore, we have to ask for the average number of particles N found at a distance between L and L + dL. For the one-dimensional case one finds the number N of particles at the distance between x and x + dx from the origin as:

$$N = c(x, t)\,dx = \frac{n_0}{\sqrt{\dfrac{4\pi vl}{3}t}}\, \exp\left(-\frac{3x^2}{4vlt}\right) dx \,. \tag{3.7}$$

This is a Gaussian distribution, which is given here without proof. Since it represents a number divided by a volume, it is proportional to the average concentration of the particles observed. Of course, these numbers are average numbers and fluctuations are present. In real chemical solutions, however, the particle numbers are so high that the fluctuations are of no concern and the interpretation as concentrations is adequate. Figure 3.3 shows the concentrations as they have evolved after increasing times.

It is clearly seen that the initially well-localized concentration of particles is spread out as time increases. The half-width of the corresponding distribution is given by:

$$x_{1/2}^2 = 0.924 \cdot v \cdot l \cdot t \,. \tag{3.8}$$

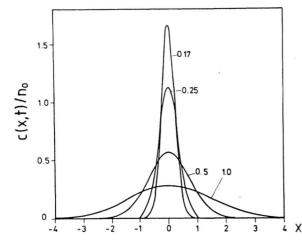

Fig. 3.3. Concentrations evolving from a point source at $t = 0$ for various times. The distribution of particles spreads out with increasing time. *Numbers* on curves are values of $\sqrt{Dt}$.

$$D = \frac{v \cdot l}{3} \text{ (cf. also Eq. 3.16)}$$

From this figure it is also easy to visualize that random walk establishes a radial mass transfer from the initial centre into the bulk of the fluid. The amount of this transfer is related to the product $v \cdot l$.

3.2 The Laws of Diffusion

To calculate the flux, i.e. the number of particles per unit area and time, crossing a given surface, we imagine a unit surface perpendicular to the x-axis in Fig. 3.4. During the time τ, all the particles with average velocity $v_x = v$, which are within the distance l to the left of the plane, can pass this plane from left to right and correspondingly, all the particles within distance l to the right of the plane with velocity $v = -v_x$ can pass from right to left.

If $c(x)$ is the concentration of particles, then the total number of particles passing through the plane at $x = 0$ is given by the difference of these two numbers. From the right we have:

$$N_r = \tfrac{1}{6}c(x + l)\cdot l . \tag{3.9a}$$

N_r is the number of particles passing during time τ and $c(x + l)$ is the concentration at position $x + l$. The factor $\tfrac{1}{6}$ results from the fact that on the average, $\tfrac{1}{6}$ of the particles has a velocity along the positive x-axis, since all directions are equivalent. Correspondingly, we have for the number N_l from the left:

$$N_l = \tfrac{1}{6}c(x - l)\cdot l . \tag{3.9b}$$

The total flux, i.e. the number of particles crossing the unit area per time unit is:

$$\frac{(N_l - N_r)}{\tau} = F_x = \frac{1}{6}\cdot\frac{l}{\tau}[c(x - l) - c(x + l)] . \tag{3.10a}$$

Since l is a very small distance, this can be expanded into a Taylor series:

$$F_x = -\frac{1}{3}\cdot\frac{l^2}{\tau}\cdot\frac{\partial c(x)}{\partial x} = -D\frac{\partial c(x)}{\partial x} . \tag{3.10b}$$

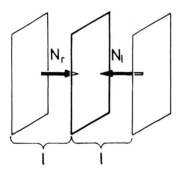

Fig. 3.4. The flux of particles through the middle area results from particles within a distance of the average step length l

This is Fick's first law of diffusion, stating that the mass flux is driven by local changes in concentration. The diffusion coefficient D is related to the statistical parameters:

$$D = \frac{1}{3} \cdot \frac{1}{\tau} = \frac{1}{3} v \cdot l \,, \tag{3.10c}$$

which also determine the random walk distance after time t in Eq. (3.6).

In an analogous way we can calculate the flux in the y and z direction, thus obtaining as a general result:

$$F = -D \,\mathrm{grad}\, c(x, y, z) \tag{3.10d}$$

for the three-dimensional case of a stagnant medium.

If, however, the medium flows with velocity v we have to add a flux resulting from advective transport. Thus, the total flux is given:

$$F = -D \,\mathrm{grad}\, c(x, y, z) + v \cdot c(x, y, z) \,. \tag{3.11}$$

Once we know the fluxes entering into a volume through its surfaces we can derive an equation which gives the spatial and time dependence of the concentration. In Fig. 3.5 we regard a small volume ΔV represented as a cube with Δx, Δy, Δz. The change of concentration inside this cube results from the fluxes into the cube and those out of it and from the rate of production of particles inside the cube:

$$\frac{\partial c}{\partial t} \Delta x \Delta y \Delta z = \Delta y \Delta z \cdot \left(-D \left[\frac{\partial c(x)}{\partial x} - \frac{\partial c(x + \Delta x)}{\partial x} \right] \right)$$

$$+ \Delta x \Delta z \left(-D \left[\frac{\partial c(y)}{\partial y} - \frac{\partial c(y + \Delta y)}{\partial y} \right] \right)$$

$$+ \Delta y \Delta x \left(-D \left[\frac{\partial c(z)}{\partial z} - \frac{\partial c(z + \Delta z)}{\partial z} \right] \right)$$

$$+ v_x \cdot [c(x) - c(x + \Delta x)] \Delta y \Delta z$$

$$+ v_y [c(y) - c(y + \Delta y) \Delta x \cdot \Delta z$$

$$+ v_z \cdot [c(z) - c(z + \Delta z)] \Delta x \cdot \Delta y + R \Delta x \Delta y \Delta z \,. \tag{3.12}$$

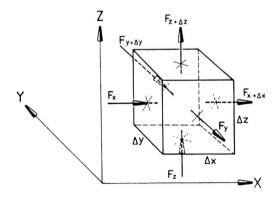

Fig. 3.5. Control volume $\Delta V = \Delta x \Delta y \Delta z$ and fluxes in and out of it

The left side of the equation represents the change of the particle number in the volume $\Delta x \Delta y \Delta z = \Delta V$ in time t. The first term in the brackets on the right side represents the difference of the particle flow into the volume and that out of it via diffusion. The terms containing v_x, v_y and v_z result from convective flow of the liquid into the faces of the cube. R represents a production rate of particles in the volume due to chemical reactions. The change of particle number in the volume ΔV during a time Δt, represented on the left side of the equation, has to be equal to the total number of particles flowing into that volume during time minus the number of particles flowing out of the volume, plus the amount produced by the chemical reaction, represented by R. Dividing Eq. (3.12) by $\Delta x \Delta y \Delta z$ and replacing the differences by the derivatives, we obtain Fick's second equation:

$$\frac{\partial c(x, y, z, t)}{\partial t} + v \operatorname{grad} c(x, y, z, t) = D \Delta c(x, y, z, t) + R . \tag{3.13}$$

3.2.1. Some Basic Solutions

In karst areas we encounter dissolution processes of limestone under a variety of boundary conditions. Dissolution occurs in water-filled joints of very narrow aperture as well as in situations where dissolved material diffuses into large water bodies, e.g. in big rooms underground.

In all these cases the diffusional transport of dissolved ionic species into the bulk of the liquid is one of the processes determining the total dissolution rate. It is therefore necessary to know the behaviour of diffusion of these situations in time and space. Therefore, in this section we discuss some basic solutions of Eq. (3.13) and their properties. For simplification, the discussion is restricted to solutions of Eq. (3.13), in its one-dimensional form and to the case of a stagnant solvent without any chemical reactions. Therefore, $v = 0$ and $R = 0$. We have therefore to deal with:

$$\frac{\partial c(x, t)}{\partial t} = D \frac{\partial^2 c(x, t)}{\partial x^2} . \tag{3.14}$$

There is ample literature dealing with solutions of Eq. (3.14) (Carslaw and Jaeger 1959, Crank 1975, Luikov 1968). We therefore will not give derivations of the solutions, but will mainly discuss their properties.

3.2.1.1 Point Source

We assume that at time $t = 0$ the concentration $c(t, x)$ of a diffusion species is limited to an infinitely small space dx at $x = 0$. Thus, we have with n_0 equal to the total number of particles:

$$c(0, x) = \begin{cases} n_0/dx & \text{for} \quad -\frac{dx}{2} \le x \le \frac{dx}{2} \\ 0 & \text{elsewhere .} \end{cases} \tag{3.15}$$

The solution for times $t > 0$ is given by a Gauss distribution:

$$c(x, t) = \frac{n_0}{\sqrt{4\pi Dt}} \cdot \exp\left(-\frac{x^2}{4Dt}\right) \tag{3.16}$$

Figure 3.3 illustrates this situation, which was also obtained from the random walk problem (cf. Eq. 3.7). At $t = 0$ we have a sharply defined concentration close to $x = 0$, which spreads out as time increases. The total area of each concentration distribution represents the total number of particles. Since this number is conserved, it is equal to the area of the initial rectangular distribution. The half-width of each distribution is given by:

$$x_{1/2}^2 = 2.76 \cdot D \cdot t. \tag{3.17}$$

This is related to the average distance L_t the particles have moved away from the origin. Comparing this to the results of our random walk consideration in section 3.1, we see that the solution of the equation of diffusion (Eq. 3.14) exactly reflects the results of random walk.

3.2.1.2 Diffusion into a Semi-Infinite Body

We assume a large plane wall of solid bordered by an infinite body of water to its right. At $t = 0$ the concentration of the dissolved species is c_0 at the border, at $x = 0$ and zero elsewhere. During the entire process of dissolution, we assume the concentration to be constant $c = c_0$ at $x = 0$. This is the situation, which occurs when chemical surface reaction rates are so fast that equilibrium at the surface between the solid and the species in the solution is achieved very quickly. The solution $c(x, t)$ of Eq. (3.14) is given by:

$$c(x, t) = c_0 \left[1 - \text{erf}\left(\frac{x}{2\sqrt{Dt}}\right) \right], \tag{3.18}$$

where the Gauss error function $\text{erf}(U)$ is given by:

$$\text{erf}(u) = \frac{2}{\sqrt{\pi}} \cdot \int_0^u \exp(-u^2) du. \tag{3.18a}$$

Figure 3.6 shows the concentration for several times $t > 0$. The distance $x_{1/2}$, where the concentration has dropped to $c_0/2$, increases with time and is given by:

$$x_{1/2} = \sqrt{D \cdot t} \tag{3.19}$$

which again reflects the behaviour of random walk.

To find the mass flux transported from the wall into the liquid according to Fick's first law, Eq. (3.10), we have to calculate the first derivative of $c(x, t)$ with respect to x at $x = 0$.

$$F = -D\frac{\partial c(x, t)}{\partial x/x = 0} = \frac{Dc_0}{\sqrt{\pi Dt}}. \tag{3.20}$$

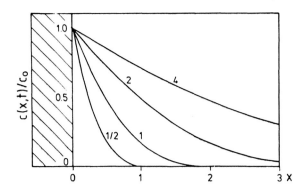

Fig. 3.6. Diffusion into a semi-infinite body of solution: concentrations evolving in time. At $t \geq 0$ the concentration at the wall is constant c_0. At $t = 0$ the concentration for $x > 0$ is 0. *Numbers* on curves give values of $2\sqrt{Dt}$

This shows that the flux from the wall decreases with time. The reason for this is that the concentration close to the wall increases with time. From the view of random walk now the number of particles moving towards the wall increases with increasing concentration close to the wall and compensates for the particles moving into the bulk. This reduces the total flux into the bulk.

3.2.1.3   Diffusion in Confined Conduits

The case in which water flows in joints in the saturated zone can be modelled by water flowing between two parallel planes, constituted by the limestone rock. Their distance is $2 \cdot a$. We assume that at $t = 0$ and for all times later, the concentration of dissolved species at $x = \pm a$ is in equilibrium with the wall, $c(\pm a, 0) = c_{eq}$. Elsewhere it is at a constant level at $t = 0$; $c(x, 0) = c_0$. The concentration $c(x, t)$ as it develops in time can be given as a series:

$$\frac{c(x, t) - c_0}{c_{eq} - c_0} = 1 - \sum_{n=1}^{\infty} \frac{2(-1)^{n+1}}{\mu_n} \cos\left(\mu_n \frac{x}{a}\right) \cdot \exp\left(-\mu_n^2 \frac{D \cdot t}{a^2}\right);$$

$$\mu_n = \frac{2n - 1}{2} \cdot \pi. \tag{3.21}$$

It is important to note that for times $t > 0.4 \cdot a^2/D$ only the first term in the sum contributes, since the exponentials for $n > 1$ drop off rapidly in comparison. For times $t > 4 \cdot a^2/D$ the system has come to an equilibrium with constant concentration $c(x, t) = c_{eq}$ everywhere. Figure 3.7 shows the concentration for various times given in multiples of decay time T_d defined below. This is in contrast to the previous cases, where the solid is bordered by an infinite body of water and a stationary state cannot be achieved. In the case of water bodies with finite dimensions, however, the system approaches equilibrium with the exponential decay time T_d, which in any case is proportional to the square of a length, characteristic for the extension of the water body, and is inversely proportional to the diffusion coefficient. In this case:

$$T_d = \frac{4a^2}{\pi^2 D}. \tag{3.22}$$

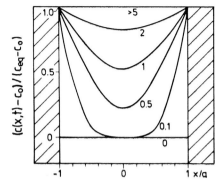

Fig. 3.7. Diffusion into a solution confined by two parallel planes with distance 2a. For $t \geq 0$ the concentration at the walls, $x = \pm a$, is constant c_{eq}. At $t = 0$ the concentration in the solution is c_0. The *numbers* on the curves give the time in units of $T_d = 4a^2/\pi^2 D$

For $t > 0.4a^2/D$ the mass flux from the walls is given by the derivative of $c(x, t)$:

$$F(\pm a) = \pm(c_{eq} - c_0) \cdot \frac{2D}{a} \exp\left(-\frac{\pi^2 Dt}{4a^2}\right). \tag{3.23}$$

Note that F is a vector directed away from the wall. With increasing dimension of the waterbody, diffusional mass transport decreases.

Therefore, also in cases where mass transport is controlled by several competing processes, such as chemical kinetics on the surface and in the solution, with increasing dimension of the waterbody, transport by diffusion will be lowered. On the other hand, since chemical kinetics are independent of the dimension of the waterbody, there always will be, with increasing dimension, a limit beyond which diffusion becomes the slowest process and controls mass transport entirely.

So far we have discussed boundary conditions, where the concentration at the walls is at equilibrium with the dissolving solid. Under many geological conditions of dissolution of minerals there are different boundary conditions. If the reaction rate of a dissolving mineral depends by some rate law on the chemical composition of the solution, a flux is prescribed at the solid surface. In the case of only slowly changing concentrations in the solution, this flux remains practically constant in a pseudostationary state. This is exactly what happens in the dissolution of limestone (cf. Chap. 7).

Therefore, the boundary conditions are:

$$\frac{\partial c}{\partial x/\pm a} = \pm \frac{F}{D}, \tag{3.24}$$

where the value F prescribes the flux from the walls.

Note that instead of concentration, now its derivative with respect to x is prescribed at $x = \pm a$. With this boundary condition the solution reads:

$$c(x, t) - c_0 = \frac{F}{a} \cdot t + \frac{F \cdot a}{D} \cdot \frac{3x^2 - a^2}{6a^2} - \frac{F \cdot a}{D} \sum_{n=1}^{\infty} \frac{2 \cdot (-1)^n}{\mu_n^2} \cdot \cos\left(\frac{\mu_n x}{a}\right)$$

$$\cdot \exp\left(-\frac{\mu_n^2 D \cdot t}{a^2}\right) \tag{3.25}$$

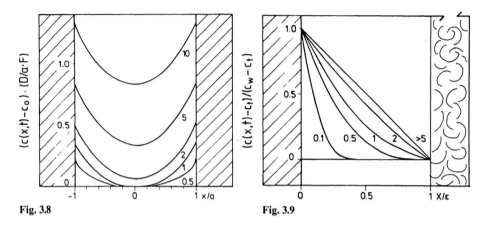

Fig. 3.8

Fig. 3.9

Fig. 3.8. Diffusion into a solution confined by two parallel planes with distance $2a$: concentration evolving in time. Flux F from the walls is constant. At $t = 0$ the initial concentration is c_0. *Numbers* on curves give the time in units of $T_d = a^2/\pi^2 D$

Fig. 3.9. Diffusion through a layer with thickness ε: concentration evolving in time. At $x = 0$ the concentration is c_w and at $x = \varepsilon$ it is c_t for $t \geq 0$. Initially at $t = 0$ the concentration in the layer is c_t. *Numbers* on the curves give the time in units of $T_d = \varepsilon^2/\pi^2 D$

where

$$\mu_n = n\pi, \qquad T_d = \frac{a^2}{\pi^2 D}. \tag{3.25a}$$

The first term in Eq. (3.25) results from the fact that due to the constant flux from the wall, the concentration increases linearly in time. The second term denotes a stationary parabolic concentration profile, which is established after approximately $t > 4T_d$. T_d is the decay time of the time-dependent third term. Figure 3.8 illustrates the concentration distributions for various times given in multiples of T_d.

To complete the series of important examples, we discuss finally a very common case.

If water flows turbulently on limestone, there is a small layer of liquid between the surface of the rock and the turbulent core, where mass transport is effected by molecular diffusion. In the turbulent core, however, due to the chaotic movement of water parcels there is perfect mixing between them and the concentration is constant with respect to space.

Mass transport is therefore determined by molecular diffusion through a layer of thickness ε, with prescribed concentrations c_w at the wall and c_t at the boundary with the turbulent core. The concentration $c(x, t)$ is then given by:

$$\frac{c(x, t) - c_t}{c_w - c_t} = 1 - \frac{x}{\varepsilon} + \sum_{n=1}^{\infty} \frac{2 \cdot (-1)^{n+1}}{\mu_n} \sin\left(\mu_n \frac{x - \varepsilon}{\varepsilon}\right) \cdot \exp\left(-\mu_n^2 \frac{Dt}{\varepsilon^2}\right);$$

$$\mu_n = n\pi; \qquad T_d = \frac{\varepsilon^2}{\pi^2 D}. \tag{3.26}$$

Thus, a linear concentration profile is established which is approached with the decay time T_d. This stationary profile is given by:

$$c(x, \infty) = -\frac{x}{\varepsilon}(c_w - c_t) + c_w .\tag{3.26a}$$

Figure 3.9 shows the concentrations in the layer for various times in multiples of T_d. The flux transported through the layer in the stationary state is:

$$F = \frac{D}{\varepsilon}(c_w - c_t) .\tag{3.27}$$

3.3   Diffusive and Advective Mass Transport

It is most common in geological processes that dissolution of minerals occurs in moving water bodies. This can happen in laminar flow in narrow joints and partings of the rock. If the dimensions of flow conduits are in the order of a few centimetres, usually turbulent flow occurs. Therefore, to describe mass transport, one has to consider also the terms of advective mass transport in Eq. (3.13):

$$\frac{\partial c}{\partial t} + v\,\mathrm{grad}\,c = D\Delta c .\tag{3.13}$$

In the case of laminar or turbulent flow, one usually replaces v, which depends on the spatial coordinates by its average value, thus assuming plug flow with constant velocity v.

It is easy to obtain the solution to this equation, if the solution in a stagnant medium, i.e. $v = 0$, is known. To illustrate this, we observe the solution from a frame of reference, which is fixed to a small volume of the liquid, thus moving with velocity v relative to a fixed frame. To visualize this, we may imagine a river flowing with velocity v. Into this river we inject, as a point source, a given amount of a dye. For an observer moving with a boat, which drifts down the river with the speed v, the dye spreads according to the stagnant solution of Eq. (3.16). An observer standing on the bank of the river, however, sees the same spreading but simultaneously the dye moves away from him with velocity v.

In general, the solution in the fixed frame of reference is related to that in a stagnant medium, i.e. $v = 0$, by replacing the spatial coordinates r in the stagnant solution by:

$$r' = r - vt .\tag{3.28}$$

Thus, the spreading of the point source and its propagation in the river for the one-dimensional case can be directly obtained from Eq. (3.16) as:

$$c(x, t) = \frac{n_0}{\sqrt{4\pi Dt}} \cdot \exp\left(-\frac{(x - vt)^2}{4Dt}\right) .\tag{3.29}$$

Thus, the advective term in Eq. (3.13) just describes this coordinate transformation, provided v is a constant everywhere.

3.3.1   Diffusion in Turbulent Flow

In laminar flow each small element of volume dV moves on a streamline during its propagation in the flow and its velocity can be predicted at any moment. There is no interference with other elements of volume, and accordingly no mixing of the liquids contained in it, into neighbouring elements (cf. Chap. 5). Mass transport between two neighbouring elements is only possible by molecular diffusion.

In turbulent flow, however, large statistical fluctuations of velocity occur at any point. There are large variations of the velocities both parallel and perpendicular to the average flow velocities. Due to their stochastic character they cannot be predicted, i.e. each particle moves on a chaotic path. These fluctuations cause mixture between small volume elements which are sufficiently close to each other. The average distance a volume element can travel until losing its individuality by mixing with other volume elements, is called mixing length.

Let us now consider a specific particle located in such a volume element, for instance a cation dissolved from a mineral. This cation moves with the volume element in average a distance L on a chaotic pathway which results from the fluctuations of velocity. The average velocity of this statistical motion is U. After the particle has moved this distance L, a new volume element around the cation can be defined, which is driven with average velocity U in another direction for the average distance L. This is a random walk process. It is superposed to the random walk process of molecular diffusion, which occurs on a much smaller spatial scale. Therefore, turbulence increases mass transport by diffusion and as a consequence the constant of diffusion in Eq. (3.23) has to be increased according to Eq. (3.10a):

$$D = \tfrac{1}{3} v \cdot l + \tfrac{1}{3} U \cdot L = D_m + D_e \, . \tag{3.30}$$

Here, the first term gives the contribution due to molecular diffusion and the second term, resulting from turbulent diffusion, is called eddy diffusion. Since the scale of L is larger than l by many orders of magnitude, D_e also exceeds D_m by many orders of magnitude. Estimations of D_e for turbulent flow in pipes are given by Skelland (1974) and Bird et al. (1960). The fact that mass transport is critically determined by the flow conditions of the solvent is of utmost importance in geological processes. Therefore, consideration of flow conditions is an absolute must, when regarding dissolution or precipitation processes in real cases.

3.3.2   Hydrodynamic Dispersion

Water flowing in porous media, e.g. sand or porous rocks, on a macroscopic scale follows streamlines, which can be derived from the hydraulic heads in the aquifer according to Darcy's law (cf. Chap. 5). On a microscopic scale, however, due to the statistical structure of the media, the individual particles follow a tortuous pathway through the interstices between the grains. Thus, in a situation, where on a macroscopic scale, flow velocity is uniform everywhere, on a microscopic scale a water particle follows a random pathway through the labyrinth of open spaces between the grains with varying velocities. If a dye is injected into such an aquifer at a given point for a short time, the dye molecules are transported by advection along the streamline and simultaneously, due to the random motion of the flow on a microscopic scale, they are dispersed and spread out. Experiments show that the amount of spreading, transverse to the flow, is different from that longitudinal to it. This effect of spreading is called hydrodynamic dispersion and is discussed in detail in Bear (1972).

Three mechanisms are operative in hydrodynamic dispersion. The first occurs in the individual pore channels, where flow is laminar in most cases. In these channels the molecules travelling in the centre have higher velocities than those close to the walls. This induces a spread longitudinal to the flow. A second process contributing mainly to longitudinal dispersion is due to the fact that different pore channels are of differing width and therefore flow velocities vary accordingly. The third mechanism comes from the branching of the pore channels. This resembles most closely the random walk, since each molecule travels in average freely a way along the channel until by branching, it is forced statistically in a new direction. This mechanism is the main contribution to transverse dispersion.

Figure 3.10a illustrates the porous medium as a channel network with statistically distributed directions, widths and lengths of the channels. The random path taken by an individual molecule on its travel through the medium is also shown. Figure 3.10b shows how velocity distribution in one individual channel changes due to a varying cross-section. These velocity distributions also cause dispersion, since particles in the middle of the channel are transported faster than those at the edges.

The coefficient of diffusion due to transverse dispersion is related to the parameters of random walk, the average channel length and the average velocity v. A length characteristic for the porous medium is the average grain diameter d. Thus,

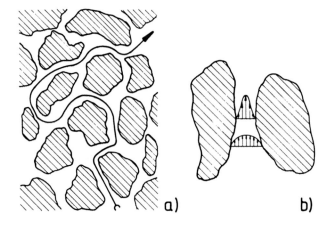

Fig. 3.10. a Random walk in a porous medium; **b** distribution of velocity in a single channel between two grains

with this length representing the average distance in random walk, the diffusion constant due to transverse dispersion is written as:

$$D_{dis}^{trans} = D_m \left(1 + \frac{v \cdot D}{D_m}\right) = D_m(1 + P_e).$$ (3.31)

This relation is found experimentally to be true for Peclet numbers P_e between 1 and 10. The Peclet number is given by $v \cdot d/D$ and compares transverse coefficient diffusion to molecular diffusion coefficient (Bear 1972).

Longitudinal dispersion is related to transverse dispersion for $Pe > 0.5$ by an experimentally determined relation (Blackwell et al. 1959):

$$D_{dis}^{long} = 8.8 \, D_{dis}^{trans} \cdot P_e^{1.17}.$$ (3.31a)

So far we have discussed mass transport of an injected substance. But hydrodynamic dispersion also governs mass transport in the case of dissolution of minerals from the grains composing the porous medium.

If water flows in an individual pore channel of length l and width w, dissolving material from the grains, a concentration gradient perpendicular to the pore walls builds up within a time T_d. This time is given, depending on the boundary conditions of dissolution, either by Eq. (3.22) or (3.25):

$$T_d \approx \frac{w^2}{\pi^2 D_m}.$$ (3.32)

This concentration gradient builds up due to molecular diffusion in an individual parcel of liquid, the residence time of which in the channel is $t = l/v$. After this time the solution leaves the channel and mixing occurs. The ratio T_d/t now gives the number of mixing processes, which occur due to confluences from different channels at each interconnecting point within time t. For $T_d/t \gg 1$ mixing destroys the gradients forming in the pores by molecular diffusion. On a macroscopic scale, concentration gradients are then determined by mechanical dispersion with high coefficient of diffusion D_e. In the case $T_d/t \ll 1$ dispersion can be neglected and molecular diffusion determines the dissolution process. Note that the ratio t/T_d is approximately equal to the Peclet number. As in the case of turbulent flow, also in porous flow hydrodynamic conditions determine the coefficient of diffusion and must therefore be regarded in every geological consideration of dissolution processes.

3.4 Diffusion Coefficients and Their Magnitudes

3.4.1 Molecular Diffusion

In most geological problems one has to deal with diffusion of small molecules. These may be either neutral species, such as gases N_2, O_2, CO_2 and H_2S, or ionic species, such as Ca^{2+}, HCO_3^-, CO_3^{2-}, H^+, OH^- and many others. All these small molecules have diffusion coefficients of comparable magnitude in the order of 10^{-5} cm^{2}s^{-1}.

Table 3.1. Molecular coefficients of diffusion for various species encountered in karst water

Ion	$D_m(10^{-6}$ cm^2 s^{-1}) Temperature (°C)			[a]
	0	18	25	
H$^+$	56.1	81.7	93.1	
OH$^-$	25.6	44.9	52.7	
Ca^{2+}	3.73	6.73	7.93	
Mg^{2+}	3.56	5.94	7.05	
CO$_3^{2-}$	4.39	7.80	9.55	
HCO$_3^-$	—	—	11.8	
SO$_4^{2-}$	5.00	8.90	10.7	
Cl$^-$	10.1	17.1	20.3	
Na$^+$	6.27	11.3	13.3	
CO$_2$	5.6	16.0	20.1	[b]

[a] Li and Gregory (1974). [b] Landolt-Börnstein (1969)

Table 3.1 lists the numbers for a variety of ionic species in the temperature region between 0°C and 25°C, showing that the diffusion coefficients increase with temperature. The two small ions H$^+$, OH$^-$ show coefficients which are larger by a factor of between 5 and 10, whereas some metal ions are smaller by about a factor of 2 from the average (Li and Gregory 1974).

Thus, for instance, if CaCl$_2$ dissolves in water one could expect the Cl$^-$ species to diffuse more quickly into the solution than the Ca^{2+} species. If this, however, did happen there would be a separation of charges. These induce electric fields which retard Cl$^-$ and enhance Ca^{2+} motion in such a way that charge neutrality of the solution is maintained. Thus, the diffusional motion of both ions is tied together. It can be described by a common coefficient of diffusion, which is an average between the two individual ones.

For a binary electrolyte $(A^{+z_+})v + (B^{-z_-})v^-$, where $z\pm$ denotes the ionic charges and $v\pm$ the stoichiometric coefficients, the joint coefficient of diffusion is given by:

$$D_j = \frac{(z_+ + z_-)D_+ D_-}{z_+ D_+ + z_- D_-} .$$ (3.33)

This formula was first derived by Nernst. Its derivation can be taken from standard text books, e.g. Jordan (1979). It should be mentioned that the coefficients of diffusion depend also on the concentrations of ions encountered. In the realm of groundwater, however, this is of minor influence and can be neglected.

For gases in an aqueous solution the constants of diffusion are in the region between 1×10^{-5} to 2×10^{-5} cm^{2}s^{-1} (Lerman 1979).

3.4.2 Eddy Diffusion

In turbulent flow we find an increase in the effective constant of diffusion when moving away from the wall into the turbulent core. Bird et al. (1960) gave a crude method to calculate the constant of eddy diffusion from the velocity distribution encountered in a straight tube of radius R (cf. Chap. 5).

For hydraulic gradients in the order of 10^{-2} one finds that eddy diffusion constants D_e are enhanced by a factor of n in comparison to the molecular D_m.

$$D_e = nD_m , \tag{3.34}$$

with n between $10^2 - 10^5$. In the centre of the tube, n is crudely estimated by:

$$n = \sqrt{\left(\frac{R}{2}\right)^3} \cdot \sqrt{\frac{g \cdot J}{v^2}} \approx 10^3 \sqrt{R^3 J} , \tag{3.35}$$

where R is the radius of the tube, g earth's acceleration 981 cm s^{-2}, v is the kinematic viscosity and J the hydraulic gradient (cf. Chap. 5).

3.4.3 Hydrodynamic Dispersion

The magnitudes of the effective diffusion coefficients are determined by the Peclet number (cf. Eqs. 3.27, 3.28).

The Peclet number is determined by the average grain diameters composing the medium and the average velocity of a particle along its statistical way. Thus, a wide variety of diffusion enhancement is possible.

4 Chemical Kinetics

In the dissolution of limestone, transport of the ions from the surface of the solid into the bulk of the solution by diffusion is only one step. When the chemical processes at the surface are so fast that ions removed by diffusion into the bulk are immediately replaced by those released from the solid, i.e. the surface concentration of Ca^{2+} is at saturation, then diffusion determines the dissolution rate entirely. If, however, this is not so, then chemical rate laws, i.e. chemical kinetics, have to be considered when dealing with dissolution rates. In $CaCO_3$ dissolution there are two chemical processes which are sufficiently slow to play an important role in determining dissolution rates, i.e. the amount of $CaCO_3$ removed from the solid per unit area and time.

One is a heterogeneous process at the surface, where the chemical composition of the solution at the surface determines the rates. The other process is homogeneous and converts CO_2 in the bulk of the solution into H^+ and HCO_3^-. The protons thus delivered are needed to react with CO_3^{2-} released from $CaCO_3$ into HCO_3^-. Thus, accumulation of CO_3^{2-} is prevented and the ionic activity $(Ca^{2+}) \cdot (CO_3^{2-})$ product is kept sufficiently low for further dissolution to proceed as long as H^+ can be delivered. In this chapter we will provide the basic knowledge on chemical kinetics, which is necessary to understand the role it plays in dissolution of limestone.

4.1 Rate Laws of Elementary and Overall Reactions

4.1.1 Elementary Reactions

The conversion of CO_2 in an aqueous solution by hydration into H_2CO_3:

$$H_2O + CO_2 \rightleftharpoons H_2CO_3 \tag{4.1}$$

is a process, in which a CO_2 molecule approaches an H_2O molecule sufficiently close and in a geometrically appropriate way so that the two molecules fit sterically. Then the reaction takes place. Such a process where the reaction partners are the true species, written down in the reaction equation, is called an elementary reaction. Since H_2O, CO_2 and H_2CO_3 finally reach chemical equilibrium, it is obvious that at equilibrium each elementary reaction (forward reaction) has to be balanced by a corresponding back reaction, which also is an elementary reaction. In our case H_2CO_3 decomposes into H_2O and CO_2. If this back reaction is extremely slow in comparison to the forward reaction, chemical equilibrium is shifted to the side of the reaction products, i.e. the reaction appears to proceed only in one direction.

In general, we may write an elementary reaction as:

$$n_A \cdot A + n_B B \xrightarrow{k_f} n_P P + n_Q Q \,. \qquad (4.2)$$

This means that by a collision of n_A molecules of the kind A and n_B molecules of kind B, reaction products P and Q are formed. Since in each process the molecules have to approach at least to a limiting reaction distance, the number of reactions per second is related to the probability that a sufficient number of molecules of each kind is present at the same time in a small volume in the dimension of the reaction distance, e.g. a sphere with the diameter of this distance.

The probability of finding one molecule in a given volume is proportional to the concentration of the species. The probability of finding n molecules simultaneously in this volume is the n'th power of the probability of finding only one. Thus, the number of elementary reactions per unit volume and time is given by:

$$k_f \cdot [A]^{n_A} \cdot [B]^{n_B} = R_f \,. \qquad (4.3)$$

k_f is a factor called the rate constant. The stoichiometric factors n appear as exponents, since the probability of finding at the same time two molecules of A is proportional to $[A]^2$, or of finding at the same time one of A and one of B is proportional to $[A] \cdot [B]$ and so on.

From Eq. (4.3) one can derive the change of concentrations in time. In each individual reaction process n_A molecules of species A and n_B molecules of species B are removed from the solution. Therefore,

$$-\frac{d[A]}{dt} = n_A k_f \cdot [A]^{n_A} \cdot [B]^{n_B} = n_A \cdot R_f \,;$$

$$-\frac{d[B]}{dt} = n_B k_f \cdot [A]^{n_A} \cdot [B]^{n_B} = n_B \cdot R_f \,. \qquad (4.4)$$

The negative sign is due to the fact that the number of molecules decreases. The sum of the exponents $(n_A + n_B + \cdots)$ is termed the order of the reaction. In each individual process correspondingly reaction products are formed and we have also:

$$\frac{d[P]}{dt} = n_P k_f \cdot [A]^{n_A} \cdot [B]^{n_B} = n_P R_f \,;$$

$$\frac{dQ}{dt} = n_Q k_f [A]^{n_A} \cdot [B]^{n_B} = n_Q R_f \,. \qquad (4.5)$$

It is important to note that the reaction rates are determined by the concentrations of the species and not by the activities.

Reaction (4.2) is balanced by the back reaction:

$$n_P Q + n_P P \xrightarrow{k_b} n_A A + n_B B \,. \qquad (4.6)$$

The change in concentrations from this reaction is analogously:

$$-\frac{dP}{dt} = n_P k_b [P]^{n_P} \cdot [Q]^{n_Q} = n_P R_b \, ;$$

$$-\frac{dQ}{dt} = n_Q k_b [P]^{n_P} \cdot [Q]^{n_Q} = n_Q R_b \, ;$$

$$\frac{dA}{dt} = n_A k_B [P]^{n_P} \cdot [Q]^{n_Q} = n_A R_b \, ;$$

$$\frac{dB}{dt} = n_B k_B [P]^{n_P} \cdot [Q]^{k_Q} = n_B R_b \, . \tag{4.7}$$

Thus, the total rates are the combination of the forward and backward contributions:

$$\frac{1}{n_A} \frac{d[A]}{dt} = \frac{1}{n_B} \frac{d[B]}{dt} = -k_f [A]^{n_A} \cdot [B]^{n_B} + k_b [P]^{n_P} \cdot [Q]^{n_Q} \, ;$$

$$\frac{1}{n_P} \frac{[dP]}{dt} = \frac{1}{n_Q} \frac{d[Q]}{dt} = +k_f [A]^{n_A} \cdot [B]^{n_B} - k_b [P]^{n_P} \cdot [Q]^{n_Q} \, . \tag{4.8}$$

The equations can be extended correspondingly if more reaction and product species exist.

In case of equilibrium, forward and backward reactions cancel each other and all the derivatives become zero. From this one obtains:

$$\frac{k_b}{k_f} = K = \frac{[A]^{n_A} \cdot [B]^{n_B}}{[Q]^{n_Q} \cdot [P]^{n_Q}} \tag{4.9}$$

which is the mass action law, written in concentrations.

4.1.2 Overall Reactions

Experimental investigations on the kinetics of chemical reactions:

$$n_A A + n_B B \xrightarrow{k_f} n_Q Q + n_P P \, , \tag{4.10}$$

often reveal rate laws of the type:

$$\frac{1}{n_A} \frac{d[A]}{dt} = \frac{1}{n_B} \frac{d[B]}{dt} = -n_P \frac{d[P]}{dt} = -n_Q \frac{d[Q]}{dt}$$

$$= k_f' [A]^{n_A'} \cdot [B]^{n_B'} \cdot [P]^{n_P'} \cdot [Q]^{n_Q'} \, , \tag{4.11}$$

where the n' are no longer related to the stoichiometric coefficients but can take any value, positive or negative. These rate equations are no longer stoichiometrically true as those of elementary reactions. The reason for this is that Eq. (4.10) describes an overall reaction, which comprises intermediate reactants and products, which do not occur explicitly in the overall reaction equation.

Systems CO_2-H_2O and $CaCO_3-CO_2-H_2O$

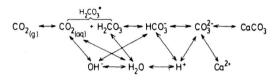

Gas   Solution   Solid

Fig. 4.1. Chemical reactions during dissolution of $CaCO_3$ according to the system CO_2-H_2O (Usdowski 1982)

The generally used reaction equation for dissolution of limestone:

$$CaCO_3 + H_2O + CO_2 \rightarrow Ca^{2+} + 2\,HCO_3^-$$

is such an overall reaction, since it comprises three reactions not appearing explicitly in this equation.

Figure 4.1 gives an illustration of these processes. CO_2^g enters from the gas phase into the liquid-forming aqueous CO_2^{aq} which converts into carbonic acid H_2CO_3, dissociating into HCO_3^- and CO_3^{2-}, thereby delivering H^+ which reacts with the carbonate ions released from the solid to HCO_3^-. Furthermore, a parallel reaction converts CO_2^{aq} into HCO_3^- by removing OH^- from the solution.

Thus, overall reactions can be decomposed in a series of elementary reactions, which might be consecutive and parallel as illustrated schematically by the reaction scheme:

$$A \rightarrow B \rightarrow C \rightarrow D \underset{F \longrightarrow P}{\overset{E \longrightarrow P}{\diagdown}} \qquad (4.12)$$

where all the reactions are now elementary reactions with individual rate laws. The decomposition of an overall reaction into elementary reaction steps reveals the real reaction mechanism. The rate laws can then be written as combinations of those of the elementary reactions. When all these reactions proceed with comparable rates, very complicated expressions can result. In some cases, however, only one of the elementary processes determines the rates. This is then called the rate-determining reaction. In consecutive reactions it is the process with the slowest reaction, which represents a bottleneck. This is the same situation as on traffic lanes where the narrowest part of the roads determines the number of cars which can pass in a given time interval.

In the case of several parallel reactions the situation is opposite, since the fastest process now determines the rates. To illustrate how rate equations of the type in Eq. (4.11) result, Lasaga (1981) has given the example of ozone decomposition:

$$2O_3 \rightarrow 3O_2 . \qquad (4.13)$$

This reaction has a mechanism where the first step is a reversible reaction:

$$(1) \qquad O_3 \underset{k_1^-}{\overset{k_1^+}{\rightleftharpoons}} O_2 + O , \qquad (4.14)$$

decomposing ozone into atomic and molecular oxygen. This is followed by the slow reaction:

(2) $O + O_3 \xrightarrow{k_2} 2O_2$,   (4.15)

where the back reaction is extremely slow and can be neglected.

The rate laws of the elementary reactions are:

$$\frac{d[O]}{dt} = k_1^+[O_3] - k_1^-[O_2][O] - k_2[O_3][O] ;$$

$$\frac{d[O_2]}{dt} = k_1^+[O_3] - k_1^-[O_2][O] - 2k_2[O][O_3] .$$ (4.16)

Since k_2 is very small, one can assume that the concentration of oxygen atoms reaches a steady state, such that the number of atoms removed by reaction (2) is equal to those delivered by reaction (1). The concentration $[O]_s$ in this steady state can be obtained by replacing $[O]$ by $[O]_s$ and observing that the derivative is zero:

$$[O]_s = \frac{k_1^+[O_3]}{k_1^-[O_2] + k_2[O_3]} \approx \frac{k_1^+[O_3]}{k_1^-[O_2]} ,$$ (4.17)

since

$$k_1^-[O_2] \gg k_2[O_3] .$$

Inserting this for $[O]$ in Eq. (4.16) one finds:

$$\frac{d[O_2]}{dt} = \frac{2k_1^+k_2}{k_1^-} \frac{[O_3]^2}{[O_2]} = k'[O_3]^2[O_2]^{-1} ,$$ (4.18)

in agreement with experimental observations. The overall rate constant k' is revealed as being composed of the rate constants of the elementary reactions.

So far we have considered overall reactions proceeding in only one direction. Since, however, each of the elementary reactions is reversible, i.e. has a back reaction of sufficiently large rates, in general also the overall reaction shows a back reaction, with corresponding rate equations. The condition for equilibrium is the balance of both reactions. In contrast to the elementary reactions, in general no mass action equation can be derived from the overall reaction, since the exponents in overall reaction rates are not stoichiometrically true.

There is, however, a very important principle of detailed balance stating that at equilibrium of the overall reaction each elementary reaction is at equilibrium also. From this follows that at equilibrium each forward rate is balanced by the opposing backward rate in all the individual elementary reactions acting in the overall process. This principle of detailed balancing provides a link between thermodynamics and kinetics, if the reaction mechanism is known. It relates, according to Eq. (4.9), one of the two kinetic constants for each elementary reaction to the corresponding equilibrium constant.

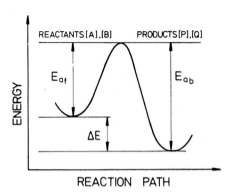

Fig. 4.2. Potential barrier associated with the activation energies of forward and backward reactions

4.1.3 Temperature Dependence of Rate Constants

To accomplish an elementary chemical reaction the molecules involved have to approach a certain distance. This has to be done in such a way that more complicated molecules also have to fit sterically into each other. The approach must therefore be in a special pathway, denoted by a reaction coordinate. Note that the following arguments relate to an elementary reaction exclusively.

Chemical reactions involve the rearrangement of molecules by distorting molecular structures or breaking bonds. After the reaction a relaxation to the final product takes place. The first step requires an activation energy E_{af}. The relaxation to the final product releases an energy E_{ab}.

This is illustrated in Fig. 4.2, which shows the energy as a function of the reaction coordinate. Thus, the forward reactants need at least the energy E_{af} to overcome the activation barrier. The back reaction needs the energy E_{ab} to form the original reactants from the products. The probability for a particle to have an energy higher than E is given by Boltzmann statistics to be proportional to $\exp(-E/RT)$, where T in °K ist the temperature and R the gas constant. The reaction rate is then proportional to a collision factor A, giving the number of collisions per time appropriate for a reaction, and the number of reactants (or products in the back reaction) with energy $E > E_{af}$ (E_{ab} in the back reaction). A is usually slightly dependent on T; $A = A(T)$:

$$R_f = A_f(T) \cdot \exp\left(-\frac{E_{af}}{RT}\right) \cdot [A]^{n_A} \cdot [B]^{n_B}$$

$$R_b = A_b(T) \cdot \exp\left(-\frac{E_{ab}}{RT}\right) \cdot [Q]^{n_Q} \cdot [P]^{n_P} . \tag{4.19}$$

From this we find the rate constants k to be:

$$k_f = A_f(T) \cdot \exp\left(-\frac{E_{af}}{RT}\right) ;$$

$$k_b = A_b(T) \cdot \exp\left(-\frac{E_{ab}}{RT}\right) . \tag{4.20}$$

This is an Arrhenius-type equation. Since in an elementary reaction k_f and k_b are related to the equilibrium constant, we have:

$$K = \frac{k_f}{k_b} = \frac{A_f(T)}{A_b(T)} \cdot \exp\left(-\frac{(E_{af} - E_{ab})}{RT}\right) = K_0 \exp\left(-\frac{\Delta E}{RT}\right). \tag{4.21}$$

This is the general equation for the temperature dependence of equilibrium constants. The energy difference ΔE is the energy released in the reaction and is nearly equal to the enthalpy change in most reactions.

The size of the activation energy determines the temperature dependence of the reaction rates and furthermore provides information on the type of reaction mechanism.

Reactions may occur by a diffusion controlled mechanism. In this case there is an immediate reaction once the reactants have met, without the necessity of overcoming barriers. Therefore, the rate constants are proportional to the flux of reactant A towards reactant B, i.e. into a sphere with B in the centre and with a radius equal to the distance which is necessary for the reaction to take place. This flux is proportional to the coefficients of diffusion. These coefficients in general also show an Arrhenius-type temperature dependence with low activation energies ($\Delta E < 20$ kJ mol^{-1}). This is also true for heterogeneous reactions, where diffusional fluxes determine dissolution or precipitation rates.

In the case of chemically controlled reactions (surface controlled in heterogeneous reactions), the activation energies are higher, as for instance in dissolution of calcite, which shows activation energies $\Delta E \approx 32$ kJ mol^{-1} for two of three parallel, concurrent elementary reactions (Plummer et al. 1978). These reactions are important at pH > 6 and will be discussed in Chapter 6. At low pH mainly attack of hydrogen ions effects dissolution of $CaCO_3$, as is the case of dissolution of calcite in HCl. Here, one finds activation energies of $\Delta E = 8$ kJ mol^{-1}, showing that diffusional transport of H^+ to the calcite surface determines the reaction rate (Plummer et al. 1978, Sjöberg 1983).

4.2   Approaching Equilibrium

In this section we give the differential equations and their solutions in time for a few common types of reactions. These solutions describe the time dependence of the concentration of reactants and products, when the reaction proceeds towards equilibrium. From these solutions a mean time can be derived, which gives a measure for the time until the reaction has come to equilibrium. We will not give the details of the mathematics. These are given in detail by Capellos and Bielski (1972).

In the following we discriminate between reversible and irreversible reactions. Reactions are irreversible when the rate constant of the back reaction is extremely small, i.e. the back reaction practically does not occur. Reversible reactions are those, in which both forward and back rates are non-zero and cancel at equilibrium in a dynamical way.

4.2.1 Irreversible Reactions

4.2.1.1 First-Order Reactions

These reactions are of type $A \rightarrow B$ with rates:

$$\frac{d[A]}{dt} = -k \cdot [A] \, . \tag{4.22}$$

Integration yields:

$$[A](t) = [A]_0 \exp(-kt) \, , \tag{4.23}$$

where $[A]_0$ is the concentration of A at $t = 0$. If we assume B not to be present at $t = 0$, i.e. $[B]_0 = 0$, one obtains:

$$[B](t) = [A]_0(1 - \exp(-kt)) \, . \tag{4.24}$$

After the reaction has been completed A ist totally converted into B.
From Eq. (4.23) one defines the mean lifetime or time constant:

$$T_d = \frac{1}{k} \, . \tag{4.25}$$

After this time the concentration of A has fallen to $1/e = 0.368$ of its initial value, i.e. $[A](T_d) = 0.368 [A]_0$.
One also defines a half-lifetime $T_{1/2}$, which elapses until 50% of the initial A has been converted to B:

$$T_{1/2} = 0.693 \cdot T_d \, . \tag{4.26}$$

4.2.1.2 Consecutive First-Order Reactions

These reactions are of the type:

$$A \xrightarrow{k_1} B \xrightarrow{k_2} C \xrightarrow{k_3} D \, .$$

D is the final product, B and C are intermediates.
We assume $[A](0) = [A]_0$ at time $t = 0$ and all the other species to be zero at $t = 0$. The differential equations read:

$$\frac{d[A]}{dt} = -k_1[A] \, ;$$

$$\frac{d[B]}{dt} = -k_2[B] + k_1[A] \, ;$$

$$\frac{d[C]}{dt} = -k_3[C] + k_2[B] \, ;$$

$$\frac{d[D]}{dt} = +k_3[C] \, . \tag{4.27}$$

The solutions read:

$$[A] = [A]_0 \cdot \exp(-k_1 t) \, ;$$

$$[D] = [A]_0 \cdot \left[1 - \frac{k_2 k_3 \exp(-k_1 t)}{(k_2 - k_1)(k_3 - k_1)} - \frac{k_1 k_3 \exp(-k_2 t)}{(k_1 - k_2)(k_3 - k_2)} \right.$$

$$\left. - \frac{k_1 k_2 \exp(-k_3 t)}{(k_1 - k_3)(k_2 - k_3)} \right] . \tag{4.28}$$

The solutions for the intermediates are of no interest here, since we are only interested in the time dependence of the final products. The time constant of the $[A](t)$ is given by k_1. The products, however, follow a complicated behaviour, made up of three exponentials with time constants determined by the rate constants k_1, k_2 and k_3. When one of these constants is small compared to the others, e.g. $k_2 \ll k_3$, k_1, Eq. (4.28) is approximated by:

$$[D] = A_0 [1 - \exp(-k_2 t)] , \tag{4.29}$$

since the two exponentials with k_1, k_3 decay rapidly in comparison to the exponential with k_2. Thus, the slowest reaction becomes rate-limiting in consecutive reactions. This is true also for n stage reactions, with n consecutive steps.

4.2.1.3 Parallel First-Order Reactions

We consider reactions occurring simultaneously of type:

$$A \xrightarrow{k_1} B \, ;$$

$$A \xrightarrow{k_2} C \, ;$$

$$A \xrightarrow{k_3} D \, .$$

The differential equations are:

$$\frac{d[A]}{dt} = -(k_1 + k_2 + k_3)[A] = -k[A] \, ;$$

$$\frac{d[B]}{dt} = k_1 [A] \, ;$$

$$\frac{d[C]}{dt} = k_2 [A] \, ;$$

$$\frac{d[D]}{dt} = k_3 [A] \, . \tag{4.30}$$

The solutions can be obtained by direct integration;

$$[A] = [A]_0 \cdot \exp(-kt) \, ;$$

$$[B] = [A]_0 \cdot \frac{k_1}{k} \cdot [1 - \exp(-kt)] \, ;$$

$$[C] = [A]_0 \cdot \frac{k_2}{k} \cdot [1 - \exp(-kt)] \,;$$

$$[D] = [A_0] \cdot \frac{k_2}{k} \cdot [1 - \exp(-kt)] \,. \tag{4.31}$$

In this case all products and the reactants change with one common exponential. In contrast to consecutive reactions, irreversible parallel reactions are determined by the largest rate constant, since the total reaction constant k is determined by the sum of the individual reaction constants.

4.2.1.4   n-th Order Reactions

Often one encounters empirical rate laws of the type:

$$-\frac{d[A]}{dt} = k[A]^n \,. \tag{4.32}$$

This is the case in many processes of dissolution or precipitation, where [A] means $(c_A^{eq} - c_A) : c_A$ concentration in the solution and c_A^{eq} the equilibrium concentration. The solution of Eq. (4.32) is obtained by direct integration as:

$$\frac{1}{[A]^{n-1}} - \frac{1}{[A]_0^{n-1}} = (n - 1)kt \,. \tag{4.33}$$

The half-lifetime of the reaction is obtained by substituting $[A] = [A]_0/2$:

$$T_{1/2} = \frac{2^{n-1} - 1}{(n - 1)k[A_0]^{n-1}} \,. \tag{4.34}$$

Note that $T_{1/2}$ depends on the initial concentration of [A]. The derivation of Eq. (4.34) is only true of $n \neq 1$.

Figure 4.3 shows the time dependence of [A] for $n = 2$, $n = 3$, $n = 5$, $n = 10$ compared to the first-order reaction. All the reactions share a numerically equal rate constant k and start with an equal concentration $[A]_0$. The reactions become slower with increasing n.

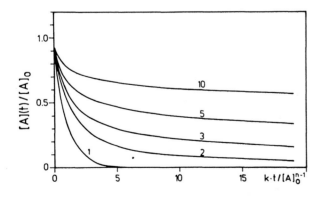

Fig. 4.3. Time dependence of concentration [A](t) for an nth-order reaction. *Numbers* on the curves denote order n of the reaction. The time is given in units of $[A]_0^{n-1}/k$. For $[A]_0 = 1$ the time scale is identical for all orders of reaction

4.2.2 Reversible Reactions

4.2.2.1 First-Order Reactions

The reaction is $A \underset{k^-}{\overset{k^+}{\rightleftharpoons}} B$. It shows the simplest case of two opposing first-order reactions. The differential equations are:

$$\frac{d[A]}{dt} = -[A] \cdot k^+ + [B] \cdot k^- \; ;$$

$$\frac{d[B]}{dt} = [A] \cdot k^+ - [B] \cdot k^- \; . \tag{4.35}$$

The initial conditions at $t = 0$ are $[A](0) = [A]_0$, $[B](0) = 0$.
    With these conditions we find the solutions:

$$[A] = [A]_0 \cdot \frac{\{k^- + k^+ \cdot \exp[-(k^+ + k^-)t]\}}{k + k} \; ;$$

$$[B] = [A]_0 \cdot \frac{k^+\{1 - \exp[-(k^+ + k^-)t]\}}{k^+ + k^-} \; . \tag{4.36}$$

Equilibrium is approached with a time constant, determined by the sum of the two rate constants. At equilibrium, i.e. in the limit $t \to \infty$, we find:

$$\frac{[A]_{eq}}{[B]_{eq}} = \frac{k^-}{k^+} \; . \tag{4.37}$$

4.2.2.2 First-and Second-Order Reactions

These reactions of the type:

$$A \underset{k^-}{\overset{k^+}{\rightleftharpoons}} B + C$$

occur in the dissociation of acids. The rate equation for $[A]$ is:

$$\frac{d[A]}{dt} = -k^+[A] + k^-[B][C] \; . \tag{4.38}$$

The initial conditions are $[A](0) = [A]_0, [B](0) = [C](0) = 0$.
    The rate equations for $[B]$ and $[C]$ are then identical, since in the reaction for all times $[B] = [C] = [X]$:

$$\frac{d[X]}{dt} = k^+([A]_0 - [X]) - k^-[X]^2 \; . \tag{4.39a}$$

For $[A]$ we have written:

$$[A] = [A]_0 - [B] = [A]_0 - [X] \; . \tag{4.39b}$$

At equilibrium the derivatives are zero and we have:

$$k^-[X]_{eq}^2 = k^+([A_0] - [X]_{eq}),$$ (4.40)

where $[X]_{eq}$ is the equilibrium value of $[B]$ and $[C]$.

The solution for $[X]$ reads then:

$$k^+t = \frac{[X]_{eq}}{2[A]_0 - [X]_{eq}} \ln \frac{[A]_0[X]_{eq} + [A]_0[X] - [X][X]_{eq}}{[A]_0([X]_{eq} - [X])}$$ (4.41)

with

$$[X]_{eq} = -\frac{k^+}{2k^-} + \sqrt{\left(\frac{k^+}{2k^-}\right)^2 + \frac{k^+}{k^-}[A]_0}.$$ (4.42)

4.3   The Kinetics of the Reaction $H_2O + CO_2 \rightleftharpoons H^+ + HCO_3^-$

Conversion of carbon dioxide into hydrogen ions and bicarbonate is the first step in weathering processes of limestone, since the hydrogen ions are the aggressive reactants in the dissolution of $CaCO_3$. The conversion process depends heavily on the pH of the solution and is slowest in the region of pH $\simeq 7.5$. This is the realm of karst water. In average the time constants for the reaction are about 10 s in this region. From these values one can easily realize that the kinetics of this process may be rate determining in the removal of limestone.

The overall reaction:

$$H_2O + CO_2 \rightleftharpoons HCO_3^- + H^+$$

is composed of two parallel reactions. At pH < 8 the predominant pathway is reaction I:

$$CO_2 + H_2O \underset{k_I^-}{\overset{k_I^+}{\rightleftharpoons}} H_2CO_3,$$ (I)

which is followed by an instantaneous reaction:

$$H_2CO_3 \rightleftharpoons HCO_3^- + H^+.$$

The rate constants in this last reaction are so large that equilibrium is established in a very short time, i.e. less than 1 ms. Therefore, during the conversion process, where the first reaction is rate-limiting, equilibrium exists such that:

$$[H_2CO_3] = [H^+][HCO_3^-]\gamma_H\gamma_{HCO_3}/K_{H_2CO_3}.$$ (2.3)

The rate equations for the reaction are given by:

$$\frac{d[CO_2]}{dt} = -k_I^+[H_2O][CO_2] + k_I^-[H_2CO_3]$$

$$= -k_I^+[CO_2] + k_I^- K_{H_2CO_3}^{-1}\gamma_H\gamma_{HCO_3}[H^+][HCO_3^-];$$

$$\frac{d[HCO_3^-]}{dt} = k_I^+[CO_2] - k_I^- K_{H_2CO_3}^{-1}\gamma_H\gamma_{HCO_3}[H^+][HCO_3^-].$$ (4.43)

Note that in both equations by definition the activities of H_2O are equal to 1.
From this the overall reaction I can be written as:

$$H_2O + CO_2 \underset{k_a}{\overset{k_1^+}{\rightleftharpoons}} H^+ + HCO_3^- \, ,$$

with the rate constant

$$k_a = k_1^- K_{H_2CO_3} \gamma_H \gamma_{HCO_3^-} \, . \tag{4.44}$$

A second parallel reaction contributes significantly above pH = 8. This reaction II
is:

$$CO_2 + OH^- \underset{k_2^-}{\overset{k_2^+}{\rightleftharpoons}} HCO_3^- \, , \tag{II}$$

with rate equations:

$$\frac{d[CO_2]}{dt} = -k_2^+[CO_2][OH^-] + k_2^-[HCO_3^-] \, ;$$

$$\frac{d[HCO_3^-]}{dt} = k_2^+[CO_2][OH^-] - k_2^-[HCO_3^-] \, . \tag{4.45}$$

Adding Eqs. (4.43) and (4.45) we obtain the total rate equations:

$$\frac{d[CO_2]}{dt} = -(k_1^+ + k_2^+[OH^-])[CO_2] + (k_a[H^+] + k_2^-)[HCO_3^-]$$

$$= -\frac{d[HCO_3^-]}{dt} \, . \tag{4.46}$$

Thus, the total reaction is:

$$H_2O + CO_2 \underset{k^-}{\overset{k^+}{\rightleftharpoons}} HCO_3^- + H^+ \, ,$$

with pH-dependent rate constants:

$$k^+ = k_1^+ + k_2^+[OH^-] \, , \qquad k^- = k_a[H^+] + k_2^- \, . \tag{4.47}$$

The numerical value of k_1^+, k_1^-, k_2^+ and k_2^- and their temperature dependence have
been investigated by many authors. They are reviewed by Kern (1960) and by
Usdowski (1982). Table 4.1 lists the temperature dependence and the values of these
constants and gives some analytical expressions.

Figure 4.4 gives the log k^+ as a function of pH. Below pH = 8, essentially
reaction I contributes. The rise in log k^+ beyond pH = 8 is due to increasing
$[OH^-]$, which increases the rates of reaction II. The right-hand ordinate of Fig. 4.4
relates to values of k^- also shown as a function of pH. At low pH the back reaction
is dominant, thus shifting equilibrium to H_2CO_3.

Table 4.1. Reaction rate constants for $CO_2 \rightleftharpoons HCO_3^- + H^+$ conversion as a function of temperature (°K) (Usdowski 1982)

°C	$-\log k_1^+$ s^{-1}	$\log k_1^-$ s^{-1}	$\log k_2^+$ $mol^{-1}\,s^{-1}$	$-\log k_2^-$ s^{-1}
0	2.68	0.32	3.04	5.33
5	2.41	0.55	3.23	4.99
10	2.17	0.78	3.41	4.66
15	1.55	1.01	3.59	4.34
20	1.76	1.22	3.76	4.02
25	1.58	1.43	3.93	3.72

$\log k_1^+ = 329.850 - 110.54 \log T - 17265.4/T$
$\log k_1^- = 13.558 - 3617.1/T$
$\log k_2^+ = 13.635 - 2985/T$
$\log k_2^- = 14.09 - 5308/T;\ T = °K;\ \log = \log_{10}$

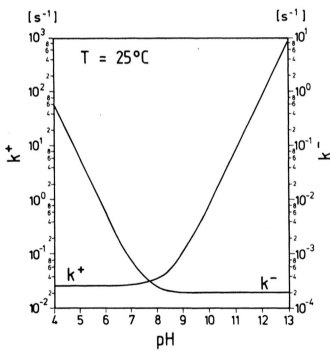

Fig. 4.4. Rate constants k^+ and k^- for the overall reaction $CO_2 \rightleftharpoons H^+ + HCO_3^-$ as a function of pH

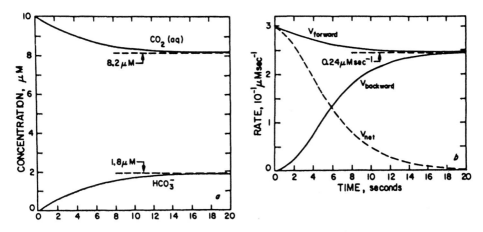

Fig. 4.5. Concentrations of CO_2^{aq} and HCO_3^- as a function of time (*left hand part*) in the reaction $CO_2^{aq} + H_2O \rightarrow HCO_3^- + H^+$ under conditions of a closed system. The *right hand part* shows forward, backward and netto reaction rates during the reaction (Stumm and Morgan 1981)

4.3.1 Examples

To obtain some insight into the kinetics of CO_2 conversion and how fast they proceed, we discuss first the following situation. Into pure water with pH = 7 a certain amount of CO_2 is dissolved instantaneously and then the system is closed. We ask for the process of equilibration between CO_2 and HCO_3^-. One important point is that for each CO_2 converted, one HCO_3^- and one H^+ is produced. Thus, the pH value of the solution changes with time. Initially, pH is at 7. With conversion of CO_2 the pH drops. In the region of pH < 7, however, only the parallel reaction I contributes, which is a first- and second-order reversible reaction of Section 4.2.2.2 with $[H^+] = [HCO_3^-]$. The rate equations are given by Eq. (4.43).

Figure 4.5 shows the time evolution of CO_2 and HCO_3^- and also the reaction rates as calculated from Eqs. (4.43) and (4.41). The initial concentration of dissolved $[CO_2^{aq}] = 1 \times 10^{-5}$ mol l^{-1}. Equilibrium is established within 20 s.

As a second example we take the case where CO_2^{aq} is dissolved in a buffered solution with constant pH. In this case the concentrations of H_2CO_3 and HCO_3^- are related to the total amount of dissolved carbonate species $\Sigma[C] = [H_2CO_3] + [HCO_3^-] + [CO_3^{2-}]$:

$$[H_2CO_3] = \frac{\Sigma[C]}{1 + \dfrac{K_{H_2CO_3}}{\gamma_H \cdot \gamma_{HCO_3}[H^+]} + \dfrac{K_{H_2CO_3}K_2}{\gamma_H^2 \cdot \gamma_{CO_3}[H^+]^2}} = \alpha_0' \Sigma[C];$$

$$[HCO_3^-] = \frac{\Sigma[C]}{1 + \dfrac{[H^+]\gamma_H\gamma_{HCO_3}}{K_{H_2CO_3}} + \dfrac{K_2\gamma_{HCO_3}}{\gamma_H\gamma_{CO_3}[H^+]}} = \alpha_1' \Sigma[C]. \qquad (4.48)$$

These equations are analogous to Eqs. (2.10) to (2.13). The only change is substituting $[H_2CO_3^*]$ by $[H_2CO_3]$ and K_1 by $K_{H_2CO_3}$.

The total forward reaction is then:

$$\frac{d[CO_2]}{dt} = -k^+[CO_2] \tag{4.49}$$

and the back reaction gives:

$$\frac{d[CO_2]}{dt} = (k_1^- \alpha_1' + k_2^- \alpha_0') \Sigma[C] = k' \Sigma[C] . \tag{4.50}$$

Provided that at $t = 0$ the initial carbonate species is exclusively CO_2 we have:

$$[CO_2]_0 = [CO_2] + \Sigma[C] . \tag{4.51}$$

The total reaction rate is then obtained by adding forward and back reaction rates as:

$$\frac{d[CO_2]}{dt} = -(k^+ + k^-)[CO_2] + k'[CO_2]^0 . \tag{4.52}$$

This is a pseudo first-order equation (cf. Sect. 4.2.2.1).

Equilibrium is approached by an exponential with time constant:

$$T_d = \frac{1}{k^+ + k'} . \tag{4.53}$$

Figure 4.6 shows the half-lifetime and the time to attain 99% of the equilibrium values as a function of pH.

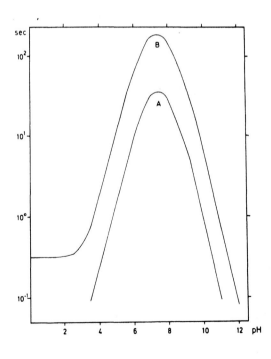

Fig. 4.6. Time for 50% (*curve A*) and 99% (*curve B*) equilibration in the reaction converting CO_2 into carbonate species $\Sigma[C]$ (Usdowski 1982)

The process is slowest at pH $\approx$ 7.5 as we stated before.

The third example is important in calcareous solutions. In this case a large concentration of HCO_3^- due to the calcite dissolved is present. Therefore, in a good approximation we can assume HCO_3^- to be constant if not too large amounts of CO_2 are added to the solution.

The reaction rate is then given:

$$\frac{d[CO_2]}{dt} = -k^+[CO_2] + k^-[HCO_3^-] . \tag{4.54}$$

Since $[HCO_3^-]$ is constant and in such a buffered solution the change of pH on addition of CO_2 is moderate. Therefore, k^- in the last term in Eq. (4.54) can be considered as constant. Thus, the reaction is pseudofirst-order and proceeds exponentially with a time constant:

$$T_d = 1/k^+ . \tag{4.55}$$

The value of k^+ can be taken from Fig. 4.4. Below pH $= 7$, $T_d = 38$ s, above pH $= 8$ there is a strong increase and at pH $= 10$, $T_d = 1$ s.

4.4 Mixed Kinetics

So far we have discussed two extreme cases. In the first one, mass transport is entirely determined by diffusion, whereas in the other one chemical rate laws control the processes exclusively. In many processes, however, both diffusion and chemical reactions, are of comparable effectiveness, and we have the case where diffusion and rate laws simultaneously control what happens.

To obtain some insight into this regime of mixed kinetics we consider the following situation as illustrated in Fig. 4.7. The concentration of a molecular species, dissolved from the surface of a solid, has concentration c_0 at the solid-liquid interphase at $x = 0$. There is a layer of thickness ε where diffusion is controlled by the molecular diffusion coefficient D. Outside this layer the liquid is in turbulent motion and the coefficient of eddy diffusion is high, such that no concentration gradients can exist and everywhere in this region $c = c_B$. For simplicity we assume

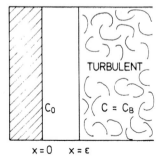

Fig. 4.7. Boundary conditions for transport of chemically reacting species dissolved from a solid at $x = 0$ (see text)

that the volume of the turbulent region is extremely large, such that the solution is infinitely diluted and $c_B = 0$. Now the molecules in the liquid react in a homogeneous, first-order chemical reaction with reaction rate:

$$\frac{dc}{dt} = -kc \,. \tag{4.56}$$

Mass transport of the species by molecular diffusion from the surface across the layer into the turbulent region is described by Fick's second law, Eq. (3.13):

$$\frac{\partial c}{\partial t} = D\frac{\partial^2 c}{\partial x^2} - kc \,. \tag{4.57}$$

In the stationary state the time derivative is zero and we have:

$$D\frac{\partial^2 c}{\partial x^2} = kc \,. \tag{4.58}$$

The solution to this equation reads:

$$c(x) = A\exp(-x/\lambda) + B\exp(+x/\lambda) \,; \qquad \lambda = \sqrt{D/k} \,; \tag{4.59}$$

and A and B are calculated from the boundary conditions:

$$A = \frac{-c_0\exp(\varepsilon/\lambda)}{\exp(-\varepsilon/\lambda) - \exp(\varepsilon/\lambda)} \,; \qquad B = \frac{c_0\exp(-\varepsilon/\lambda)}{\exp(-\varepsilon/\lambda) - \exp(\varepsilon/\lambda)} \,. \tag{4.59a}$$

Figure 4.8 illustrates this solution for three cases. In the first case, $\lambda \ll \varepsilon$. This is shown for $\lambda = 0.1\varepsilon$ and $\lambda = 0.2\varepsilon$. There is a steep descent of the curves which is determined by an exponential decay resulting from the first term in Eq. (4.59). Thus, the molecules after penetrating the distance λ, called the diffusion length, have undergone a chemical reaction. Therefore, λ is the penetration distance, which molecules can diffuse in average until a chemical reaction occurs. The larger the rate constant k, the shorter is the distance. So $\lambda \gg \varepsilon$ is the case of fast chemical reaction and the entire reaction proceeds within a small layer with thickness λ.

If the reaction is slow, i.e. k is small, we have the other limiting case, where $\lambda \gg \varepsilon$. The linear curve in Fig. 4.8 results for $\lambda \geq 3\varepsilon$. Here, mass transport is entirely

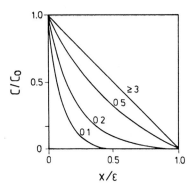

Fig. **4.8.** Stationary concentrations for mass transport by diffusion and a simultaneous, first-order chemical reaction for the boundary conditions as described in Fig. 4.7. *Numbers* on the curves give values of λ/ε (see text)

by diffusion, since the reaction is so slow that all molecules can cross the layer ε without reacting. In this limiting case. Eq. (4.59) can be approximated to a high degree of accuracy to:

$$c(x) = c_0 \cdot \left(1 - \frac{x}{\varepsilon} \right).$$
(4.60)

A solution identical to Eq. (4.60) is obtained from Fick's second law without a chemical reaction, i.e. $k = 0$ (cf. Eq. 3.26).

The intermediate case, where $\lambda \approx \varepsilon$ is illustrated by $\lambda = 0.5\varepsilon$. Most of the molecules are transported into the bulk by diffusion: There is, however, a chance of reaction for a sufficiently large part of them. This is the reason why the curve bends below the linear one.

The flux from the surface at $x = 0$ can be calculated by differentiation of Eq. (4.59) as:

$$-D\frac{\partial c}{\partial x/x = 0} = \frac{D}{\lambda} c_0 \frac{1 + \exp(-2\varepsilon/\lambda)}{1 - \exp(-2\varepsilon/\lambda)} = F_c .$$
(4.61)

In the case $\lambda \gg \varepsilon$, i.e. pure diffusional control, the flux is:

$$F_d = \frac{D}{\varepsilon} c_0 .$$
(4.62)

The ratio of flux F_c with a chemical reaction to the flux F_d without a reaction is given by:

$$f_e = \frac{F_c}{F_d} = \frac{\varepsilon}{\lambda} \cdot \frac{1 + \exp(-2\varepsilon/\lambda)}{1 - \exp(-2\varepsilon/\lambda)} .$$
(4.63)

Thus, the presence of a fast chemical reaction enhances mass transport from the wall of a solid by factor f_e. This is often called chemically enhanced diffusion.

One interesting point is that this chemically enhanced flux no longer depends on the thickness ε of the water layer, as soon ε exceeds $\varepsilon > 3\lambda$. Figure 4.9 shows the flux F given by Eq. (4.61) as a function of the thickness ε of the layer. The flux approaches a constant value for $\varepsilon > 2\lambda$. We will encounter this situation as an important feature in the dissolution of calcite (cf. Chap. 7).

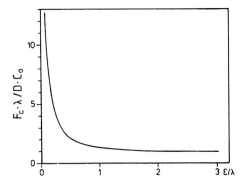

Fig. 4.9. Chemically enhanced flux F_c (Eq. 4.61) as a function of ε/λ. Note that at large $\varepsilon \gg \lambda$ the flux becomes independent of ε and is determined entirely by the depth λ of the reaction zone and the constant of diffusion D

In our example we have implicitly assumed that the concentration c_0 at $x = 0$ remains constant, irregardless of the extent to which mass transport is enhanced from the solid by a chemical, homogeneous reaction in the solution. In other words, the reaction releasing the molecules from the solid is infinitely fast and each molecule removed from the surface, whether by diffusion into the solution or by chemical reaction, is replaced immediately. Very often this is not the case and the heterogeneous reaction at the surface plays an important role. Again, we consider the geometry shown in Fig. 4.7. But now we assume that there is no homogeneous reaction in the solution. Instead, the heterogeneous reaction at the surface of the solid has finite rates. The flux thus resulting from the wall is given by:

$$F_c = k(c_{eq} - c_0) , \tag{4.64}$$

where c_{eq} is the concentration at which equilibrium is established, and no further dissolution is possible, i.e. $F_c = 0$.

The flux removed from the surface by diffusion into the bulk is given by using Eq. (3.27) which reads then:

$$F_d = +\frac{D}{\varepsilon}(c_0 - c_B) = k^d(c_0 - c_B) . \tag{3.27}$$

Since in the dissolution process the amount of substance released from the solid phase equals the amount removed from the surface by diffusion, we have:

$$F_c = F_d . \tag{4.65}$$

From this condition we can calculate the concentration c_0 at $x = 0$, which depends on the reaction rate constant k:

$$c_0 = \frac{k c_{eq} + k^d c_B}{k + k^d} . \tag{4.66}$$

Inserting this into Eq. (4.64) one obtains

$$F = \frac{k \cdot k^d}{k + k^d} \cdot (c_{eq} - c_B) = \alpha(c_{eq} - c_B) . \tag{4.67}$$

Thus, the flux established depends simultaneously on the heterogeneous reaction rate at the surface, i.e. k and c_{eq}, and on the transfer coefficient k^d resulting from diffusion with diffusion coefficient D across the layer of width ε.

In the case of a very fast reaction at the surface, i.e. $k \gg k^d$, the mass transfer coefficient α is determined entirely by diffusion:

$$\alpha = k^d . \tag{4.68}$$

In the other extreme limit, i.e. a very slow heterogeneous reaction with $k \ll k^d$, we have:

$$\alpha = k \quad \text{and} \quad c_0 = c_B . \tag{4.69}$$

Here, the heterogeneous reaction is rate-determining and the diffusional flux is extremely low. Therefore, the force, driving diffusion, i.e. the concentration gradient, is small also and c_B must be lower than c_{eq} only by a very small amount. In the extreme case of $k^d = 0$, no dissolution takes place at all ($F = 0$) and $c_B = c_{eq}$ which also comprises $F = 0$.

One further remark concerning Eq. (4.67) should be given. The flux F is given by the concentrations c_B and c_{eq}. Both concentrations can be measured and one has a law, which can be directly verified experimentally. In principle, this law relates to an overall reaction, since the intermediate steps are hidden in the constant α, which is also determined experimentally.

In many experiments one finds dissolution rates from solids which can be expressed by empirical laws:

$$F = \gamma \cdot (c_{eq} - c_B)^n , \tag{4.70}$$

where $n > 0$. Values of n up to $n = 4$ have been observed for instance in dissolution experiments on calcite (Plummer and Wigley 1976). Empirical laws of this type usually result from complex mechanisms operating simultaneously.

There may be parallel heterogeneous reactions at the surface with differing reaction orders and also homogeneous reactions in the solution, for instance the reaction of CO_2 into HCO_3^- in the dissolution of calcite.

To really understand therefore dissolution processes one has to disentangle all these mechanisms.

5 Hydrodynamics of Flow

Flow of groundwater in karst areas covers a wide field of hydrodynamic conditions. There is diffuse flow through partings in the rocks, such as joints or beddings, which initially, i.e. prior to widening by dissolution, might have apertures as small as 2×10^{-3} cm (Davis 1968). To describe the hydrodynamic properties of these pathways of flow, one may visualize these partings either as two parallel planes with a fixed distance, or as a two-dimensional porous medium. In any case flow is laminar and flow velocities are small, in the order of 10^{-3} cm s^{-1}. Thus, diffuse flow is characterized by a long retention time of karst water in the rock. The springs fed by this kind of water, which has sufficient time to equilibrate with respect temperature and chemical composition, usually show only small variations in temperature and are close to calcite saturation. Due to the long retention time of water and the large storage volume of the aquifer, they react slowly to flood pulses.

In contrast, flow also takes place in pipelike conduits, which have been created by solutional activity. If the diameter of these conduits exceeds 1 cm, flow in these pipes becomes turbulent and flow velocities are high in comparison to the laminar flow regime. Springs fed by these conduits are termed conduit springs. Since the retention time of water in conduit aquifers is short, these springs show large variations in temperature and chemical composition. They react very quickly to flood pulses, since the runoff from a pipe system reacts practically instantaneously to changes in pressure heads. Examples of both types of springs and their classification were first given in much detail by Shuster and White (1971).

Figure 5.1 illustrates these two end-member flow systems. Of course, there are many intermediate cases possible, e.g. where springs are fed from feeders of both kinds simultaneously.

The average velocities of karst waters from the above considerations are expected to be in the range between 10^{-3} cm s^{-1} up to 1 m s^{-1}.

Milanovic (1981) reported investigations of average velocities in karst water flow in the Dinaric karst. From a total of 281 dye tests the histogram shown in Fig. 5.2 was obtained. The diagram shows the most probable velocity to be in the region of a few cm s^{-1}, whereas velocities above 20 cm s^{-1} are scarce. This indicates that in this highly karstified area most of the flow occurs in relatively small conduits with diameters in the order of a few centimetres. As we have discussed already in Chapter 3, mass transport of dissolved limestone away from the walls of the rock depends critically on the hydrodynamic conditions. Therefore, the understanding of karstification processes must also be based on the knowledge of the basic principles of hydrodynamics of flow. These will be discussed in the following sections.

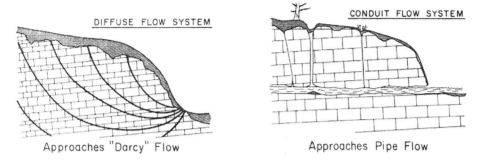

Fig. 5.1. Schematic representation of flow systems in karst. Diffuse flow system results from flow through narrow partings. Conduit flow systems are due to pipe flow in cave passages (Shuster and White 1971)

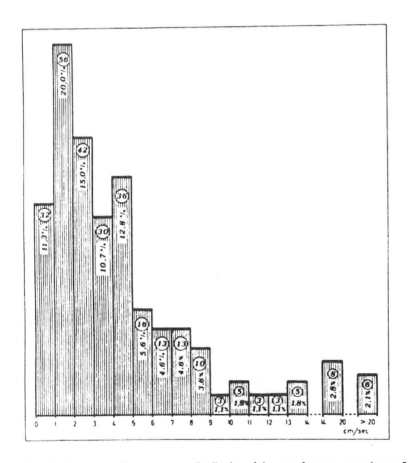

Fig. 5.2. Histogram of the percentage distribution of the most frequent groundwater flow velocities in the Dinaric karst (Milanovic 1981)

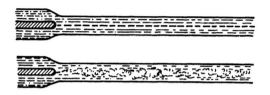

Fig. 5.3. Trace of injected dye in laminar flow (*upper part*) exhibits a streamline (*dotted line*). In turbulent flow (*lower part*) eddies occur

5.1 Laminar and Turbulent Flow

The different types of flow can be visualized by Reynolds' experiment. From a large reservoir of water, flow through a circular pipe is maintained as shown in Fig. 5.3. Dye is injected into the centre of the flow tube. At small flow velocities a streamline coloured by the dye appears. Each parcel of fluid has its predetermined path to follow, which therefore is marked as a streamline by the dye injected. If one separates two different, individual small parcels of water, even though they may deform, their pathways never cross each other and mixing of the fluid contained in the parcels is impossible.

If one increases the flow velocity, the previously well-defined path, marked by the dye, starts to blur and eddies occur. At still higher velocities within a short distance from the inlet, complete mixing results. In this case we have fully turbulent flow. In turbulent flow pathways of individual parcels of fluids become chaotic, and after a short distance of travel, crossing to other pathways occurs with complete mixing of the contents of different water parcels.

The reason for this behaviour lies in the action of two forces of different origin on a volume element. One is the friction between neighbouring water layers represented by the viscosity. The other is the force of inertia, which is related to the acceleration of volume elements along their pathways. At small velocities accelerations and therefore forces of inertia are small compared to those exerted by viscosity. Thus, fluctuations which are capable of driving a small volume of water away from its path, determined by the equation of motion, are damped and flow is along individual streamlines.

If the velocity increases, then the forces of inertia also increase, which eventually rule out those of viscosity. Then small water parcels are driven out of their streamlines and penetrate into other ones. Thus, flow becomes turbulent and large fluctuations in velocity arise. The criterion to decide whether flow will be laminar or turbulent is given by Reynolds number, N_{Re}. This dimensionless number is given by:

$$N_{Re} = \frac{v \cdot d}{\nu} ; \qquad (5.1)$$

where v is a characteristic velocity, e.g. the average flow velocity in the tube, d is a characteristic geometric dimension, i.e. the diameter of the tube, and ν the kinematic viscosity. In smooth tubes for $N_{Re} < 2000$ flow is laminar. For $N_{Re} \geq 2000$ turbulent flow is gradually established, which becomes fully turbulent at $N_{Re} = 10\,000$.

5.1.1 The Law of Bernoulli

Karst water aquifers to a considerable extent can be described as a complex, interconnected system of conduits with varying dimensions in which accordingly laminar or turbulent flow exists. This flow is driven by height differences between the input and the output. The most important equation which governs the hydraulics of such a system is the law of Bernoulli, which relates flow velocities in a conduit to pressure and elevations.

Let us consider the motion of volume element dV of the liquid. From Newton's equation of motion we have:

$$dV \cdot \rho \frac{dv}{dt} = (G + P + Z) \cdot dV, \qquad (5.2)$$

where ρ is the density, G is the force per volume due to gravitation, P that due to the pressure p and Z due to friction by viscosity. The work done, after the element has travelled a distance ds, is obtained by multiplying Eq. (5.2) by ds. The work dW is done on that element dV by the forces on the right-hand side of Eq. (5.2). Since the liquid is incompressible, no work for deformation of dV is performed. From Eq. (5.2):

$$dW = (G + P + Z)dsdV = \rho \cdot dV \cdot \frac{dv}{dt} ds = \rho dV(dv \cdot v), \qquad (5.3)$$

since $ds/dt = v$ and therefore $\frac{dv}{dt} \cdot ds = v \cdot dv$.

Using $d\left(\frac{v^2}{2}\right) = v \cdot dv$ we obtain:

$$dW = (-\rho g \cdot dh - dp + Zds)dV = \rho \cdot d\left(\frac{v^2}{2}\right) \cdot dV. \qquad (5.4)$$

The right-hand side of this equation gives the change in kinetic energy. The first term in parentheses is due to the change of potential energy given by $-\rho gdh$; g is the acceleration due to earth's gravity. This represents the loss of energy when a unit element of volume drops by the height dh, related to ds. The second term (dp) represents the change in hydrostatic pressure p along ds and gives the work done by pressure forces. The negative sign denotes the fact that the kinetic energy of an element of volume increases with decreasing pressure. Finally, the last term Zds is work done by friction, which is lost as mechanical energy and converted into heat. Rearranging Eq. (5.4) yields:

$$d\left(\rho gh + p + \rho \frac{v^2}{2}\right) = Zds = dh_f, \qquad (5.5)$$

which by integration gives:

$$h + \frac{p}{\rho g} + \frac{v^2}{2g} + h_f = \text{const};$$

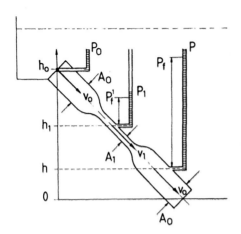

Fig. 5.4. Bernouilli's law illustrated in a tube with varying cross-section. Without frictional losses $h + p/\rho g$ at the inlet and the outlet are equal due to equal flow velocities. The *arrows* indicate the pressure p_f if frictional loss is existent. In the narrow part of the tube hydrostatic pressure is low due to the larger velocity $v_1 > v_0$; p_f^1 indicates the hydrostatic pressure with frictional loss

$$h_f = -\int \frac{Z}{\rho g} \cdot ds .$$ (5.6)

This is the usual form of the general Bernoulli equation. Its meaning is visualized in Fig. 5.4. A fluid flows from height h_0 through a tube of varying diameter. We assume entrance and outlet diameters to be equal. The flow velocity is v_0 at the inlet and also at the outlet. At h_0 the hydrodynamic pressure is p_0.

1. We first assume friction to be negligible, i.e. $h_f = 0$. Then Eq. (5.6) states that everywhere in the tube:

$$h + \frac{p}{\rho g} + \frac{v^2}{2g} = h_0 + \frac{p_0}{\rho g} + \frac{v_0^2}{2g} .$$ (5.7)

By measuring the hydrostatic pressure as indicated in Fig. 5.4, one finds:

$$\frac{p}{\rho g} = (h_0 - h) + p_0 + \frac{v_0^2 - v^2}{2g} .$$ (5.8)

Thus, hydrostatic pressure increases at the dispense of height loss and decreases with increasing velocity as is the case in the place with narrow diameter, where according to mass conservation the equation of continuity holds:

$$v_0 A_0 = v_1 A_1 ,$$ (5.9)

where v_0 is the velocity in that part of the tube with the cross-sectional area A_0 and v_1 is the velocity where the cross-sectional area is A_1.

2. If friction is present, Eq. (5.5) tells us:

$$dh_f = \frac{Zds}{\rho g} = d\left(h + \frac{p}{\rho g} + \frac{v^2}{2g} \right).$$ (5.10)

Thus, there is a continuous loss of energy as the fluid moves along the streamline. This can be seen in Fig. 5.4, where the arrows indicate the hydrostatic pressure height $p/\rho g$, if friction is present.

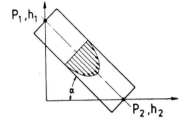

Fig. 5.5. Velocity distribution in a circular or flat rectangular channel in laminar flow

5.1.2 Laminar Flow

We consider a fluid flowing down a circular tube of length l and radius R, the entrance of which is at height h_1 and pressure p_1. The outlet is at h_2 and p_2. Laminar flow in this tube is characterized by a parabolic velocity distribution which is given by:

$$v(r) = (R^2 - r^2)\frac{p_1 - p_2 + \rho g(h_1 - h_2)}{4\eta l} = \frac{(R^2 - r^2) \cdot \Delta P}{4\eta l}$$ (5.11)

ΔP is a generalized pressure comprising the difference in hydrostatic pressures at both ends of the tube and the head $h_1 - h_2$, and η is the viscosity.

This equation states that at the wall of the tube, i.e. $r = R$, the velocity is zero, as expected due to adhesive forces exerted by the wall to the molecules of the liquid. The velocity distribution is shown by the arrows in Fig. 5.5.

The average velocity can be obtained by integration:

$$\bar{v} = \frac{R^2}{8\eta} \cdot \frac{\Delta P}{l} = \frac{R^2 \rho g}{8\eta} \cdot \frac{\Delta P}{\rho g \cdot l} = \frac{\rho g R^2}{8\eta} \cdot J$$ (5.12)

J is called the hydraulic gradient.

The volume rate of flow Q is then:

$$Q = \pi R^2 \bar{v} = \frac{\pi R^4}{8\eta} \cdot J.$$ (5.13)

The energy head loss h_f resulting from friction can be calculated by use of Eq. (5.10). Regarding the fact that v is constant over the whole length of the tube, one finds:

$$h_f = (h_1 - h_2) + \frac{p_1 - p_2}{\rho g},$$ (5.14)

which in combination with Eq. (5.12) yields:

$$h_f = \frac{8\eta l}{\rho g R^2} \cdot \bar{v}.$$ (5.15)

Laminar flow in karst systems very often is along joint partings which can be modelled by narrow, parallel slits with distance d. In this case one obtains a velocity distribution:

$$v = \frac{(h_1 - h_2)\rho g + p_1 - p_2}{2\eta l}\left(\frac{d^2}{4} - z^2\right),$$

(5.16)

where the coordinate is perpendicular to the confining plane and $z = 0$ is at the centre of the slit. The average flow velocity is:

$$\bar{v} = \frac{\rho g d^2}{12\eta}\cdot\left(\frac{(h_1 - h_2) + (p_1 - p_2)/\rho g}{1}\right) = \frac{\rho g d^2}{12\eta}\cdot J .$$

(5.17)

The volume rate is:

$$Q = \bar{v}W\cdot d = \frac{\rho g W d^3}{12\eta}\cdot J ,$$

(5.18)

where W is the width of the slit. It should be noted that the expressions above are related to the case of a narrow slit, i.e. $W \gg d$.

The head loss by friction h_f is given by:

$$h_f = \frac{12\cdot\eta\cdot l}{\rho g d^2}\cdot\bar{v} .$$

(5.19)

Flow rates for tubes of non-circular cross-section are calculated by multiplying Eqs. (5.17) and (5.18) by a geometrical correction factor M (Beek and Mutzall 1975). Figure 5.6 shows this correction factor for various shapes.

Both geometries show commonly the characteristics of laminar flow in general:

1. Average flow velocity or volumetric flow rate depend linearly on the driving force J called the hydraulic gradient;

2. Energy loss due to friction depends linearly on the average flow velocity.

There is an analogy between laminar flow in conduits and flow of electric current in resistors due to statement (1) and the continuity equation. The latter expresses the fact that at each point of any network of conduits the rate of flow to the point equals that away from it. Thus, any laminar flow network can be simulated by a corresponding network of resistors and voltages.

5.1.3 Turbulent Flow

In turbulent flow velocity fluctuations of the individual atoms occur. There is, however, a time-averaged velocity of the flow which describes the transport of the liquid. Velocity distributions for turbulent flow in conduits are therefore to be described in terms of this time-averaged velocity. In the following we consider turbulent flow in a circular conduit. In the centre of this circular tube there will be clearly a time-averaged flow velocity. Random fluctuations move particles on chaotic path lines along the average velocity. At the wall of the tube, because of adhesive forces, the flow velocity is zero and there are also no fluctuations. With increasing distance from the wall, the velocity and the fluctuations increase until both velocity and fluctuations reach a maximum.

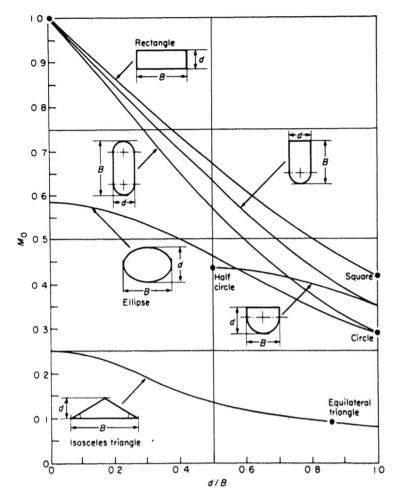

Fig. 5.6. Correction factors M for various geometries of conduits for laminar flow velocities (see text) (Beek and Mutzall 1975)

A crude model of the velocity distribution and fluctuations is presented in Fig. 5.7. Close to the wall one has a viscous sublayer, where the motion of the liquid is described as in laminar flow. In this region turbulent fluctuations are small and random motion is almost entirely due to Brownian motion of the molecules. Mass transport is due to molecular diffusion. Next to this sublayer there is a transition buffer zone, where velocity and fluctuations increase until the fully turbulent core is reached. Diffusion by turbulent random motion increases correspondingly which is characterized by an increase in the effective coefficient of diffusion. In the turbulent case, flow is fully turbulent, i.e. turbulent random motion determines momentum and mass transport.

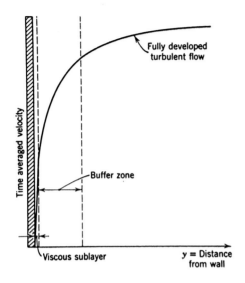

Fig. 5.7. Time averaged velocity in dependence of distance from the wall of a tube. Random fluctuations are small in the viscous sublayer, but increase in the buffer zone until they are fully developed in the core (Bird et al. 1960)

The velocity distribution of turbulent flow in a circular pipe cannot be calculated from first principles. One has to use semi-empirical methods (Bird et al. 1960).

Near the wall the velocity distribution can be expressed by:

$$v^+ = s^+, \qquad 0 \leq s^+ \leq 5, \tag{5.20}$$

where v^+ and s^+ are dimensionless and are related to the velocity v and the distance s from the wall by:

$$v^+ = \frac{v}{v^*}, \qquad s^+ = sv^* \frac{\rho}{\eta}, \qquad v^* = \sqrt{\frac{\tau_0}{\rho}}, \tag{5.21}$$

where τ_0 is the shear stress exerted to the wall of the tube by turbulent friction. It is related to the force acting onto the tube of length l and radius R by:

$$F = \tau_0 \cdot 2\pi R l. \tag{5.22}$$

Thus, the region $s^+ \leq 5$ defines the laminar viscous sublayer where the velocity increases linearly with distance from the wall. The buffer zone extends from $5 \leq s^+ \leq 26$ and v^+ is given by rather complicated analytical expressions, which will not be discussed here. They can be found in standard text books, e.g. Stephenson (1984). For $s^+ \geq 26$ the velocity profile in the turbulent core is given by:

$$v^+ = \frac{1}{0.36} \ln s^+ + 3.8. \tag{5.23}$$

Figure 5.8 plots v^+ versus s^+. The points are experimental data and show good agreement to the given expressions. Note that s^+ is plotted logarithmically.

To obtain the mean flow velocity averaged over the diameter of the tube, one has to perform an integration of the velocity distribution with respect to s^+. Its

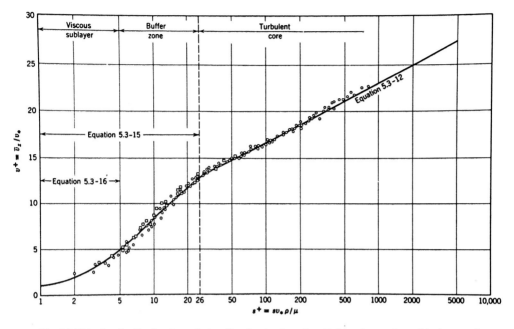

Fig. 5.8. Velocity distribution for turbulent flow in circular tubes. *Full line* is a semi-empirical expression. *Points* are from the experiment (Bird et al. 1960)

result is given by

$$\bar{v}^2 = \frac{8\tau_0}{\rho \lambda_f},$$ (5.24)

where λ_f is a friction factor, depending on the Reynolds number N_{Re} only. This is a very important result as it gives possibilities to determine λ_f as a function of N_{Re} experimentally.

Since in stationary flow the force exerted by friction along the wall must be balanced by the force due to gravity and pressure in the same direction, we have:

$$\{\rho g(h_1 - h_2) + p_1 - p_2\} \pi R^2 = 2\pi R \tau_0 \cdot l.$$ (5.25)

From this we obtain:

$$\tau_0 = \frac{R}{2} \cdot \frac{\Delta p}{l}.$$ (5.26)

Using Eq. (5.24) one finally has:

$$\bar{v}^2 = \frac{h}{l} \cdot \frac{2gd_t}{\lambda_f} \cdot J,$$ (5.27)

where d_t is the diameter of the tube, h is the hydraulic head along the tube length and is given by:

$$\frac{\Delta P}{\rho g} = h_1 - h_2 + \frac{p_1 - p_2}{\rho g} = h \, . \tag{5.28}$$

This is the Darcy-Weisbach equation, which states that in contrast to laminar flow, the average velocity is proportional to the square root of the driving force Δp. For a straight circular tube Eq. (5.28) states also (cf. Eq. 5.10) that the friction loss h_f is equal to the head h.

The friction factor λ_f is obtained from Eqs. (5.23) and (5.24) and thus depends on the assumptions made on τ_0.

For smooth pipes a good semi-empirical approximation is (Stephenson 1984):

$$\frac{1}{\sqrt{\lambda_f}} = 2 \log (N_{Re} \cdot \sqrt{\lambda_f}) - 0.8 \, . \tag{5.29}$$

To obtain an expression for the width ε of the laminar sublayer we use Eqs. (5.26) and (5.21) to find:

$$\varepsilon = \frac{5 \cdot \sqrt{2\eta}}{\rho \sqrt{Rg \cdot h/l}} \, . \tag{5.30}$$

So far we have considered hydrodynamically smooth pipes. If the wall of the pipe is rough, we can define roughness by the ratio k/d_t, where k is the average height of the roughness projections. In this case λ_f can be written (Stephenson 1984) as:

$$\frac{1}{\sqrt{\lambda_f}} = -2 \log \frac{k}{d_t} + 1.14 \tag{5.31}$$

for large values of k/d_t.

Expressions (5.29) and (5.30) can be combined to give the Colebrook-White formula, covering both smooth and rough pipe walls and also the transition zone:

$$\frac{1}{\sqrt{\lambda_f}} = -2 \log \left(\frac{k}{3.71d_t} + \frac{2.51}{N_{Re}\sqrt{\lambda_f}} \right) \, . \tag{5.32}$$

From the Darcy-Weisbach equation, we can write:

$$N_{Re} \cdot \sqrt{\lambda_f} = \sqrt{\frac{2gh}{l}} \cdot \frac{d_t^{3/2} \rho}{\eta} \, . \tag{5.33}$$

Inserting Eq. (5.33) into (5.32) and then Eq. (5.32) into (5.24), one obtains an explicit expression for the average flow velocity v:

$$\bar{v} = \sqrt{\frac{2ghd_t}{l}} \left[-2 \log \left(\frac{k}{3.7d_t} + \frac{2.51v}{\sqrt{2gd_t^3 \cdot h/l}} \right) \right] \, . \tag{5.34}$$

On the basis of this equation simple explicit flow-head loss graphs can be constructed, which are published by the Hydraulics Research Station at Wallingford (1969). Figure 5.9 shows such a graph calculated for smooth pipes (k = 0) and rough pipes with k ≠ 0. Flow velocities v are given as a function of the hydraulic gradient for various tube diameters d_t.

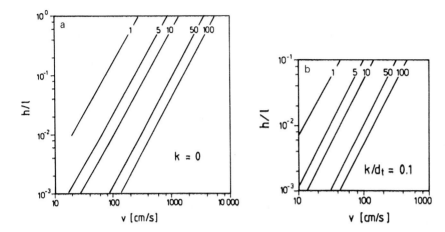

Fig. 5.9a,b. Flow velocities in straight circular tubes for various diameters d_t (*numbers* on the curves) in turbulent flow as a function of the hydraulic gradient. **a** For smooth tubes, $k = 0$; **b** for rough tubes, $k/d_t = 0.1$

If the conduit is not circular, a general expression for the Darcy-Weisbach equation can be given by use of the hydraulic radius R_h, which is introduced empirically by:

$$R_h = \frac{A}{P_m},$$ (5.35)

where A is the cross-section of the water stream and P_m the wetted perimeter of the conduit. Then, by replacing d_t by $4R_h$ in all the corresponding equations, one can easily derive the corresponding expressions for general conduit shapes.

Note that this empiricism does not hold for the case of laminar flow. Furthermore expression (5.34) is valid only for $N_{Re} > N_{Re}^{cr}$. N_{Re}^{cr} is the critical value, where turbulence occurs.

In turbulent flow the average flow velocity and correspondingly the volume flow no longer depend linearly on the hydraulic head h, but to a reasonably good approximation (cf. Fig. 5.9 and Eq. 5.34) on the square root of h. Thus, turbulent flow, in contrast to laminar flow, cannot be modelled by electrical analogues.

So far the results are valid only for straight tubes. In karst systems conduit tubes show changes of cross-section and bends. These contribute to the head loss by friction.

In this case the friction loss (cf. Eq. 5.27):

$$h = \frac{\bar{v}^2}{2g} \cdot \frac{1 \cdot \lambda_f}{d_t}$$ (5.36)

is augmented by head losses from changes in the tube geometry and one obtains:

$$h = \frac{\bar{v}^2}{2g} \cdot \frac{1 \cdot \lambda_f}{d_t} + \sum_i \frac{\bar{v}^2}{2g} \cdot K_i.$$ (5.37)

The loss coefficients K_i are due to contraction or expansions of the cross-sectional area, or due to bending of the conduit and due to exit or entrance losses. They exhibit values in the order of one and can be taken for special cases from the standard literature (Stephenson 1984; Bird et al. 1960). The effect of these head losses is to reduce the flow velocity compared to geometrically straight tubes.

Bögli (1980) has analyzed the influence of such losses on cave passages in the Hölloch, Switzerland. He finds for this special case geometrical losses amounting to 13% of those resulting from the friction at the walls.

The Darcy-Weisbach equation can also be used for open channel flow by using the hydraulic radius, i.e. the ratio of the cross-sectional area of the flowing liquid to the wetted perimeter of the channel.

Finally, it should be noted that there are a variety of other semi-empirical formulas for turbulent flow in tubes and open channel flow, such as those by Manning, Chezy and Hazen-Williams.

They all are related to the Darcy-Weisbach equation and the corresponding friction factors can all be expressed in terms of N_{Re} and λ_f (Stephenson 1984).

5.1.4 An Example of Hydraulic Characteristics of Karst Aquifers

One might ask whether the flow equations given for rather idealized conditions can be used to describe flow in real karst conduits which are of utmost complexity.

Milanovic (1981) has given a beautiful example that this is really the case. He investigated the relationship between the discharge capacity of the Ombla Spring,

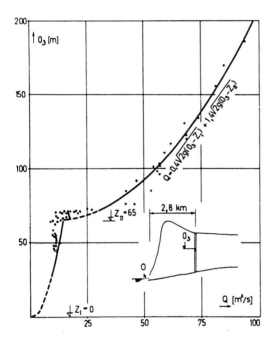

Fig. 5.10. Discharge of the Ombla Spring as a function of water level elevation O_3 in a borehole (Milanovic 1981)

a huge karst spring near Dubrovnik, Yugoslavia, and the water level in boreholes located in the watershed of the spring. There is a close correlation with coefficient $r_c = 0.96$ between these quantities. This was interpreted by Hogdin and Ivetic (1976) cited in Milanovic (1981). As soon as the water level in the borehole reaches an elevation above 65 m, a rapid increase in discharge of the spring occurs. This shows the existence of a karst channel connecting to the spring, and draining the aquifer, once it has been filled above that elevation. Figure 5.10 shows the discharge Q of the spring as a function of the elevation O_3 of the water level in the borehole. The full line has been calculated by the equations:

$$Q = 0.4 \cdot \sqrt{2gO_3} \quad \text{for} \quad O_3 \le 65\text{m} \, ;$$

$$Q = 0.4 \cdot \sqrt{2gO_3} + 1.4\sqrt{2g(O_3 - 65)} \quad \text{for} \quad O_3 > 65\text{m} \, . \tag{5.38}$$

The first term is due to a complex conduit system draining the lower part of the aquifers, whereas the second term becomes operative only for $h \ge 65$ m. The fact that the discharge depends on the square root of the elevations shows the characteristics of turbulent flow in conduits and open channels. More examples showing these characteristics are discussed by Bonacci (1987).

5.1.5 Flow Through Porous Media

The first experiment to analyze flow in porous media, e.g. sands, was carried out by Darcy in 1856. The experimental set-up is shown in Fig. 5.11. A circular cylinder of cross-section A and length L is filled with sand. The inlet is at height Z_1. It is kept under constant pressure p_1 by a water column above it. The corresponding

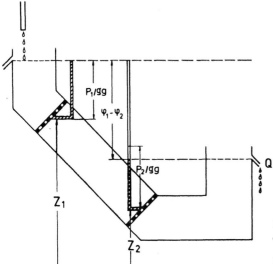

Fig. 5.11. Illustration of Darcy's law. The driving force to flow Q is the hydraulic potential $\varphi_1 - \varphi_2 = \dfrac{p_1 - p_2}{\rho g} + z_1 - z_2$

data at the outlet are described by Z_2 and p_2 respectively. The flow Q through the cylinder is found experimentally to be given by:

$$Q = \frac{A \cdot K}{L} \cdot \left(Z_1 - Z_2 + \frac{p_1 - p_2}{\rho g} \right) = AKJ , \qquad (5.39)$$

where K is a constant and is called the hydraulic conductivity. It depends on the properties of both the fluid and the porous medium. Note that this equation is analogous to Eqs. (5.12) and (5.18), which describe laminar flow in tubes and in narrow slits, with the proportionality constant determined by viscosity and the geometrical data of the conduit.

To transform Eq. (5.39) to the average velocity of the liquid in the medium, we define a specific discharge:

$$v = \frac{Q}{A} = K \cdot J = K \cdot \frac{\Delta P}{\rho g} . \qquad (5.40)$$

Assuming further that Darcy's findings can be generalized to be true also locally in any part of the cylinder, we can write:

$$v = \frac{K}{\rho g} \cdot \frac{dP}{dl} = K \cdot J . \qquad (5.41)$$

This equation is limited to one-dimensional flow. In a three-dimensional medium the generalization is:

$$v = KJ = K \cdot grad \left(\frac{p}{\rho g} + h \right) = -K \, grad \, \varphi . \qquad (5.42)$$

The hydraulic gradient **J** is now a vector and flow velocity **v** is parallel to it. Usually one defines a potential φ, which we will use from now on. Equation (5.42) remains also valid for inhomogeneous hydraulic conductivities varying spatially, i.e. $K = K(x, y, z)$.

For incompressible liquids the equation of continuity in three dimensions in a confined aquifer reads:

$$\frac{\partial v_x}{\partial x} + \frac{\partial v_y}{\partial y} + \frac{\partial v_z}{\partial z} = 0 . \qquad (5.43)$$

Substituting v by Eq. (5.42) one obtains:

$$grad \, (K \, grad \, \varphi) = 0 , \qquad \varphi = -\frac{P}{\rho g} - h . \qquad (5.44)$$

In the case of homogeneous media, where K is constant, this simplifies to the Laplace equation:

$$\Delta(K\varphi) = \Delta\Phi = 0 , \qquad K\varphi = \Phi . \qquad (5.45)$$

Since we are interested mainly in two-dimensional flow problems, the equation to be discussed in terms of the generalized potential is:

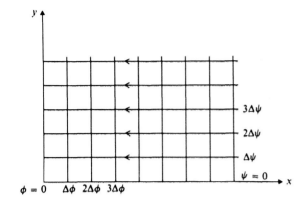

Fig. 5.12. Flow net for homogeneous flow along the x-axis in the x-y plane (see text)

$$\frac{\partial^2 \Phi}{\partial x^2} + \frac{\partial^2 \Phi}{\partial y^2} = 0 \,. \tag{5.46}$$

This equation can be solved numerically once the boundary conditions are known (Kinzelbach 1986). In many cases analytical solutions are also given (Bear 1979).

The simplest solution is that of homogeneous, uniform flow with velocity v in an infinite plane:

$$\Phi = -v_x \cdot x - v_y y \,. \tag{5.47a}$$

The lines of equal potential are shown in Fig. 5.12 for $v_y = 0$.

One now can define a function ψ, which is also a solution of Eq. (5.46):

$$\psi = -v_x y + v_y x \,. \tag{5.47b}$$

Lines of equal ψ are called streamlines and are also shown in Fig. 5.12. There is a generally valid relation between the stream functions $\psi(x, y)$ and the potentials $\Phi(x, y)$:

$$v_y = \frac{\partial \psi}{\partial x} = -\frac{\partial \Phi}{\partial y} \,;$$

$$v_x = -\frac{\partial \psi}{\partial y} = -\frac{\partial \Phi}{\partial x} \,. \tag{5.48}$$

Thus, by use of Eq. (5.44) ψ can be calculated if Φ is known.

There are two important properties of streamlines:

1. Streamlines show the average path that a particle moves in the flow. This can be seen in the following way. If a particle has moved a small distance defined by $ds = (dx, dy)$ with velocity $v = (v_x, v_y)$, these two vectors are parallel. Therefore,

$$v \times ds = 0 \,. \tag{5.49}$$

Inserting Eq. (5.48) for v one finds:

$$\frac{\partial \psi}{\partial x} dx + \frac{\partial \psi}{\partial y} dy = d\psi = 0 \,. \tag{5.50}$$

This states that the stream function does not change its value along the path of the particle. Therefore, lines of constant ψ are path lines of the particle.

2. Streamlines and equipotential lines are perpendicular to each other. In our example of homogeneous flow along a plane one can visualize lines of equal potential as lines of equal height at the inclining plane. The flow of water then will be in the direction of steepest descent, which is the shortest distance between two adjacent lines of equal height. Therefore, its velocity will be perpendicular to the lines of equal height.

This can be also generally expressed mathematically by:

$$\text{grad } \Phi \cdot \text{grad } \psi = 0 , \qquad (5.51)$$

as can be seen immediately from Eq. (5.48). Since the gradient of any function is a vector pointing perpendicular to its isolines, Eq. (5.51) expresses isolines of Φ to be perpendicular to those of ψ and vice versa.

A net constructed from equipotential lines and streamlines is called a flow net. In its simplest form it is shown by Fig. 5.12.

The properties (1) and (2), which are valid generally as long as the medium is isotropic and homogeneous, can be easily visualized from this figure.

Any two streamlines with ψ_1 and ψ_2 confine an area in the flow net, which is called a stream tube. Water flowing in this tube cannot cross the boundaries defined by these two streamlines. The discharge Q delivered by a stream tube can be calculated by integration of v over a cross-section of the tube, and by using Eq. (5.48) one obtains:

$$\Delta Q = \psi_2 - \psi_1 . \qquad (5.52)$$

If one constructs flow nets where the difference $\Delta\Phi$ of the potential between neighbouring equipotential lines is equal to the difference $\Delta\psi$ of the stream function between neighbouring streamlines, one obtains quadratic flow nets. All these properties are easily seen in Fig. 5.12. To illustrate these properties also on flow nets for different geometries, Fig. 5.13 shows the flow net of an infinite, confined aquifer with a recharge well of radius r_w at the origin. This recharging well imposes the boundary condition that for $r < r_w$ the potential is kept constant. Because of symmetry the equipotential lines are circles centred at $r = 0$. The streamlines are straight lines originating from the well. For mathematical details of this pumping or recharging well problem the reader is referred to text books, e.g. Bear (1979).

All the properties of flow nets, which have been discussed, can also be seen in Fig. 5.13.

Since the Laplace equation (5.45) is linear, any of two different solutions can be added to give a new solution with corresponding boundary conditions. If we superimpose a recharging or pumping well to a homogeneous, uniform base flow, the flow net shown in Fig. 5.14 results.

The base flow is from right to left. The well is pumping and flow enters the well. By reversing all the arrows we have the opposite where the well is recharging at the same rate into an aquifer with base flow also reversed. The flow nets remain unchanged.

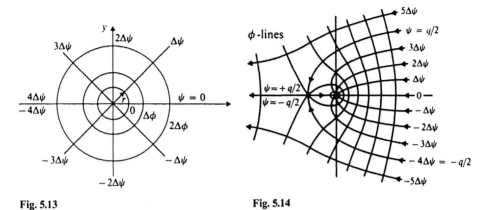

Fig. 5.13   **Fig. 5.14**

Fig. 5.13. Flow net for radial flow from a well centred at r = 0. Flow lines are the *straight lines*. Lines of equipotential are *circles* (see text)

Fig. 5.14. Superposition of homogeneous base flow and radial flow into a sink. Note the water divide at lines $\psi = q/2$ and $\psi = -q/2$; q is the discharge into the sink

There is one very important general point, which occurs whenever flow originates from different pumping or recharging sources. In Fig. 5.14 the flow net can be divided into two regions. In the region between $\psi = -4\varDelta\psi$ to $\psi = 4\varDelta\psi$ all the streamlines enter the pumping well. In the case of a recharging well the arrows are reversed and correspondingly all those lines leave the well. Thus, in the latter case all the water recharged into the aquifer occupies a domain bordered by these two streamlines, which constitute a water divide to the water resulting from base flow. Thus, these two waters of different origin never mix, if one neglects molecular diffusion and dispersion.

This statement is generally true for any combinations of wells recharging into an aquifer with any combination of outlets, due to the fact that each volume element of fluid maintains its identity in laminar flow along its path and mixing with neighbouring elements is excluded. In general, each recharging inlet defines its own flow domain comprising some appropriate outlets. Flow domains of different inlets are separated by water divides.

This is illustrated in Fig. 5.15a (see page 98), which shows these domains in a confined aquifer with four inlets and one outlet. This figure has been drawn as a result of an experiment in which, by using dyes, these flow domains could be viewed in a hydraulic filtration model, consisting of a confined tabular body of sand (Ewers 1978). Similarly, other geometries of inputs and outlets are shown in Fig. 5.15b and c.

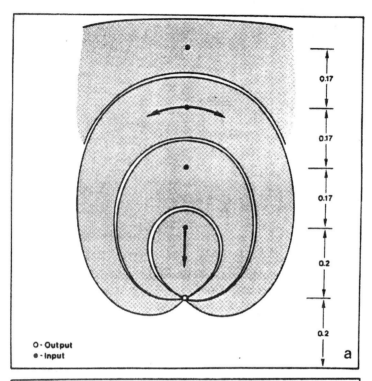

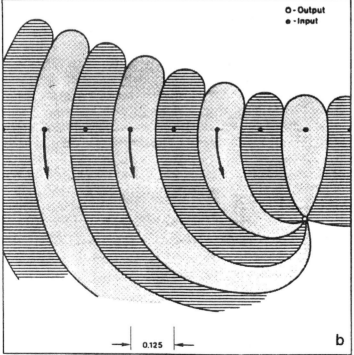

Fig. 5.15a,b

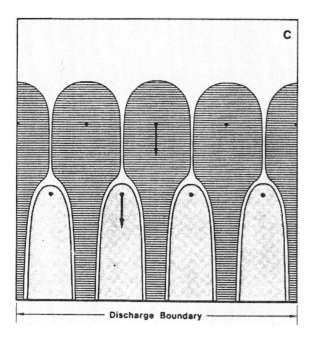

Discharge Boundary

Fig. 5.15a–c. Flow domains in various input-output geometries. a Input and output situated in one linear array; b linear array of inputs, output is separate; c two linear arrays of inputs discharge to an extended output boundary. Water of neighbouring flow domains (differently marked) cannot cross the boundaries between the domains (Ewers 1982)

Part II
Principles of Dissolution and Precipitation of CaCO$_3$

6 Dissolution and Precipitation of Calcite: The Chemistry of the Heterogeneous Surface

As already stated dissolution of calcite is a complex process comprising three different, simultaneously acting mechanisms:

1. Diffusion of the reactants, such as H^+, H_2CO_3 and CO_2, towards the calcite surface and also diffusion of the dissolved products Ca^{2+}, CO_3^{2-} and HCO_3^- into the bulk of the fluid.
2. Conversion of CO_2 into the aggressive reactants H^+ and H_2CO_3 by the reaction $H_2O + CO_2^{aq} = H^+ + HCO_3^-$.
3. The release of Ca^{2+} and CO_3^{2-} from the solid is a heterogeneous chemical process and depends in some way on the activities of the reactants and products at the surface. Figure 6.1 gives a schematic diagram of these processes.

The release of Ca^{2+} from the surface is accomplished by three simultaneously acting elementary reactions (cf. Chap. 1). The first is the attack of H^+ to $CaCO_3$ which can be visualized in the reaction of calcite with strong acids:

$$CaCO_3 + H^+ \rightleftharpoons HCO_3^- + Ca^{2+} .$$

The second is the direct reaction of $CaCO_3$ with undissociated carbonic acid:

$$CaCO_3 + H_2CO_3 \rightleftharpoons 2HCO_3^- + Ca^{2+} ,$$

and the third is physical dissolution in water, where Ca^{2+} and CO_3^{2-} are released:

$$CaCO_3 + H_2O \rightleftharpoons Ca^{2+} + CO_3^{2-} + H_2O \rightleftharpoons Ca^{2+} + HCO_3^- + OH^- .$$

To calculate dissolution rates at the surface one has to know the rate equations, which in some way are functions of the activities $(H^+)_s$, $(HCO_3^-)_s$, $(CO_3^{2-})_s$, $(Ca^{2+})_s$ and $(H_2CO_3^*)_s$ at the surface. These activities, however, are dependent on the rate of the surface reactions, on the rate of transport of reactants and products and on the rate of production of H^+ and H_2CO_3 by conversion of CO_2. Finally stoichiometry in an $H_2O-CO_2-CaCO_3$ system (cf. Eq. 2.31, Fig. 2.12), which is valid in the realm of karst waters, states that for each Ca^{2+} released from the solid, the corresponding CO_3^{2-} has to react with one H_2CO_3, or $HCO_3^- + H^+$ respectively, forming HCO_3^-.

If dissolution proceeds sufficiently slow, as is always the case in karst environments and in practically any laboratory experiment, then a quasi-stationary state is established. The condition for the surface concentrations is such that the rate of Ca^{2+} released from the surface has to be equal to the rate of Ca^{2+} transported into the bulk, and also to the rate of CO_2 converted into H^+ and HCO_3^-. Therefore, the

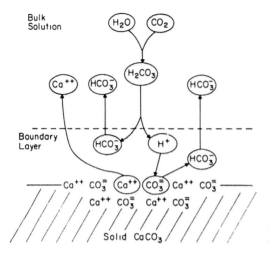

Fig. 6.1. Schematic diagram of mass transport processes and chemical reactions in the dissolution of calcite (White 1977)

actual dissolution rate is determined by a complex, concerted action of all three mechanisms. Changing one of them changes the dissolution rate.

To obtain information on the rate laws of the heterogeneous reactions at the calcite surface by dissolution experiments requires exact knowledge on the transport properties in the experiment performed and one also has to know the exact conditions of CO_2 conversion.

There is a tremendous amount of literature on calcite dissolution experiments performed with various materials (synthetic calcite, natural calcites and limestone) under various conditions (pure H_2O–CO_2 system, seawater and artificial seawater) using various experimental techniques. In this chapter we will focus on these experiments which supply information on the heterogeneous surface reactions. With this knowledge and als using the theory of mass transport (cf. Chap. 3) and the data on CO_2 conversion (cf. Sect. 4.3), one is able to model dissolution rates in situations which are related to karst environments (cf. Chap. 7).

6.1 Experimental Methods of Calcite Dissolution

To design an experiment giving information on surface reactions, one has to provide two conditions: (1) transport of all the species into the bulk solution has to be very effective and (2) CO_2 conversion has to be fast, such that the surface reactions, as the slowest processes, determine dissolution rates.

To enhance mass transport one usually stirs the solution, surrounding the calcite surfaces, by an impeller, thus creating some kind of turbulent flow, where mass transport into the bulk is determined by high eddy diffusion. In any kind of turbulent flow around surfaces, a hydrodynamic boundary layer exists adjacent to the surface, where mass transport is determined by molecular diffusion. In this diffusion boundary layer, flow is determined by molecular viscosity, since eddies cannot penetrate into this layer (cf. Sects. 3.4 and 5.1.3). The thickness of this layer determines a region of high resistance to mass transport and it is therefore desirable to keep it as small as possible. To obtain this, two types of experimental setups are used, which will be described in some detail.

6.1.1 Rotating Disc System

For certain systems a quantitative treatment of mass transfer is possible, which is not generally the case. In the case of a flat circular disc, rotating with angular velocity ω around an axis perpendicular to its plane in an infinitely large body of fluid, Levich (1962) has solved the flow equations for viscous, incompressible flow. A detailed treatment is given also by Pleskov and Filinovskii (1975).

Figure 6.2 shows the flow lines. The disc rotates around the y-axis. There is flow from the bulk towards the disc, shown by the flow lines in Fig. 6.2a. As the flow approaches the disc the flow lines bend over and finally proceed parallel to the surface of the disc. The radial dependence on the flow lines is shown in Fig. 6.2b.

From the boundary condition that the relative velocity between the surface of the disc and the fluid in touch with it has to be zero, one can see that a thin layer of fluid is dragged with the disc. Mass transport across this layer is governed by molecular diffusion. Once dissolved molecules have penetrated through this layer they are transported away by the laminar flow along the flow lines parallel to the surface. Thus, the laminar flow field acts like one large eddy. Figure 6.3 illustrates this situation.

A practical experimental setup is shown in Fig. 6.4. A piece of the dissolving material is inserted into the lower surface of an appropriately shaped cylindrical, symmetrical body, which rotates in a vessel with a diameter, which is large compared to that of the disc. It is of utmost importance that the surface of the disc is extremely smooth so as not to disturb the flow field. Rotational speeds up to 10 000 rpm are used and care has to be taken to avoid wobbling of the disc. Detailed descriptions of rotating disc systems have been given by Cornet et al. (1969), Gregory and Riddiford (1956) and Riddiford (1966).

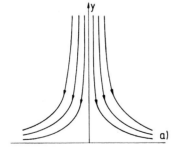 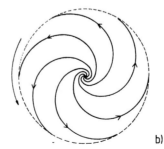

Fig. 6.2a,b. Flow lines at a rotating disc. **a** Projection of flow lines in a plane perpendicular to the disc. The axis of rotation is identical with the y-axis. **b** Flow lines projected into the plane of the disc

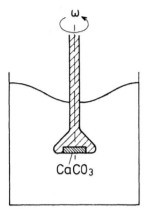

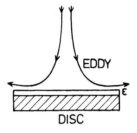

Fig. 6.3. Illustration of laminar boundary layer of thickness ε. Molecules, having penetrated this layer by molecular diffusion, are transported quickly along the flow lines which act as one large eddy

◀ Fig. 6.4. Rotating disc. A body of cylindrical symmetry is immersed into the solvent contained in a comparatively large vessel. The dissolving solid is mounted into the lower surface such that a very smooth surface is established

The mass transfer from the rotating disc has been calculated in detail by Pleskov and Filinovskii (1975). From the results of this calculation, a diffusion boundary layer of thickness ε can be defined:

$$\varepsilon = 0.45 \left(\frac{v}{D}\right)^{-1/3} \cdot \varepsilon_h , \tag{6.1}$$

where ε_h is the thickness of the hydrodynamic boundary layer, which is determined from the flow characteristics and is given by:

$$\varepsilon_h = 3.6 \left(\frac{v}{\omega}\right)^{1/2} . \tag{6.2}$$

ω is the angular velocity of the disc and v the kinematic viscosity.

Note that the diffusion boundary layer and the hydrodynamic boundary layer thickness ε and ε_h are related by the Schmidt number:

$$N_{Sc} = \left(\frac{v}{D}\right) . \tag{6.3}$$

This is a general relation in all types of hydrodynamic transport problems (Beek and Mutzall 1975).

To calculate the mass flux from a rotating disc we use the results of the situation elucidated in Chapter 3 (Fig. 3.9), which is completely analogous with the geometry of Fig. 6.2. Assuming the concentration of the dissolving substance at the surface of the disc as c_w and that in the bulk as c_t, the flux across the boundary according to Eq. (3.27) is:

$$F_d = \frac{D}{\varepsilon}(c_w - c_t) = k^d(c_w - c_t) . \tag{3.27}$$

It should be noted here that this formulation is called a Nernst model, first proposed by Nernst (1904). It assumes that mass transport by molecular diffusion is essentially across a Nernst layer of thickness ε, which has to be defined for each special case.

For a first-order heterogeneous reaction at the surface, the flux F_c from the surface can be written as (cf. Eq. 4.64):

$$F_c = k(c_{eq} - c_w), \tag{4.64}$$

where c_{eq} is the concentration at saturation. By equating $F_d = F_c$ we obtain c_w and correspondingly the flux F as discussed in Section 4.4:

$$F = \frac{k^d \cdot k}{k^d + k}(c_{eq} - c_t) = k(c_{eq} - c_t). \tag{4.67}$$

Inserting

$$k^d = \frac{D}{\varepsilon} = \frac{D^{2/3}\omega^{1/2}}{1.61\nu^{1/6}}, \tag{6.4}$$

one obtains:

$$\frac{1}{F} = \left(\frac{1.61\nu^{1/6}}{D^{2/3}\omega^{1/2}} + \frac{1}{k}\right) \cdot \frac{1}{c_{eq} - c_t}. \tag{6.5}$$

Thus, by plotting the reciprocal rate $1/F$ versus $\omega^{-1/2}$ one obtains a straight line. From the slope of this line the diffusion coefficient D can be found, whereas the intercept with the $1/F$-axis gives the chemical rate constant k. Note that the mass flux increases with increasing rotational speed ω. This is due to the fact that the diffusion boundary layer becomes thinner with increasing ω, as is reflected by Eqs. (6.1) and (6.2).

This behaviour, indeed, has been found by Rickard and Sjöberg (1983) for the initial dissolution rates of $CaCO_3$ in pure 0.7-M $KCL-H_2O$ solutions at pH 8.4. Figure 6.5 shows a plot of reciprocal rates versus $\omega^{-1/2}$ for Carrara marble and

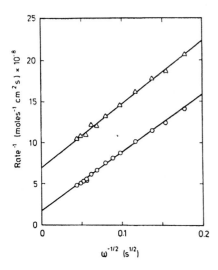

Fig. 6.5. Reciprocal rate $1/F$ versus $\omega^{-1/2}$ for Carrara marble (o) and Iceland spar (∆) in 0.7 M KCl solutions at pH = 8.4 and 25°C (Rickard and Sjöberg 1983)

Iceland spar. The experimental data fit well to the straight lines predicted by Eq. (6.5).

Note that this behaviour is only to be expected for first–order chemical kinetics at the surface. A much more complex situation arises for higher order kinetics. Nevertheless, the rotating disc method has its merits due to the defined hydrodynamics of the system.

It also has disadvantages due to the small surface of the material to be dissolved and the large volume of liquid, which has to be used to exactly define the hydrodynamics. As a result of this, the approach of the concentrations towards equilibrium is extremely slow and the experiment is very time-consuming.

6.1.2 Batch Experiments

To overcome this difficulty the area of the dissolving material should be increased. This can be easily accomplished by using small crystals with diameters in the range between a few microns up to a few hundred microns with surface areas of $1 \text{ m}^2 \text{ g}^{-1}$ calcite down to $0.1 \text{ m}^2 \text{ g}^{-1}$. A sufficient amount of material is suspended into a solution stirred by an impeller at sufficiently high rates to keep all particles in suspension. Thus, one can increase the surface area by a factor of 10 to 100 in comparison to rotating disc experiments and accordingly reduce the time necessary to perform the experiment. The hydrodynamics of the system, however, is poorly defined and it is difficult to interpret the data in terms of surface reaction kinetics. In many experiments, therefore, one changes stirring rates and observes dissolution as a function of stirring rates. If dissolution rates become independent of the stirring rates, one concludes that the system is governed by surface reactions (Plummer and Wigley 1976). This conclusion, however, need not necessarily be correct. The independence of stirring rate may also result when hydrodynamic conditions remain unaffected by changing the stirring rates (Sjöberg and Rickard 1983). This will be discussed in some detail later. As a crude indicator to decide whether a process is transport or chemically controlled often the stirring coefficient n is used. In many cases the dissolution rates depend on ω^n. For $n > 0.5$ transport control dominates, otherwise chemical control determines the process (Bircumshaw and Riddiford 1952).

In the discussion of mass transport properties two cases have to be considered. If the particles are small with diameters less then 10μ they move with the liquid such that the relative velocity between the particle and the liquid is zero. In this case mass transport of dissolving material has been calculated by Nielsen (1964) for the case of spherically shaped particles as:

$$F = \frac{D}{r}(c_W - c_B), \tag{6.6}$$

where r is the radius of the particle and c_W the concentration at its surface, c_B is the concentration in the bulk of the solution. For particles which are larger, r in Eq. (6.5) has to be replaced by ε, the thickness of a diffusion boundary layer surrounding the suspended particle. Since ε depends on the relative velocity between fluid and particle, which for larger particles is no longer zero, it remains undefined in most experiments and, if at all, only the order of magnitude can be given.

Plummer and Wigley (1976) have estimated ε for particles of an average size between 59 to 630 μ by calculating the Sherwood number from the hydrodynamics of the experiment. From the Sherwood number, which gives the ratio of the particle diameter to ε, the thickness of the diffusion boundary layer is found to be in the order of 10 μ.

In spite of the poor definition of the magnitude ε of the boundary layer, stirred batch experiments can be useful in extracting information on dissolution kinetics for the two limiting cases. If the surface reaction rates are sufficiently low, such that:

$$k^d = D/\varepsilon \gg k \,,$$

one can extract surface reaction constants with some confidence from Eq. (4.67). In the other limit:

$$k^d = D/\varepsilon \ll k \,,$$

no information on k can be obtained. If, however, several parallel reactions occur on the surface, as is the case with calcite, information can be obtained on the relative velocity of different reactions. Thus, at low pH values (pH < 5) one finds that calcite dissolution depends heavily on the rate of stirring, whereas at high pH > 6 this dependence is drastically reduced. This is a hint that the reaction:

$$CaCO_3 + H^+ \rightleftharpoons Ca^{2+} + HCO_3^-$$

is a very fast reaction, which dominates at low pH. At high pH the reaction:

$$CaCO_3 + H_2O \rightleftharpoons Ca^{2+} + HCO_3^- + OH^-$$

becomes dominant as a comparatively slow reaction. This will be discussed in detail in Section 6.2.

To show the importance of the influence of hydrodynamic conditions on stirred batch setups Sjöberg and Rickard (1983) have performed those experiments on the same material in different reaction vessels. Figure 6.6 shows calcite dissolution rates

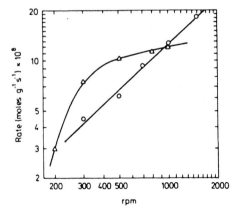

Fig. 6.6. Log rate per unit surface area versus rotational velocity of impeller stirrer for runs with calcite crystal fragments (125–250 μ) in 0.7 M KCl solutions at pH 8.4 and 25°C. Runs performed in round-bottomed vessels without baffles ($\triangle$) shows a high stirring dependence at low stirring rates which diminishes at higher rates, when the particles become fully suspended. For runs performed in the rotating disc vessel with baffles ($\circ$), the dissolution rate is slower except at very high stirring rates (Sjöberg and Rikkard 1983)

in dependence on the rotational speed of the impeller in a rounded, unbaffled vessel in comparison to those with a baffled vessel. The baffled vessel data show a high dependence on stirring rate with $F \alpha \omega^{0.9}$, indicating transport control. The unbaffled vessel shows a bending curve, which seems to indicate surface control above $\omega >$ 500 rpm. From the fact that the rates are lower than in the baffled vessel, however, one has to conclude that transport controls dissolution rates. Thus, the dependence of the boundary layer thickness ε on rotational speed is heavily influenced by the boundary conditions imposed by the surrounding vessel. From these findings a combination of experiments using rotating disc experiments and stirred batch experiments seems to be the appropriate way to eludicate the details of dissolution processes.

A final comment should be made on the special case of calcite dissolution. In the introduction to this chapter we have stated that CO_2 conversion may be a rate-limiting reaction. Therefore, in all experiments one has to be aware of its importance. The amount of CO_2 converted per time, in any case is proportional to the volume V of the solution. It also must be equal to the amount of calcite dissolving, which depends on the surface A of calcite present in solution. Thus, one can increase the amount of CO_2 converted by increasing the volume. As will be shown later in Chapter 7 a ratio of $V/A \geq 1$ cm is sufficient under all experimental conditions to exclude CO_2 conversion as a rate-limiting reaction.

6.1.3   Measurement of Dissolution Rates

To determine the dissolution rates experimentally one has to measure the concentration of Ca^{2+} as a function of time during the experimental run. From the equation:

$$V \cdot \frac{d[Ca^{2+}]}{dt} = A \cdot F , \tag{6.7}$$

the rate F can be determined, if the surface area A of the calcite and the volume V of the solution are known. The time derivative of the Ca^{2+} concentration is determined by measuring $[Ca^{2+}]$ (t) at sufficiently close time intervals and calculating the time derivative by a difference formula (Plummer and Wigley 1976).

Figure 6.7 shows a typical reaction vessel for stirred batch experiments. The solution is stirred in a closed vessel, which is kept at a constant temperature by use of a thermostat.

To keep the solution at a constant CO_2 concentration a CO_2-containing atmosphere of known p_{CO_2} is bubbled through the solution at a flow rate of a few litres per minute. Several electrodes to measure pH, conductivity, etc. are mounted into the lid.

Finally inlets, for either supplying reagents, such as acids, into the solution or to withdraw a sample of solution for analysis, are provided. The reaction vessel for rotating disc experiments is similar, replacing the impeller by the rotating disc. The Ca^{2+} concentration can be measured directly by using Ca^{2+}-sensitive electrodes (Compton and Daly 1984) or by monitoring conductivity (Baumann et al. 1985). In

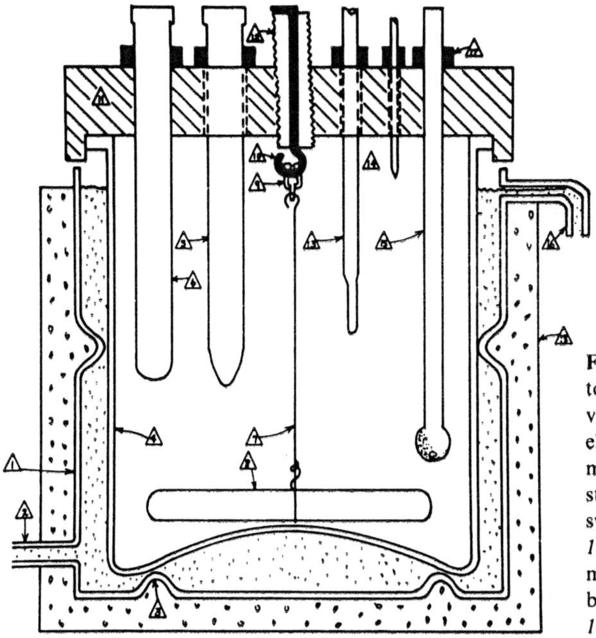

REACTION VESSEL

Fig. 6.7. Diagram of reaction vessel (not to scale). *1* Water bath; *2* water inlet; *3* vessel support; *4* reaction chamber; *5* pH electrode; *6* calomel electrode; *7* platinum magnetic stirrer support wire; *8* magnetic stirrer bar; *9* Teflon swivel; *10* platinum swivel support wire; *11* Teflon vessel top; *12* Teflon stirrer adjuster; *13* thermometer; *14* acid or base inlet; *15* glass bubbler; *16* water outlet; *17* tygon collar; *18* foam insulation (Morse 1974)

most experiments, however, pH is monitored during the experiment. If p_{CO_2} and pH are known, computer programs such as WATEQ can be used to calculate the concentration of the corresponding species in the solution, e.g. Ca^{2+}, HCO_3^-, CO_3^{2-}, etc. Thus, the chemical composition of the solution is known at any time in the experimental run.

There are two ways of performing the experiments. In the "free-drift" experiment, one prepares an H_2O–CO_2 system of known composition and adds a sufficient amount of $CaCO_3$. The system then reacts to equilibrium at a fixed p_{CO_2}. This is the condition of the open system, discussed in Chapter 2. These experimental conditions are close to the conditions of calcite dissolution in the unsaturated zone of calcareous soils.

The second way is the pH-stat method, first used by Morse (1974). Here, one keeps pH fixed during dissolution. This is achieved by adding diluted hydrochloric acid by use of an automatic burette, which is controlled by the desired pH, preset in the experiment. This method extends the region of chemical compositions of the solution above those which can be obtained by the free-drift experiments, since in addition to p_{CO_2}, pH can now be chosen as an independent parameter. Since in the pH-stat experiment the overall reaction is:

$$CaCO_3 + 2H^+ \rightarrow Ca^{2+} + CO_3^{2-} + 2H^+ \rightarrow Ca^{2+} + H_2CO_3,$$

the rate of dissolved $CaCO_3$ can be obtained from the rate of addition of acid via the stoichiometry of the reaction above. Such pH-stat experiments are especially suited to investigations relating to dissolution of calcite in seawater. In this geological situation the composition of the solution surrounding the particle stays constant, and it is easily possible to simulate this natural environment by the pH-stat method.

Further details on pH-stat experiments are given by Sjöberg (1978) and Plummer et al. (1978).

6.2   The Kinetics of Calcite Dissolution

Since dissolution kinetics are of high geological interest in many natural environments, an increasing interest has focussed on this problem since the early 1970s. In this section we will review some of this huge body of literature concerning calcite. Two approaches have been taken. In the first one, by using pure material, i.e. synthetic $CaCO_3$ or pure Iceland spar, the aim of the experiments has been to obtain information on the chemical kinetics of the surface reactions. The final aim of these investigations is to find a mechanistic reaction model. Since reaction models are based on reversible, elementary chemical reactions they also give information on the precipitation properties of the substance under consideration. Therefore, information on these models can also be obtained by observing the kinetics of precipitation. The results from experiments of this kind are therefore corroborative to data of dissolution experiments.

The second approach focusses mainly on natural materials, such as biogenic calcites of varying magnesium content. Therefore, many experiments on such materials have been performed in seawater or artificial seawater. The approach is an empirical one. The results are expressed in empirical formulas with the saturation state Ω as a parameter.

Three review articles on this subject have been published up to now. The first by Plummer et al. (1979) is concerned primarily with the mechanistic interpretation of reaction kinetics in systems composed of pure $CaCO_3$, H_2O, CO_2 and a diluted, strong acid or base to establish a pH from 4 up to 10. The second by Morse and Berner (1979) deals mainly with calcite and aragonite dissolution in seawater and applies the results to marine sediments. The latest review by Morse (1983) reported also on calcite precipitation and devoted much attention to natural systems. Furthermore, it addressed the question of inhibition of dissolution and precipitation by substances such as Mg^{2+}, orthophosphate, heavy metals and organic material.

6.2.1   Three Regions of Calcite Dissolution

An important advance in eludicating the interplay of diffusional mass transport and surface kinetics in calcite was achieved by Berner and Morse (1974). They performed experiments at 25°C and 1 atm pressure on fine-grained synthetic calcite in natural and artificial seawater and in an $NaCl–CaCl_2$ solution of the same ionic strength and calcium concentration as seawater at various pressures of CO_2. They used pH-

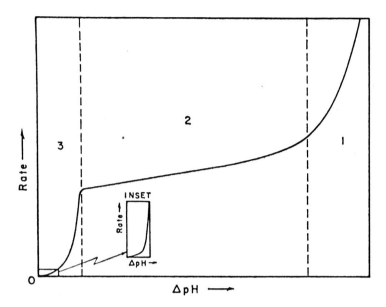

Fig. 6.8. Schematic plot of calcite dissolution rate versus ΔpH. The *inset* is a magnified portion of the region near equilibrium (Berner and Morse 1974)

stat techniques to measure the initial rates upon addition of $CaCO_3$ particles to the seawater or pseudo seawater solutions.

Figure 6.8 shows a schematic plot of their results. The dissolution rates are plotted against ΔpH. This parameter is defined as the difference of the pH of the undersaturated solution in the experiment and the pH at saturation with calcite. Note that in this type of experiment the saturated solution with respect to Ca^{2+} and p_{CO_2} has the same composition as the undersaturated one at the start of the run (Berner and Wilde 1972). In this case, i.e. only with this special definition, one has the relation:

$$\Delta pH = -\tfrac{1}{2}\log\Omega . \tag{6.8}$$

Note that therefore all rates in Figs. 6.8, 6.9, 6.10 and 6.11 refer to one common calcium concentration, and pH is the variable parameter.

In Fig. 6.8 three regions of undersaturation can be clearly defined, representing different types of reaction control. In region 1 at low pH of the solution (i.e. high $[H^+]$) there is a strong increase in dissolution rates with increasing ΔpH. In region 2 all data plot as a linear function of ΔpH. There is a sharp discontinuity separating region 3 from region 2 by a steep drop of rates when approaching equilibrium pH, i.e. ΔpH = 0. Close to equilibrium a region of inhibited dissolution is present (see insert).

Figure 6.9 presents a plot of the actual experimental data on a logarithmic scale for various p_{CO_2}. In contrast to the extreme change of the rates by three orders of magnitude in dependence on ΔpH, there is only a relatively weak dependence on the p_{CO_2} in equilibrium with the solution.

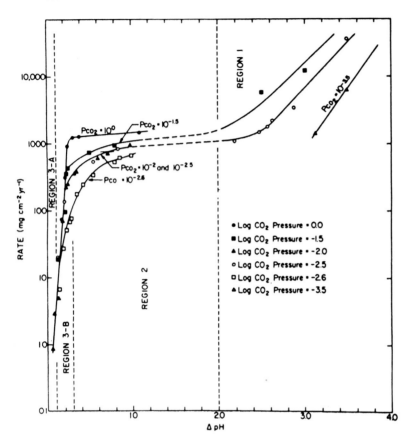

Fig. 6.9. Solution rates from the experimental data of Berner and Morse (1974) plotted as functions of pH for various pressures of CO_2 (White 1977)

To eludicate the nature of the different mechanisms, Berner and Morse (1974) compared the experimental data to those obtained from a theory of diffusion controlled dissolution.

In their experiments particles with diameters below 10 μ were used. Thus, the theory of Nielsen (1964) was employed to calculate the diffusional flux from a sphere of the corresponding particle diameter (cf. Eq. 6.6). For each species of ions:

$$F = \frac{D}{r}(c_{eq} - c_B),\tag{6.6}$$

where c_{eq} is the equilibrium concentrations at the surface and c_B is the concentration in the bulk of the solution. From the mass balance of dissolution we have:

$$F_{Ca} = F_{H_2CO_3} + F_{HCO_3} + F_{CO_3},\tag{6.9a}$$

where F is the diffusion flux of each species to and from the calcite surface. Equation

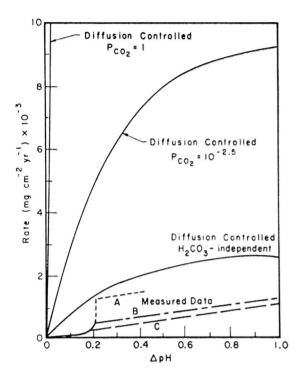

Fig. 6.10. Measured dissolution rates as functions of ΔpH compared to theoretical results from mass transport controlled by diffusion. Measured curves A: $p_{CO_2} = 1$ atm; B: $p_{CO_2} = 10^{-1.5}$ atm; C: $p_{CO_2} = 10^{-2}$ atm and $10^{-2.5}$ atm (Berner and Morse 1974)

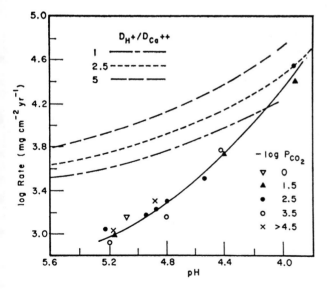

Fig. 6.11. Log of measured dissolution rates versus pH compared with the rates calculated by the H_2CO_3 independent, diffusion-controlled, theoretical model for various ratios D_H/D_{Ca} (Berner and Morse 1974)

(6.9a) is the consequence of the fact that each calcium ion released from the solid is accompanied by one carbon atom which is also delivered into the solution.

Charge balance requires:

$$2F_{Ca} + F_H - F_{HCO_3} - 2F_{CO_3} - F_{OH} = 0 \,, \tag{6.9b}$$

since no electric current is associated with the dissolution process. To calculate the fluxes one assumes that the diffusion constants of all species are equal to D, except that of the hydrogen ion D_H, which is larger. The bulk concentration in the solution can be calculated from the experimentally known calcium ion concentration, pH and p_{CO_2} by employing corresponding mass action laws, Eqs. (2.2), (2.3) and (2.4). To obtain the concentration c_{eq} at the calcite surface one employs mass action equations, Eqs. (2.2), (2.3) and (2.14). Together with Eqs. (6.8) and (6.9) and neglecting [OH$^-$] for the present experimental conditions, this constitutes a system of five equations for the five unknown equilibrium concentrations. From these, finally, all fluxes can be calculated according to Eq. (6.6). In this approach one assumes that all the species are involved in reactions at the calcite surface, thus producing a concentration gradient.

From the weak dependence of dissolution rates on p_{CO_2}, however, one could conclude that H_2CO_3 does not react with the surface and therefore its bulk and equilibrium concentrations have to be equal. This approximation has also been employed as the H_2CO_3-independent model.

Figure 6.10 compares the calculated values for diffusion controlled reactions with the experimental curves A, B and C in regions 2 and 3. Clearly, the diffusion controlled model in all cases gives rates, which are higher by one order of magnitude than the experimental data. The values obtained from the H_2CO_3-independent model are lower but also cannot explain the experimental data.

Figure 6.11 shows calculations for the H_2CO_3-independent model, employing different ratios for the diffusion constants $D = D_{Ca}$ and D_H. Only at pH ≈ 4 do the rates approach the experimental data, indicating transport control. At pH > 4.4, i.e. regions 2 and 3, the experimental data indicate a transition from transport control to chemical control at the surface.

Thus, there is evidence, at this stage of knowledge, that at low pH there is a region determined by transport control of H$^+$ ions, which dominate the chemical reaction at the surface (region 1). Regions 2 and 3, however, show dissolution rates where chemical control becomes increasingly important when approaching equilibrium.

6.2.2 The Mechanistic Model of Plummer, Wigley, and Parkhurst (PWP Model)

The most comprehensive investigation on calcite dissolution leading to a deeper understanding is that of Plummer et al. (1978). By using both methods, free-drift and pH-stat, they studied calcite dissolution on small crystals of Iceland spar (average diameter 0.06 and 0.03 cm) in the pure H_2O-CO_2-$CaCO_3$ system. They performed experiments over a wide range of p_{CO_2} (0 to 1 atm) and temperatures (5° to 60°C). The results were summarized in a mechanistic rate equation, which enables one to

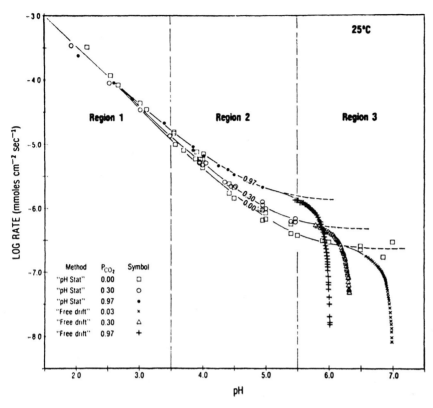

Fig. 6.12. Log rate of calcite dissolution as a function of pH. *Regions 1* and *2* are very far from equilibrium. Dependence of rates on p_{CO_2} in *region 1* is negligible, whereas dissolution increases with increasing p_{CO_2} in *region 2*. In free-drift experiments *regions 1* and *2* cover the first 25% of calcite dissolution, whereas the major reaction occurs in *region 3*, where dissolution rates drop steeply when equilibrium pH is approached (Plummer et al. 1978)

calculate the rate, once the activities of calcium, bicarbonate and hydrogen ions and H_2CO_3 at the calcite surface are known.

Figure 6.12 shows representative experimental data for pH-stat and free-drift experiments. Since in the free-drift experimental runs pH was measured as a function of time to calculate the calcium ion concentration by a computer model, pH is the appropriate parameter in dependence on which the rates are plotted. There is close agreement of the results obtained by pH-stat and free-drift experiments, provided pH, p_{CO_2} and calcium ion concentration are equal. Note that in the representation of data in Fig. 6.12 the calcium ion concentration is not constant, for varying pH as it is in the representation of Figs. 6.8 to 6.11. Therefore, a direct comparison between these two sets of data is not possible.

The data from the pH-stat experiments presented in Fig. 6.12 are all taken at compositions of the solution far from equilibrium, as can be seen by their comparatively high dissolution rates. In the free-drift experiments the solutional compositions are much closer to equilibrium during the final stage of the experimental run and the rates drop sharply as this happens. The narrow range of 0.5 pH units, however, where the free-drift curves bend down, represents 75% of calcite dissolution. Thus, in this small region the final 75% of reaction is accomplished. Note that in all experiments p_{CO_2} is kept constant by bubbling an N_2–CO_2 gas mixture with given p_{CO_2} into the solution.

Three regions of the dissolution mechanism can also be seen in Fig. 6.12. At low pH (region 1) the logarithm of the rate is linearly related to pH. In this region a strong dependence of rates on the stirring rate is observed. This indicates that transport of hydrogen ions to the surface of the calcite crystals controls the reaction in region 1.

The extension of this region depends on p_{CO_2}. At $p_{CO_2} = 1$ atm the upper limit is at pH = 3.5, whereas at $p_{CO_2} = 0$ atm a value of pH = 4.5 is found. Note also that in region 1 dissolution rates do not depend on p_{CO_2}.

This is different in region 2. Here, the rates depend on both pH and p_{CO_2}. The extension of this region depends also on p_{CO_2}. It ranges from pH = 3.5 to 5.5 at $p_{CO_2} = 1$ atm and is shifted to values of pH = 4.5 to 6.5 at $p_{CO_2} = 0.03$ atm. In contrast to region 1, where increase in stirring rate from 800 to 2300 rpm changes dissolution rates by a factor of two, dependence of stirring rate decreases with increasing pH and is practically unnoticeable at pH 5.

In region 3 data are obtained only by the free-drift runs. The curves show an inflection point, depending on p_{CO_2} of the solution. This region is characterized by a sharp drop of rates with increasing pH. No stirring dependence of dissolution rates is observed in this region. From this Plummer et al. concluded that in region 3, which comprises 75% of the reaction, dissolution rates are largely a function of the surface reaction. As long as one assumes diffusion boundary layers of a thickness not less than 0.001 cm surrounding the crystals (Dreybrodt and Buhmann 1988), this assumption can be used as a first approximation. We will discuss this topic in Chapter 7.

From a careful analysis of the data as represented by Fig. 6.12, Plummer et al. have derived rate equations. Figure 6.13 shows dissolution rates as a function of hydrogen ion activity at different p_{CO_2} and therefore consequently also at differing Ca^{2+} concentrations.

Clearly, there is a linear relationship between the rates and the activity (H^+). The slope at a constant stirring rate is independent of p_{CO_2}. The intercepts of the lines at $(H^+) = 0$, however, are dependent on p_{CO_2} and even at $p_{CO_2} = 0$ the intercept is finite.

Figure 6.14 shows these intercepts plotted as a function of p_{CO_2}. A linear relationship between rates and p_{CO_2} is observed. The intercept at $p_{CO_2} = 0$ [and (H^+) = 0 in Fig. 6.13] shows a constant rate of dissolution to be operative. These findings have been condensed into a rate equation valid in regions 1 and 2:

$$R_f = k_1(H^+) + k_2(H_2CO_3^*) + k_3 . \tag{6.10}$$

The values of k_1, k_2 and k_3 are determined from the experimental data.

This rate equation suggests the following elementary reactions at the calcite surface:

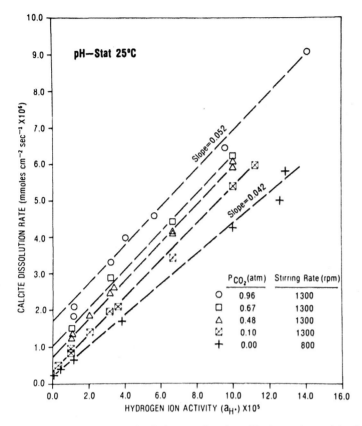

Fig. 6.13. Rate of calcite dissolution as a function of hydrogen ion activity for various p_{CO_2}. Note that at a fixed stirring rate the increase in dissolution rates upon increase of p_{CO_2} is independent of hydrogen ion activity (Plummer et al. 1978)

(I) $CaCO_3 + H^+ \rightleftharpoons Ca^{2+} + HCO_3^-$;

(II) $CaCO_3 + H_2CO_3 \rightleftharpoons Ca^{2+} + 2HCO_3^-$;

(III) $CaCO_3 + H_2O \rightleftharpoons Ca^{2+} + CO_3^{2-} + H_2O = Ca^{2+} + HCO_3^- + OH^-$.

In regions 1 and 2 of Fig. 6.12 the forward reactions are dominant. As the solution approaches equilibrium back reactions compensate for the forward reactions. At supersaturation the back reactions dominate and calcite precipitates from the solution.

The elementary reactions suggest back reaction rates of the following type (cf. Sect. 4.1):

$$R_b = R_b^I + R_b^{II} + R_b^{III} = k_4'(Ca^{2+})(HCO_3^-) + \{k_4''(HCO_3^-)\}(Ca^{2+})(HCO_3^-)$$
$$+ \{K_4''(OH^-)\}(Ca^{2+})(HCO_3^-) = k_4(Ca^{2+})(HCO_3^-) . \qquad (6.11)$$

The back rates of all three reactions have the factor $(Ca^{2+})(HCO_3^-)$ in common. To

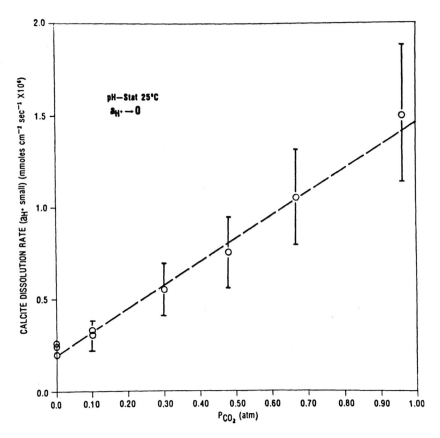

Fig. 6.14. Rate of calcite dissolution as a function of p_{CO_2} obtained from Fig. 6.13 at $a_{H^+} = 0$. (Plummer et al. 1978)

test the validity of this formulation from the experimental data one plots the observed rates R minus the term $k_1(H^+)$. Thus,

$$R - k_1(H^+) = -k_4(Ca^{2+})(HCO_3^-) + k_2(H_2CO_3^*) + k_3 . \qquad (6.11a)$$

should give a linear relation of the product $(Ca^{2+})(HCO_3^-)$ with intercepts at $k_2(H_2CO_3) + k_3$. The constant k_4 turns out to be dependent only on p_{CO_2}. Figure 6.15 shows this behaviour for various p_{CO_2}. Both the intercepts and the slope of the lines are dependent on p_{CO_2}.

The experimental findings on forward and back reactions are summarized by a rate equation (Plummer-Wigly-Parkhurst Equation):

$$R = k_1(H^+) + k_2(H_2CO_3^*) + k_3 - k_4(Ca^{2+})(HCO_3^-) . \qquad (6.12)$$

The values of k_1, k_2, k_3 are dependent only on temperature and have been determined from the experiment. Units of the activities are $mmol\ cm^{-2}\ s^{-1}$; for the rates, $mmol\ cm^{-3} = mol\ l^{-1}$.

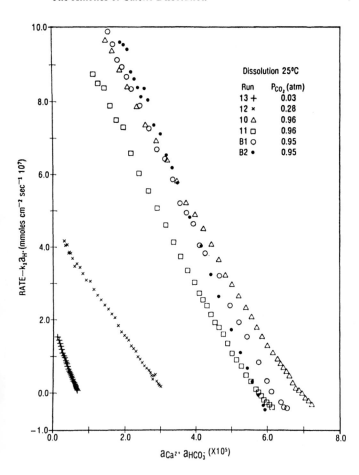

Fig. 6.15. Rate R-$k_1 \cdot (H^+)$ as a function of the activity product $(Ca^{2+})(HCO_3^-)$ for various values of p_{CO_2}. The slope of the linear relationship is $-k_4$ and the intercept at $(Ca^{2+})(HCO_3^-) = 0$ is $k_2(H_2CO_3^*) + k_3$ (Plummer et al. 1978)

At stirring rates of 1800 to 2300 rpm one finds:

$$\log k_1 = 0.198 - 444/T ,$$ (6.13)

where T is in given in Kelvin (K), units of k_1 are cm s^{-1}. The activation energy is 8.3 kJ mol^{-1} and indicates transport control as expected (cf. Sect. 4.1.3). In region 2, k_2 has been found as:

$$\log k_2 = 2.84 - 2177/T .$$ (6.14)

This corresponds to an activation energy of 42 kJ mol^{-1}, indicating control by a chemical reaction.

The rate constant k_3 shows a change in temperature dependence near 25°C. Below this temperature one finds:

$$\log k_3 = -5.86 - 317/T ,$$ (6.14a)

whereas above 25°C:

$$\log k_3 = -1.10 - 1737/T . \tag{6.14b}$$

The activation energies are 6.3 kJ mol^{-1} and 33 kJ mol^{-1} respectively.

The back rate constant k_4 depends on p_{CO_2} and temperature. This dependence has been explained by a mechanistic theory. The basis of this theory is an adsorption layer model proposed by Mullin (1972). It assumes that the reaction takes place in a thin (only a few molecules thick) adsorption layer adjacent to the crystal surface. We will denote species in this layer by the subscript s. This layer is separated from the bulk by a hydrodynamic diffusion boundary layer. Species in the bulk will be denoted by the subscript B and those at the bottom of this layer adjacent to the adsorption layer by 0. Since reaction I is fast compared to reactions II and III, which is reflected by the transport control of dissolution at low pH, (H^+) will be determined by calcite and carbonate equilibrium within the adsorption layer. The slow reaction of H_2CO_3 at this layer ensures one that $(H_2CO_3)_0 = (H_2CO_3)_s$. Thus, $(H^+)_s$ as well as $(HCO_3^-)_s$, $(Ca^{2+})_s$, $(CO_3^{2-})_s$ and $(OH^-)_s$ will be in equilibrium with $(H_2CO_3)_s$.

In terms of these species the forward reactions are now written as:

$$CaCO_3 + H_{(0)}^+ \to Ca_{(S)}^{2+} \to HCO_{3(S)}^- \tag{I}$$

$$CaCO_3 + H_2CO_{3(0)} \to Ca_{(S)}^{2+} \to 2\ HCO_{3(S)}^- \tag{II}$$

$$CaCO_3 + H_2O_{(0)} \to Ca_{(S)}^{2+} \to HCO_{3(S)}^- + OH_{(S)}^- . \tag{III}$$

The back reactions are

(I) $Ca_{(0)}^{2+} + HCO_{3(0)}^- + CO_{3(S)}^{2-} \to CaCO_3 + HCO_{3(S)}^-$

$\to CaCO_3 + H_{(S)}^+ + CO_{3(S)}^{2-} .$

In this reaction $CO_{3(S)}^-$ is a negatively charged reaction site catalyzing the reactions:

(II) $Ca_{(0)}^{2+} + HCO_{3(S)}^- + HCO_{3(0)}^- \to CaCO_3 + H_2CO_{3(S)}$

and

(III) $Ca_{(0)}^{2+} + HCO_{3(0)}^- \to CaCO_3 + OH_{(S)}^- .$

The rates associated with these reactions can now be written:

$$R_I = k_1'(H_{(0)}^+) - k_4'(Ca_{(0)}^{2+})(HCO_{3(0)}^-) ;$$

$$R_{II} = k_2'(H_2CO_{3(0)}^0) - k_4''(Ca_{(0)}^{2+})(HCO_{3(0)}^-)(HCO_{3(S)}^-) ;$$

$$R_{III} = k_3 - k_4'''(Ca_{(0)}^{2+})(HCO_{3(0)}^-)(OH_{(S)}^-) . \tag{6.15}$$

Thus, the total rate is

$$R = k_1'(H_{(0)}^+) + k_2'(H_2CO_{3(0)}^0) + k_3 - k_4(Ca_{(0)}^{2+})(HCO_{3(0)}^2) \tag{6.16}$$

with

$$k_4 = k_4' + k_4''(HCO_{3(S)}^-) + k_4'''(OH_{(S)}^-) . \tag{6.16a}$$

The activities at the bottom of the hydrodynamic diffusion boundary layer are determined by the hydrodynamic conditions of the experiment. The experimental data of Plummer et al. (1978) have been interpreted as surface controlled due to the lack of stirring dependence of the rates above pH = 5. Thus, in this region it is assumed that the values of all species at the bottom of the boundary layer (0) are close to those in the bulk (B).

To calculate k_4 from Eq. (6.16a) the principle of detailed balance is used for each of the reactions I, II and III. At equilibrium, therefore,

$$k_2(H_2CO_3^*)_{eq} = k_4''(Ca^{2+})_{eq}(HCO_3^-)_{eq} ;$$

$$k_3 = k_4'''(Ca^{2+})_{eq}(HCO_3^-)_{eq}(OH^-)_{eq} . \qquad (6.17)$$

Using the equilibrium constants K_1, K_2, K_w and K_c (cf. Eqs. 2.3, 2.4, 2.5 and 2.14) one obtains:

$$k_4'' = \frac{k_2 K_2}{K_c K_1} \qquad (6.18)$$

and

$$k_4''' = \frac{k_3 K_2}{K_c K_w} . \qquad (6.19)$$

For reaction 1 at equilibrium one finds:

$$k_1'(H^+)_{eq} = k_4'(Ca^{2+})_{eq}(HCO_3^-)_{eq} . \qquad (6.20)$$

In this equation k_1' is the chemical rate constant. Because of the transport control of the reaction at low pH, one has to be aware that $k_1' \gg k_1$, where k_1 is the transport coefficient of the transport controlled reaction. Thus, one obtains similarly:

$$k_4' = \frac{k_1' K_2}{K_c} . \qquad (6.21)$$

Combining Eqs. (6.16a) and (6.18) to Eq. (6.21) yields:

$$k_4 = \frac{K_2}{K_c}\left[k_1' + \frac{k_2}{K_1}(HCO_{3(S)}^-) + \frac{k_3}{K_w}(OH_{(S)}^-) \right] . \qquad (6.22a)$$

By using mass balance equations for $(HCO_{3(S)}^-)$ and $(OH_{(S)}^-)$ a more practical identity is derived:

$$k_4 = \frac{K_2}{K_c}\left[k_1 + \frac{1}{(H_{(S)}^+)}\left\{ k_2(H_2CO_{3(S)}^*) + k_3 \right\} \right] . \qquad (6.22b)$$

Assuming $(H_2CO_{3(S)}^*) = (H_2CO_{3(0)}^*) = (H_2CO_{3(B)}^*)$ and using for $(H_{(S)}^+)$ the corresponding value at calcite saturation related to $(H_2CO_3^*)$, the theory describes the experimental data sufficiently well.

Figure 6.16 shows the theoretical values, calculated by using k_1 instead of the unknown k_1', in dependence on p_{CO_2} at various temperatures. The experimental values of k_4 are somewhat higher than those predicted by theory. This is due to the use of k_1 instead of k_1'. The theoretical curves show only small changes on alteration of k_1. They imply, however, that k_1' is about a magnitude larger than k_1.

It is remarkable that both temperature dependence and p_{CO_2} dependence are well reproduced by the theory. This gives confidence for its applicability. One must,

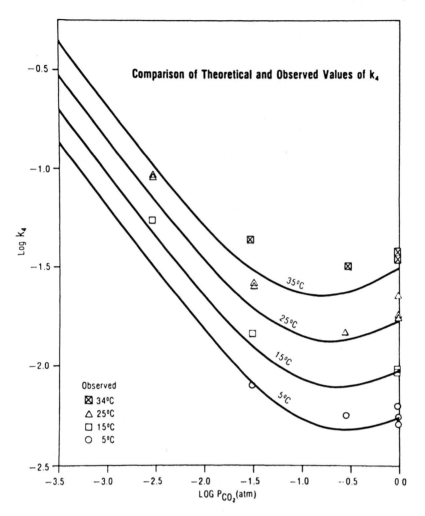

Fig. 6.16. Observed values of k_4 in comparison to theoretical values as a function of p_{CO_2} for various temperatures (Plummer et al. 1978)

however, be aware when doing so that there can be hydrodynamic conditions which do not warrant that bulk activities and those at the bottom of the boundary layer are equal. In this case a more complex transport model has to be developed (cf. Chap. 7).

Since dissolution of calcite proceeds by the simultaneous occurrence of three reactions, it is useful to discuss the relative magnitudes of the forward reaction rates at given pH and p_{CO_2} of a solution. Plummer et al. have done this by plotting onto a pH-p_{CO_2} diagram the curves, where two of the three forward rates are equal to the remaining one.

Thus, in Fig. 6.17 curve 1 represents all pH-p_{CO_2} compositions where:

$$k_1(H^+) = k_2 K_H p_{CO_2} + k_3 . \qquad\qquad (6.23a)$$

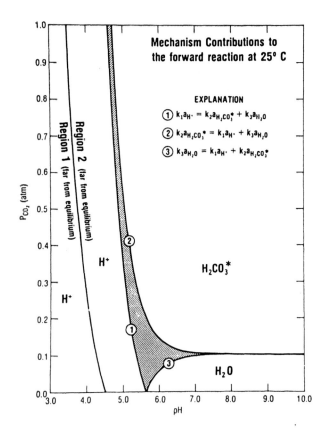

Fig. 6.17. Contributions of the reaction mechanism to the total forward rate of dissolution reaction as a function of p_{CO_2} and pH. In the *stippled area* more than one reaction contributes significantly. In the fields designated H^+, $H_2CO_3^*$ and H_2O these species determine forward reaction. Thus, at pH ≥ 7 and $p_{CO_2} \leq 0.05$ atm, i.e. the realm of karst water, the reaction $CaCO_3 + H_2O \leftrightarrow Ca^{2+} + CO_3^{2-} + H_2O$ is dominant (Plummer et al. 1978)

Accordingly, curves 2 and 3 represent the equations:

$$k_2 K_H p_{CO_2} = k_1(H^+) + k_3 \tag{6.23b}$$

and

$$k_3 = k_2 K_H p_{CO_2} + k_1(H^+). \tag{6.23c}$$

In the stippled area with curves 1, 2 and 3 as boundaries, all three reactions contribute significantly. In the region denoted with H^+ reaction I becomes increasingly dominant, when moving away from curve 1.

The curve between region 1 and region 2 (cf. Fig. 6.1) represents solution compositions, where the rates of H^+-attack of reaction I dominate those of reaction II and III by a factor of ten.

In the region denoted $H_2CO_3^*$ reaction II is dominant. In the region denoted H_2O the constant rate term k_3 determines the dissolution rates. This last region is of considerable interest, since it extends from pH > 6 and $p_{CO_2} \leq 0.1$ atm. This represents the realm of karst water and fresh water.

To visualize how the chemical evolution of the system $H_2O-CO_2-CaCO_3$ develops under open system conditions, one can easily calculate the dissolution rates as a function of the calcium ion concentration. This is done by computing the composition for an undersaturated system. Then, using the rate equation (6.16) dissolution rates are obtained. One has to be aware that in this type of calculation, one assumes that the concentrations in the bulk are identical to those at the surface, i.e. transport is so fast that the rates are controlled exclusively by the surface reaction. Furthermore, one asumes CO_2 and H_2CO_3 to be in equilibrium with each other, i.e. conversion of CO_2 into H_2CO_3 to be sufficiently fast so as not to be rate-controlling.

Figure 6.18 shows the results for an open system at 20°C and $p_{CO_2} = 0.008$ atm. The upper curve represents the dissolution rates as a function of the concentration of calcium ions already dissolved. The lower curve refers to the right coordinate scale and represents the evolution of pH. The chemical composition of this system as dissolution proceeds has been discussed in Chapter 2 and is shown in Figs. 2.9, 2.10, 2.11, 2.12. At very low calcium ion concentrations, $[Ca^{2+}]1 \times 10^{-4}$ mol l^{-1}, there is a steep decrease of dissolution rates. In this region pH is between 5.4 and 6. Reaction I contributes 2.0×10^{-7} mmol cm^{-2} s^{-1} at pH $= 5.4$ and 4.8×10^{-8} mmol cm^{-2} s^{-1} at pH $= 6$ to the dissolution rate R. The steep drop in dissolution rates at low Ca^{2+} concentration results from consumption of hydrogen ions by this reaction. Accordingly, pH drops steeply in this region. The contribution of reaction II is

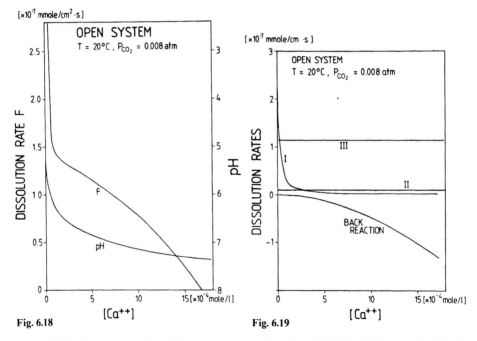

Fig. 6.18

Fig. 6.19

Fig. 6.18. Dissolution rate of calcite in the open system as a function of $[Ca^{2+}]$ which has developed in the solution during a free-drift experiment. The *scale* on the *right-hand side* refers to the pH curve, which gives the values of pH as $[Ca^{2+}]$ increases

Fig. 6.19. Contributions of forward reactions *I*, *II* and *III* to the forward rate and total rate of back reactions for the conditions as in Fig. 6.18 as a function of $[Ca^{2+}]$. As the system approaches equilibrium the back reaction rate increases and cancels forward reaction at equilibrium

constant, since under open system conditions $[H^2CO_3^*]$ is constant. It is very low $(8 \times 10^{-9}$ mmol cm^{-2} $s^{-1})$ in comparison to the contribution of reaction 3 $(k_3 = 1.12 \times 10^{-7}$ mmol cm^{-2} $s^{-1})$. The decrease in dissolution rates for calcium ion concentrations above 2×10^{-4} mmol cm^{-3} results from the back reaction which increases as the system approaches equilibrium. Figure 6.19 shows the contributions due to the three different reactions. From this one can easily see that the dominant forward reaction is due to reaction III. This is characteristic for almost all karst waters.

One important conclusion can be drawn from the shape of the rate curve in Fig. 6.18, which is representative for dissolution in open systems for a wide region of p_{CO_2} (cf. Chap. 7). In all cases it is possible to approximate these rate curves with sufficient accuracy by a linear relation for the range $[Ca^{2+}]/[Ca^{2+}]_{eq} \geq 0.2$:

$$R \approx \alpha([Ca^{2+}]_{eq} - [Ca^{2+}]), \tag{6.24}$$

where α is a constant depending on temperature and p_{CO_2}. It is in the order of 1×10^{-4} cm s^{-1}.

6.2.3 Comparison of the Mechanistic Model with Other Experiments

The first reliable experiments on dissolution of $CaCO_3$ in the pure system H_2O-CO_2 have been published by Erga and Terjesen (1956). These authors used a free-drift method and measured the calcium concentration as a function of time. The calcite particles used were of comparable size to those in the experiments of Plummer et al. (1978). In contrast to them Erga and Terjesen found a stirring dependence with a stirring coefficient of 0.22, indicating surface control (cf. Sect. 6.1.2). The experimental results are summarized in Fig. 6.20. Here, the rates are represented in dependence of the calcium concentration for various values of p_{CO_2}. This presentation is the same as that in Fig. 6.18. Plummer et al. (1979) have calculated the rates corresponding to these experimental conditions by assuming surface p_{CO_2} to be equal p_{CO_2} in the bulk. The full lines in Fig. 6.20 represent the results, scaled by a common factor of 0.5 for all the curves. This indicates either that the surface area, estimated by Erga and

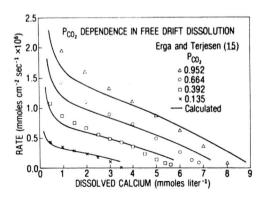

Fig. 6.20. Comparison of observed and calculated rate for four p_{CO_2} pressures in the free-drift experiment of Erga and Terjesen (1956) (Plummer et al. 1979)

Terjesen, is too high by a factor of two, or, more probably, that mixed kinetics, i.e. influence of transport control, is present. Similarly, Buhmann and Dreybrodt (1985a,b) in their dissolution experiments have also observed the experimental rates to be higher by a factor of two than those predicted by the PWP theory (cf. Chap. 7).

Nevertheless, there is quite good agreement with the experimental data, confirming the principles of the PWP model. The necessity of using a scaling factor, however, casts some doubt on the accuracy of the rate constants, which remain uncertain by a factor of two. This might well be due to the rather undefined hydrodynamic conditions in the experiment. If transport still exerts some control on the dissolution, it is to be expected that the rate constants determined from those experiments should be lower than the real ones. There is some evidence for this from recent experiments by Compton and Daly (1984). They used the rotating disc method to measure rates far from equilibrium at $[Ca^{2+}] < 1 \times 10^{-4}$ mmol cm^{-3}. From these data they found that the rate constant $k_3 = 2.1 \times 10^{-7}$ mmol cm^{-2} s^{-1} was larger than the value of 1.2×10^{-7} given by Plummer et al. (1978). Plummer et al. (1979) have also compared the predictions of their theory to the rate data on calcite dissolution in seawater or pseudo seawater obtained by Berner and Morse (1974) (cf. Fig. 6.9). Figure 6.21 plots the experimentally observed data versus calculated values. Experimental points situated at the straight line are equal to the

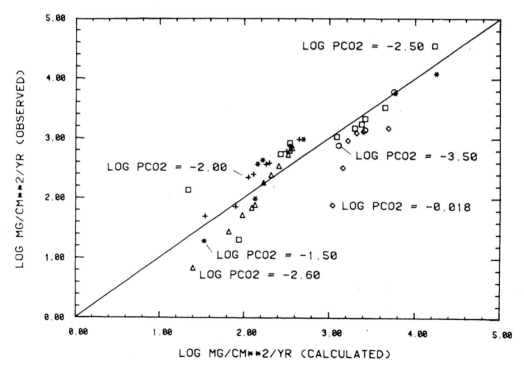

Fig. 6.21. Comparison of observed rates of Berner and Morse (1974) in pseudoseawater and seawater with rates predicted from the PWP equation (Plummer et al. 1979)

predicted ones. The accumulation of points close to this line shows that most of the points agree within a factor of two to the predicted ones.

There is a considerable amount of experimental work by Sjöberg (1976, 1978), Sjöberg and Rickard (1984a,b, 1985) and Rickard and Sjöberg (1983). These authors have investigated calcite dissolution mostly in CO_2-free solutions of high ionic strength (0.7 M KCl). They used both batch experiments (Sjöberg, 1976, 1978) and rotating disc apparatus in later work. Dissolution rates were measured using the pH-stat method in either Ca^{2+}-free solutions with KCl concentrations of 0.7 M or 0.1 M, or in solutions where $CaCl_2$ was added. In all these experiments the rates were measured as initial rates upon addition of $CaCO_3$ into a solution which was kept free of CO_2 by bubbling N_2 gas through it. After addition of calcite the system was closed to CO_2 exchange. In no case did the concentration of total carbon exceed 10^{-5} M. The experiments were performed with acid solutions with pH < 5 and also at neutral and alkaline pH.

In all experiments only forward reactions occur, since due to the low value of the carbonate concentration, the back reactions are negligibly small as $(HCO_3^-) = 0$ (cf. Eq. 6.12). The rotating disc experiments in the acid region clearly show a dependence of dissolution rates on stirring rate as $\omega^{1/2}$. From Eq. (6.5) this is only possible if k_c becomes large compared to k_T defined in Eq. (6.4). Therefore, from the $\omega^{1/2}$ stirring dependence, one concludes that transport control dominates as the rate-limiting step (Sjöberg and Rickard 1984b, 1985). The rate is a function of hydrogen ion activity and is found as:

$$R = k_T'(H^+)^{0.9} , \tag{6.25}$$

in close resemblance to the results of Plummer et al. (1978). From the temperature dependence of k_T' an apparent activation energy of 13 kJ mol^{-1} is found, close to the activation energy of k_1 obtained by Plummer et al. as 8.3 kJ mol^{-1}. This is further evidence for transport control.

At high pH dissolution rates show mixed kinetics (cf. Fig. 6.5) with comparable magnitude of transfer coefficients k_T for mass transport and k_c for the chemical reaction. The apparent rate constant k (cf. Eq. 4.67) shows an apparent activation energy of 34 kJ mol^{-1}, indicating the presence of chemical control. This value is close to that of the activation energy of k_3, which is 33 kJ mol^{-1} and dominates the forward reaction under these experimental conditions.

A direct comparison to the PWP theory is not possible since only forward rates related to k_1 and k_3 occur in all the experiments due to the absence of CO_2. The general picture as to how rates are controlled by transport or chemical reaction corroborates the results of Plummer et al. (1978, 1979). This has been summarized by Sjöberg and Rickard (1984b) as shown in Fig. 6.22, where three regions in a pH-temperature diagram are defined. At low pH transport control dominates. At intermediate pH there is a transition zone of complex kinetics. The boundary between these two regions moves to lower pH with increasing temperature.

At high pH dissolution becomes [H$^+$]-independent with mixed kinetics, where the transport control increases with increasing temperatures. At low temperatures surface chemical control gains influence.

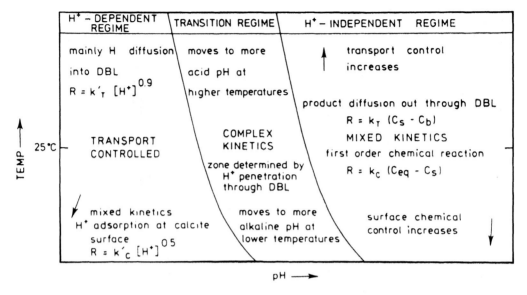

Fig. 6.22. Summary of calcite dissolution kinetics depending on the pH of the solution and temperature (Sjöberg and Rickard 1984a)

6.3 Kinetics of Calcite Precipitation

Precipitation of calcite is an important geochemical process. It determines saturation states in seawater as well as in soil water. In karst caves calcite precipitation creates the spectacular variety of speleothems.

In relation to dissolution precipitation is a process, in which the back reactions dominate. Therefore, if one assumes that both dissolution and precipitation occur due to the same elementary reactions, both dissolution and precipitation should be described by one common mechanism. The knowledge of precipitation kinetics therefore can be used to test mechanistic dissolution models.

The first major systematic studies of calcite precipitation were performed by Reddy and Nancollas (1971) and Nancollas and Reddy (1971). They used the seeded growth technique. In this method metastable, supersaturated solutions of $CaCO_3$ are prepared by adding sodium bicarbonate to $CaCl_2$ solutions. These solutions stay stable without any apparent precipitation over extended periods of time. The saturation state Ω, allowing these metastable, supersaturated solutions to exist, ranges up to $\Omega = 10$ (Reddy 1983). Upon addition of seed crystals of 10 μ size (0.3 m^2 g^{-1} surface area) precipitation occurs immediately. By monitoring pH and calcium ion concentration as a function of time, the rates of precipitation can be determined similar to dissolution experiments. By using an aqueous model one calculates the composition of the solution from the knowledge of pH and $[Ca^{2+}]$. With this method highly reproducible results were obtained. The observed rates are independent of the stirring rate, indicating surface control.

The rates were interpreted in terms of direct incorporation of Ca^{2+} and CO_3^{2-} ions into the lattice, which is equivalent to the simple elementary reaction:

$$Ca^{2+} + CO_3^{2-} \rightarrow CaCO_3 \, .$$

The rate equation then reads:

$$R = -k[(Ca^{2+})(CO_3^{2-}) - K_c] = -\frac{k}{K_c}(\Omega - 1) \, . \tag{6.26}$$

The fact that activities rather than concentrations are used can be explained on the assumption that an activated complex is involved in the elementary reaction (Inskeep and Bloom 1985).

This rate equation gives excellent agreement to the observed data in the region of $8.4 < pH < 8.8$, 1.5×10^{-4} mol $l^{-1} \le [Ca^{2+}] \le 4.5 \times 10^{-4}$ mol l^{-1}. From experiments at $10°$, $25°$ and $40°C$ the activation energy of the process was found to be 46 kJ mol^{-1} in support of a surface controlled mechanism.

All these findings were confirmed in a subsequent study by Wiechers et al. (1975). These authors extended the region of experimental conditions up to pH = 10, and also to a higher degree of supersaturation of $[Ca^{2+}]$ up to 2×10^{-3} mol l^{-1}. They also found the rates to be described by the rate equation (6.26) of Nancollas and Reddy. The activation energy was found to be 43 kJ. One shortcoming of their interpretation, however, was that the value of the rate constant k turned out to be dependent on initial pH and supersaturation.

This problem was analyzed by Sturrock et al. (1976). By using a careful analysis of the data, they showed that the rates could be described more accurately by a rate equation proposed first by Davies and Jones (1955), which reads:

$$R = -k[\{(Ca^{2+})(CO_3^{2-})\}^{1/2} - K_c^{1/2}]^2 \, . \tag{6.27}$$

In this formulation k is revealed as constant for a wide range of initial conditions. To illustrate these findings Figs. 6.23 and 6.24 show the experimental rates from different experiments plotted versus the Nancollas-Reddy equation and versus that of Davies and Jones respectively.

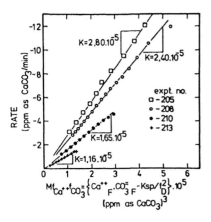

Fig. 6.23. Relationship between the rate of $CaCO_3$ precipitation and the function $[(Ca^{2+})(CO_3^-) - k_c]$ (Sturrock et al. 1976)

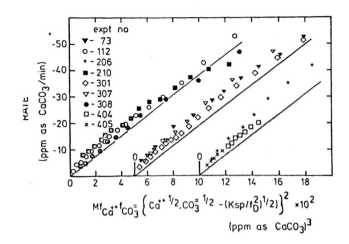

Fig. 6.24. Relationship of CaCO₃ precipitation rates and the function $\{[(Ca^{2+})(CO_3^{2-})]^{1/2} - K_c^{1/2}\}^2$ (Sturrock et al. 1976)

In both figures good linear plots are found for all data. The slopes for different runs, which represent the rate constant k, however, are different in the Nancollas-Reddy plot; Fig. 6.23 shows k to be variable with experimental conditions. In the Davies-Jones plot (Fig. 6.24) all experimental points fall on to one common linear plot, proving k to be really a constant.

To ensure that the precipitation rates depend solely on surface kinetics, Nancollas et al. (1981) provided new experimental methods by which they maintained a constant composition of the solution during the precipitation process. This was achieved by using a pH-stat technique and adding simultaneously the same amount of CaCl₂ and OH⁻, thus keeping the pH constant and at the same time replacing the amount of calcium deposited to the seed crystals. By using further a rotating disc experiment and comparing it to the seeded-growth, batch-type experiments, they found equal rate constants for both crystals and rotating disc despite their large differences in surface and probably growth-site density. This provides excellent proof that precipitation rates are purely surface-controlled in a wide range of supersaturation. Further evidence for the validity of Eq. (6.27) was established by Kazmierzak et al. (1982).

In order to analyze those data in terms of the PWP-mechanistic model, one has to recall the experimental conditions of the precipitation experiments discussed above. All these experiments have been performed in the pH region between pH = 8 to 10. CO₂ pressure is not controlled in these experiments and is low, somewhere between 10^{-3} atm down to 10^{-5} atm. In this case the PWP equation is simplified, since in this region reaction III is dominant. Therefore, neglecting the contributions of reactions I and II in Eq. (6.12) and (6.22b) one obtains:

$$R = k_3\left[1 - \frac{1}{(H_{(S)}^+)}\frac{K_2}{K_c}(HCO_3^-)(Ca^{2+})\right]. \tag{6.28}$$

Using mass action to express (HCO_3^-) in terms of (CO_3^{2-}), (H^+) and K_2 and using the saturation state Ω, one obtains:

$$R = k_3 \left[1 - \frac{(H^+)}{(H^+_{(S)})} \Omega \right].$$ (6.29)

This is very similar to the rate equation of Reddy and Nancollas, i.e. Eq. (6.26), provided $(H^+_{(S)}) \approx (H^+)$. Assuming that the concentration $(H_2CO_3^*)$ at the surface is equal to that in the bulk, the ratio of (H^+) and $(H^+_{(S)})$ varies between 0.93 to 0.97, increasing with approach to equilibrium (Plummer et al. 1979). At low p_{CO_2} a comparison of the thus calculated values from Eq. (6.29) with those from the experiment, however, show that the calculated rates are larger by a factor from 4 to 28. This discrepancy can be resolved by assuming surface $p^S_{CO_2}$, i.e. p_{CO_2} in equilibrium with $(H_2CO^*_{3(S)})$, to be larger than p_{CO_2} in the bulk. In order to obtain agreement between the experimental and calculated data, depending on saturation state Ω, $p^S_{CO_2}$ has to be assumed larger than p_{CO_2} up to a factor of four. Since $p^S_{CO_2}$ is not a measurable quantity, the comparison of experimental data so far neither proves nor disproves the PWP model.

This has stimulated further work. House (1981) performed experiments in supersaturated, dilute solutions using a free-drift seeding technique. The solutions were prepared by dissolution of a known quantity of $CaCO_3$ in an H_2O–CO_2 mixture with constant p_{CO_2}. The resultant solution is supersaturated by outgassing CO_2 with N_2 gas bubbled through the solution. Saturation states Ω up to ten are thus established. These solutions apply to the chemistry of fresh waters and karst waters and therefore the experimental data are of special interest here.

House compared the data to the Nancollas-Reddy, the Davies-Jones and the PWP-mechanistic model. He found poor agreement to the Reddy-Nancollas equation. The Davies-Jones model describes data within a wider range of conditions. Both models, however, failed to describe the data over all the experimental conditions employed. This was possible using the mechanistic model provided one assumes surface $p^S_{CO_2}$ to be larger than p_{CO_2} in the bulk. It turned out that the ratio $p^S_{CO_2}/p_{CO_2}$ is dependent on the surface area of the seeding crystals. At surface areas of 1.5 m² g⁻¹ seeding crystals, this ratio is 4 at saturation state $\Omega = 11$ and approaches 1 at equilibrium. At lower specific surface areas initial rates are smaller, for instance 1.5 at 0.5 m² g⁻¹ down to ≈ 1 at 0.2 m² g⁻¹. Thus, one might conclude that at low specific surface areas the PWP model applies without correction for $p^S_{CO_2}$.

Reddy et al. (1981) performed experiments in the region of low specific areas with seed crystals A of 0.23 m² g⁻¹ and seed crystals B with 0.7 m² g⁻¹ in a wide range of p_{CO_2}, which was kept constant during precipitation. He used pure H_2O–CO_2–$CaCO_3$ solution supersaturated with respect to calcite with constant p_{CO_2} between 0.03 and 0.3 atm. The region of the supersaturated state Ω ranged from 15 to 80. The rates were determined by a free-drift method, monitoring calcium ion concentration. Using an aqueous equilibrium model (cf. Chap. 2) the composition of the solution with respect to $[HCO_3^-]$, $[CO_3^{2-}]$ and $[H^+]$ was calculated with $[Ca^{2+}]$ and p_{CO_2} as input parameters.

Figure 6.25 shows the experimentally obtained precipitation rates for various p_{CO_2} given in percent of the N_2–CO_2 gas mixture bubbled through the precipitating

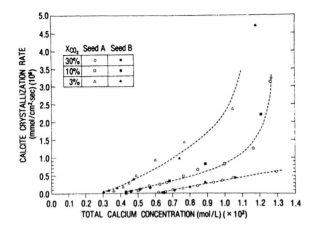

Fig. 6.25. Rate of calcite growth as a function of total calcium concentration in the solution for various p_{CO_2}, denoted as $x = \%$ CO_2 in gas mixture (Reddy et al. 1981)

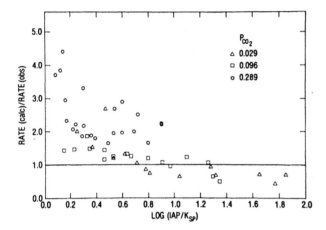

Fig. 6.26. Comparison of observed rates and those calculated according to Eq. (6.29) (Reddy et al. 1981)

solution. The rates are plotted as a function of calcium ion concentration present in the supersaturated solution.

The rates are linear at low supersaturation and are curved at higher $[Ca^{2+}]$. They are independent of specific surface area, indicating $p^S_{CO_2}$ to be independent of surface area. In view of the results found by House, this indicates $p^S_{CO_2}$ to be close to the bulk value. This is further corroborated by the independence of the rates found experimentally for stirring. By assuming $p^S_{CO_2} = p_{CO_2}$ Reddy et al. have calculated the rates from the PWP model. All the observed and calulated rates agreed within a factor of three. Figure 6.26 compares the calculated data to the observed ones by plotting their ratio versus $\log \Omega$. One finds satisfactory agreement for low p_{CO_2}. At high p_{CO_2} and low supersaturation calculated rates tend to be larger as equilibrium is approached. A trial to fit the data to expressions like $\log \Omega$ or $\log(\Omega - 1)$ as linear functions failed, thus excluding other crystallization models. From these findings one can conclude that the PWP model is appropriate to predict precipitation rates, at least in the range of chemical parameters represented by karst waters. Rates

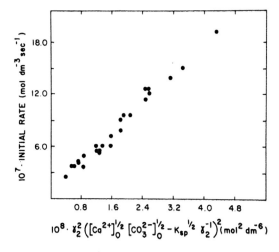

Fig. 6.27. Initial rate plot for the second form of the Davies and Jones (1955) model (Eq. 6.27) (Inskeep and Bloom 1985)

measured near equilibrium are considerably lower, not only for precipitation, but also for dissolution. This will be discussed later.

Recently, in a systematic study of precipitation kinetics Inskeep and Bloom (1985) performed pH-stat experiments, using a seeding technique with super-saturated solutions from $CaCl_2$–$KHCO_3$ mixtures. Initial $[Ca^{2+}]$ ranged from 0.7–2 mmol l^{-1}, $[HCO_3^-]$ from 4–7 mmol l^{-1} and pH from 8.25 to 8.7. In all cases p_{CO_2} was below 0.01 atm. The rates were measured as initial rates upon addition of the seed crystals.

The data were evaluated in terms of the Davies and Jones, PWP and Nancollas and Reddy models by plotting them as functions of the appropriate rate expressions to obtain straight lines in case of agreement. Figure 6.27 shows such a plot for the Davies and Jones mode. All the experimental points are reliably represented by a straight line with a high correlation factor $r = 0.978$.

A similarly high quality agreement is obtained also by using the Nancollas and Reddy model. Figure 6.28 shows the corresponding plot of the data. The good fit of both models to the experimental data might be due to the fact that the two rate expressions are mathematically very similar and yield almost identical results in the range of experimental conditions. Figure 6.29 shows a plot of the data as a function of the back rate in the PWP equation, assuming $p_{CO_2}^S = p_{CO_2}$. Although most of the data support the PWP model, some of them show deviations up to a factor of three. Trying to explain these data by correcting for $p_{CO_2}^S =$ leads to a $p_{CO_2}^S$ lower than that in the bulk. This is contrary to what one expects, since precipitation of $CaCO_3$ should release CO_2 at the surface, thus possibly increasing $p_{CO_2}^S$.

Summarizing the results of precipitation experiments, although the PWP-mechanistic model does not seem to be the one which exclusively explains the data, it is at least as appropriate as all the others. In view of the fact of its power in explaining dissolution data, however, it is the most widely applicable model, which describes precipitation and dissolution in one common theory. For geological applications it is of sufficient accuracy, since factors of two in the predictions of rates

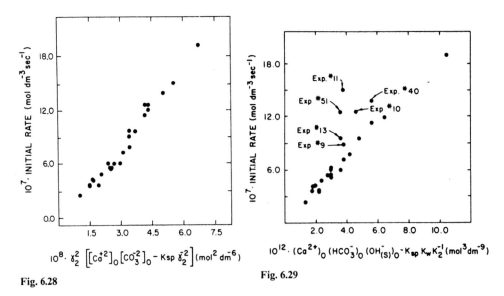

Fig. 6.28

Fig. 6.29

Fig. 6.28. Initial rate plot for the Nancollas and Reddy (1971) model, Eq. (6.26) (Inskeep and Bloom 1985)

Fig. 6.29. Initial rate plot for the PWP model, Eq. (6.12) (Inskeep and Bloom 1985)

are not of great importance, considering the many other sources of influence in a real environment.

6.4   Dissolution Close to Equilibrium

Most of the dissolution experiments which we have discussed have been performed far from equilibrium with respect to calcite. The situation close to equilibrium is obscure. There is some evidence that close to equilibrium a change of the dissolution mechanism is present, by which the rates are lowered by orders of magnitude from that expected from the predictions of the PWP model. This is of utmost importance in the initiation of karst systems, where dissolution occurs from solutions close to equilibrium. The first evidence of a change in the dissolution mechanism close to equilibrium was given by Berner and Morse (1974). White (1977) plotted all their experimental data into a summarizing representation, which is shown in Fig. 6.9. The data show a region 3A, where $\Delta pH \leq 0.1$. In this region, corresponding to a saturation state $\Omega > 0.63$, dissolution rates drop by orders of magnitude. The unit $mg\ cm^{-2}\ year^{-1}$ corresponds to $3.2 \times 10^{-10}\ mmol\ cm^{-2}\ s^{-1}$. Berner and Morse have explained this finding by an inhibition mechanism due to absorption of orthophosphate ions at the surface reaction sites. They used the theory of Burton, Cabrera and Frank (1951), which suggests that dissolution and precipitation of ionic solids proceed by attachment or release of ions from kinks at the crystalline surface.

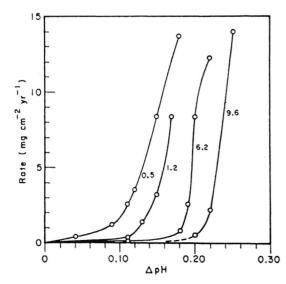

Fig. 6.30. Plots of dissolution rates of $CaCO_3$ in artificial seawater of varying dissolved phosphate content at $p_{CO_2} = 10^{-2.6}$ atm and $T = 25°C$. The *numbers* on the curves give dissolved phosphate concentrations in micromole per litre (Berner and Morse 1974)

These kinks can be favourable sites for inhibitor chemisorption. Figure 6.30 shows experimental evidence for inhibition by phosphate. The dissolution rates for calcite in seawater of varying phosphate concentrations show a region of inhibition, which extends with the concentration of phosphate. Note that concentrations of phosphate as low as micromolar values show a significant effect (Berner and Morse 1974).

Dissolution rates close to equilibrium were also reported by Plummer and Wigley (1976) in a pure H_2O–CO_2 solution at $p_{CO_2} = 1$ atm. Figure 6.31 shows the results. Dissolution rates are plotted as functions of the deficit of concentration c to the equilibrium value c_S. At $c < 0.8\ c_S$ an empirical rate law of second order is found. Above this value there is a change from second to fourth order, indicating a different mechanism. The rates drop by two orders of magnitude at $c = 0.9\ c_S$ and become too small to be measured closer to equilibrium. The units of rates are defined by mol l^{-1} s^{-1} related to the special experimental condition. This unit corresponds to 10^{-1} mmol cm^{-2} s^{-1}.

Plummer and Wigley suspect that trace inhibitors such as phosphate are responsible for the behaviour close to equilibrium. Sjöberg (1978) performed pH-stat runs to measure dissolution rates with the approach to equilibrium for different phosphate concentrations up to 7 μM. The calcium concentration of the pseudo-seawater solution was kept constant at 10^{-2} mol l^{-1}, magnesium was at 5×10^{-2} mol l^{-1}. He found a significant reduction of dissolution rates with increasing phosphate concentration. Close to nominal equilibrium the rates become too small to be measured and a new apparent equilibrium is established.

There is little systematic work on trace inhibitors close to equilibrium, but phosphate is commonly accepted as an efficient inhibitor in both dissolution and precipitation of calcite (Reddy 1977). Other inhibitors such as Cu^{2+} and Sc^{2+} have been observed by Terjesen et al. (1969). Therefore, in all experiments working with

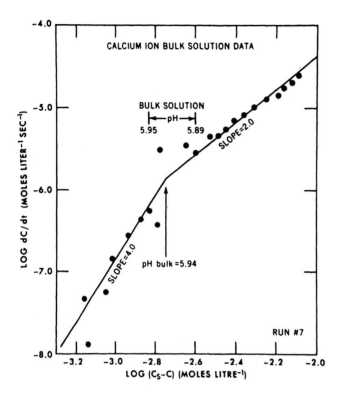

Fig. 6.31. Log rate of saturation deficit (c_s-c) of Ca^{2+} versus log rate of dissolution. There are two regions of reaction as can be seen from the different slopes (Plummer and Wigley 1976)

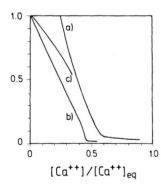

Fig. 6.32. Ratio of initially observed rate to rate observed after dissolution has proceeded to a concentration $[Ca^{2+}]$ versus the ratio of $[Ca^{2+}]/[Ca^{2+}]_{eq}$. *Curves a* and *b* show a region of inhibited dissolution for $[Ca^{2+}]/[Ca^{2+}]_{eq} > 0.5$. Curve c, see text

natural materials trace inhibitors might be present with differing influence to the dissolution close to equilibrium.

Dissolution experiments on natural samples of Iceland spar by Herman (1982), using the rotating disc method, showed a significant drop of dissolution rates at $[Ca^{2+}] \approx 0.5\,[Ca^{2+}]_{eq}$. We have performed dissolution experiments by the method described in Section 7.3.2.3. We stirred a layer of $H_2O–CO_2$ solution, 1-cm-deep on top of a marble surface and measured dissolution rates. Figure 6.32 shows the rates in units of the initial rate observed in our experiments (a) and data of Hermann (b).

Although experimental conditions with respect to p_{CO_2} are quite different, the rates behave similarly. Both show an extended zone of inhibition at undersaturation as low as approximately $0.5\ [Ca^{2+}]_{eq}$. We found also similar behaviour on samples of pure natural limestone. Rauch and White (1977) performed measurements of dissolution rates on a variety of different natural limestone samples. Curve c shows a characteristic result for a very pure limestone. From these findings one can define two regions of dissolution kinetics in natural systems. Far from equilibrium reaction mechanisms as given by Plummer et al. seem to be adequate. Close to equilibrium inhibition dominates and rates become by several orders of magnitude lower than expected from the mechanistic model.

Plummer and Wigley (1976) reported on findings of Back and Hanshaw (pers. commun.) in central Florida aquifers, where subsaturated waters with radiocarbon ages of more than 1000 years are present. Phosphate levels in these waters as high as 0.5 mg l^{-1} were found.

Michaelis et al. (1984) investigated calcite-precipitating streams in Germany. One of them, a spring at Urach, has stopped depositing calcite. The latest observation of calcite deposition was made in 1963. Measurements in 1980 and 1981 showed supersaturations up to $\Omega = 20$ without calcite precipitation. The phosphate concentration found in the water of this stream was 3 μM.

7 Modelling the Kinetics of Calcite Dissolution and Precipitation in Natural Environments of Karst Areas

7.1 Statement of the Problem

Karst evolution in its dimensions of space and time depends on the dissolution kinetics of limestone. The time needed to excavate conduits by solutional removal of limestone depends on the rates of dissolution. The length of these flow conduits is determined by the distance calcite-aggressive water can travel until it becomes saturated. This penetration distance is related to the velocity of flow and also to the kinetics of dissolution.

Thus, to understand the role of dissolution kinetics in real geological situations one has to transfer the laboratory results to situations encountered in real karst systems. This comprises the geometrical boundary conditions and also the specifications of hydrodynamic flow. Flow in joints and partings is laminar. If the spacings and hydraulic gradients are sufficiently large, however, flow will be turbulent, and mass transport of dissolved species into the bulk may become much more effective.

Very narrow partings, such as those of bedding planes prior to any enlargement by dissolution, can be visualized as two-dimensional porous media and mechanical dispersion will contribute to mass transport. In view of the manifold situations in real systems it may seem hopeless to sufficiently simplify the complex problem. Keeping in mind, however, that for geological applications theoretical predictions within a factor of approximately two are sufficiently precise, an attempt to extract one common physical core out of all these various conditions encountered in karst as a basis for a model simulating calcite dissolution and precipitation, still seems both promising and challenging. Dissolution or precipitation of calcite proceeds in three basically different circumstances, which can be classified as follows:

1. *Conditions of the open system.* There is a variety of cases where water flows on bare limestone rock with the free surface open to the atmosphere. Depending on the depth of the waterbody, flow is turbulent, as in streams and rivers, or laminar as in the case of rainwater flowing in thin layers across the rock.

The geomorphical features originating from these different situations are strikingly different. Cave rivers flowing in the vadose zone in cave passages sculpture canyons which can be of considerable dimensions, whereas rainwater flowing on bare limestone of sufficient slope creates karren. Also, the amount of calcite dissolved can be very different. In cave streams denudation rates of up to 1 mm year^{-1} have been measured (Gunn 1986, Lauritzen 1986, Palmer 1984a). In contrast the annual denudation rates on bare limestone surfaces are only about 3×10^{-3} cm (Gunn 1986, Forti 1984, Bauer 1964, Jennings 1978).

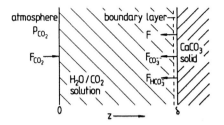

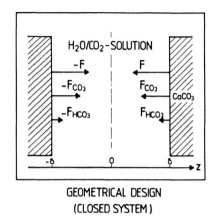

Fig. 7.1. Model of CaCO₃ dissolution in systems open to the atmosphere (Buhmann and Dreybrodt 1985a)

Fig. 7.2. Geometrical model of the CaCO₃ dissolution in a system closed to the atmosphere (Buhmann and Dreybrodt 1985b) ▶

These large differences result from the change in mass transport by diffusion due to the large changes in coefficient of diffusion when going from laminar to turbulent flow.

In spite of all these differences the common physical core can be modelled by the simple geometry shown in Fig. 7.1. We assume a plane water layer of thickness δ in contact with a plane rock surface. Due to dissolution there is a flux of Ca^{2+}, CO_3^{2-}, and HCO_3^- ions from the solid surface into the bulk. The magnitude of the diffusion coefficient in the bulk depends on the type of flow. CO_2 penetrates into the bulk by diffusion from the free surface in contact with the CO_2-containing atmosphere.

It should be noted that this model can also be applied for precipitation processes, such as calcite deposition on cave speleothems or in natural streams supersaturated with respect to calcite. This is done by simply reversing the signs of all fluxes.

2. *Conditions of the closed system.* In the phreatic zone of cave systems groundwater fills the conduits entirely and wets the exposed rock surfaces. Therefore, no exchange of CO_2 with the atmosphere is possible. Flow may occur in large conduits where it will be turbulent. But also diffuse flow in narrow partings, such as joints and beddings, is present. In this case flow tends to be laminar. Dissolution under these last conditions prevails during the initiation of karst and cave development.

A simple geometrical model to describe the conditions of calcite dissolution in the closed system is shown in Fig. 7.2. A water layer of thickness 2δ is entirely enclosed by two parallel planes of calcite. The ionic species dissolved from the limestone are transported into the bulk. The coefficient of diffusion governing this depends on the type of flow. Since the system is closed to the atmosphere, CO_2 is consumed as the dissolution proceeds.

The approximation of partings by two parallel rock surfaces is obvious, although a two-dimensional porous model may as well be applied (see below). In the intermediate and mature state of karst most of the flow is conducted via conduit tubes of circular or ellipsoidal cross-section, thus the approximation of Fig. 7.2

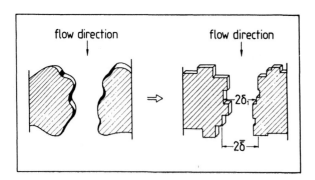

Fig. 7.3. Mathematical idealization of the irregular structure of calcite grains by parallel plane surfaces. The individual distances 2δ are approximated by a common average distance $2\bar{\delta}$ of the flow channel (Baumann et al. 1985)

seems to be questionable. It is well known, however, in transport problems that changes in geometry, such as from parallel planes to cylindrical or even spherical geometries, do not change significantly the fluxes predicted. Only a geometrical factor in the order of one is encountered. Therefore, in view of the application, the approximation in Fig. 7.2 is appropriate.

3. *Two-dimensional porous media.* As already stated narrow partings may be visualized as a two-dimensional porous medium. In this case flow is carried in the pores and interstices between the different grains of the medium. Two of these grains are shown in Fig. 7.3. Flow between these grains resembles the situation of closed system flow if one approximates the geometry according to the right-hand part of Fig. 7.3. There is one important difference. Due to hydrodynamic dispersion (Baumann et al. 1985) the coefficient of diffusion is now given by multiplying the molecular coefficient of diffusion by the Peclet number (cf. Sect. 3.3.2). In real situations this might enhance mass transport in porous media.

Thus, one can treat dissolution in saturated porous media mathematically in the same way as the closed system. One merely has to use the average grain size as the distance parameter $\bar{\delta}$ and the coefficient of diffusion resulting from hydrodynamic dispersion.

7.2 The Mass Transport Equation

In a series of papers Buhmann and Dreybrodt (1985a,b) and Baumann et al. (1985) have dealt with calcite dissolution and precipitations using the geometrical models described above. This section gives a review of this work. The aim is to establish a transport equation from which dissolution and precipitation rates can be calculated. To derive such an equation first all processes participating in the dissolution process must be known. They are (cf. Chap. 6.1) as follows:

1. The kinetics of dissolution at the phase boundary between the solvent aqueous system $CaCO_3 - H_2O - CO_2$ and the limestone rock, which depends on the chemical composition of the solvent at the phase boundary and can be derived, once this composition is known, by rate equations found experimentally and

asserted theoretically by Plummer et al. (1978, 1979). These PWP equations are at present the most recent and most reliable data on dissolution of calcite (cf. Sect. 6.2.2), at least sufficiently far from equilibrium.

2. The kinetics of conversion of CO_2, dissolved in the aqueous system, into $HCO_3^- + H^+$, which constitutes aggressive agents in the process of $CaCO_3$ dissolution. This conversion is a slow process, depending on the pH of the solution and has been reviewed in detail by Kern (1960) and Usdowski (1982). Consideration of this slow process is of utmost importance, since it can be rate-determining, especially in cases where the ratio of the solvent volume to the surface of the solid in contact with the solvent is small. This is the case when aggressive water flows in narrow joints or partings. Neglecting the process of slow CO_2 conversion leads to an overestimation of solutional rates of up to two orders of magnitude.

3. Mass transport by diffusion of the dissolved ionic species, i.e. Ca^{2+}, HCO_3^-, CO_3^{2-}, and CO_2 and H_2CO_3 from and to the phase boundaries. This transport mechanism depends on the hydrodynamic flow conditions of the solution. In laminar flow transport is determined by molecular diffusion with the coefficient of diffusion D_m. In turbulent flow eddy diffusion is dominant and the related coefficient of diffusion is by at least three orders of magnitude larger than D_m (Tien 1959, Bird et al. 1960, Skelland 1974). Therefore, hydrodynamic flow conditions are of great importance and have to be considered in any case (cf. Chap. 6).

With the geometrical conditions of Figs. 7.1 and 7.2 the transport equation is essentially two-dimensional. The z-direction of the coordinate system is perpendicular to the rock surface. Flow is parallel to the surface in the y-direction.
From Eq. (3.13) the transport equation for each ionic and molecular species i is:

$$\frac{\partial c_i(z, y, t)}{\partial t} + v_y \frac{\partial c_i(z, y, t)}{\partial y} - D \frac{\partial^2 c_i(x, y, t)}{\partial z^2} = R_i(c_1, \ldots, c_i, \ldots, c_N) . \tag{7.1}$$

In this formulation we assume that v_y is constant everywhere in the flow field, thus approximating the velocity profiles in laminar and turbulent flow by plug flow. The rate R_i designates production of the different species by chemical reactions in the solution.
Let us assume that at $t = 0$ a CO_2-H_2O solution starts to flow along the calcite surface. Then, after a transient time a stationary profile $c(z, y)$ is established and from Eq. (7.1) we obtain:

$$v_y \cdot \frac{\partial c_i(z, y)}{\partial y} - D \frac{\partial^2 c_i(z, y)}{\partial z^2} = R_i(c_1, \ldots, c_i, \ldots, c_N) . \tag{7.2}$$

If one considers a parcel of solution as shown in Fig. 7.4 travelling in the y-direction, the time t_c, in which this parcel has been in contact with the rock, is related to the distance y of travel by:

$$y_t = v_y \cdot t_c . \tag{7.3}$$

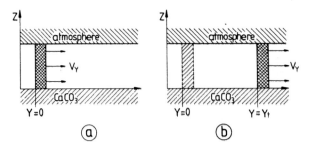

Fig. 7.4a,b. Motion of a parcel of fluid in plug flow approximation. After time t_c the parcel has moved the distance y_t. This is equivalent to a situation in which the parcel has been in contact with the calcite surface for time t_c

Introducing this into Eq. (7.2) yields:

$$\frac{\partial c_i(z, t)}{\partial t} - D \frac{\partial^2 c_i(z, t)}{\partial z^2} = R_i(c_1, \ldots, c_i, \ldots, c_N) . \tag{7.4}$$

This equation describes mass transport for a stagnant fluid, i.e. $v_y = 0$. It is a one-dimensional equation in the spatial coordinate z. From its solution the concentration profiles $c(z, t)$ can be derived. The physical reason for the equivalence of Eqs. (7.2) and (7.4) is that in plug flow, where the relative velocity between any two fluid parcels is zero, the flow does not distort distributions of the concentration. Therefore, only the time of contact to the solid determines the distribution of concentration in each parcel of fluid. This equivalence enables one to design laboratory experiments in which dissolution in karst can be simulated. This will be discussed in a subsequent section of this chapter.

To apply Eqs. (7.2) or (7.4) to a given problem the boundary conditions have to be specified. In the following we will do this and derive dissolution rates for open and closed system conditions.

7.3   Dissolution and Precipitation Kinetics in the Open System

7.3.1   Calculation of Dissolution and Precipitation Rates

We consider a stagnant water layer in contact with the rock. The boundary conditions for mass transport are shown in Fig. 7.1 for the open system. At $z = 0$, the interface atmosphere-water, there is a flux F_{CO_2} into the layer. Simultaneously, dissolution of calcite produces a flux F of Ca^{2+} ions, F_{CO_3} of CO_3^{2-} ions and F_{HCO_3} of HCO_3^- ions from the calcite surface into the bulk of the solution. For reasons of stoichiometry of the chemical reaction $CaCO_3 + H_2O + CO_2 \rightarrow CaCO_3 + H_2CO_3 \rightarrow Ca^{2+} + 2HCO_3^-$, each Ca^{2+} dissolved consumes one molecule of CO_2. This is correct for all situations of karst water compositions. Therefore, $F_{CO_2} = F$. The flux F is determined by the Plummer-Wigley-Parkhurst equation (cf. Chap. 6):

$$F = k_1 (H^+)_\delta + k_2 (H_2CO_3^*)_\delta + k_3 - k_4 (Ca^{2+})_\delta (HCO_3^-)_\delta . \tag{7.5}$$

The activities $(\ )_\delta$ are taken at $z = \delta$, the solid-liquid interface. The rate constants k_1, k_2, k_3 are only slightly dependent on temperature, whereas k_4 changes by roughly a factor of two, when going from $10°$ to $20°C$. Their values can be taken from the work of Plummer et al. The transport of all species is now given by a set of coupled differential transport equations (Bird et al. 1960), which according to Eq. (7.4) are:

$$-\frac{\partial c_i}{\partial t} + D_i \frac{\partial^2 c_i}{\partial z^2} = -R_i(c_1, \ldots, c_i, \ldots, c_N),\tag{7.6}$$

where c_i is the concentration of species i, D_i its coefficient of diffusion. R_i describes the production rate of species i by chemical reaction with other species. In our case only the reaction $H_2O + CO_2 \leftrightarrows HCO_3^- + H^+$ is slow. All the other species show very fast reactions. Therefore, they are in equilibrium with each other and the corresponding R_i are negligible. By using the reaction rates of $H_2O + CO_2 \rightleftarrows H^+ + HCO_3^-$ conversion as given in Chapter 4, Eq. (4.46), the set of transport equations for our system is given by:

$$-\frac{\partial [CO_2]}{\partial t} + D_{CO_2} \frac{\partial^2 [CO_2]}{\partial z^2} = (k_1^+ + k_2^+[OH^-])[CO_2]$$
$$- (k_a[H^+] + k_2^-)[HCO_3^-]\tag{7.7a}$$

$$-\frac{\partial [HCO_3]}{\partial t} + D\frac{\partial^2 [HCO_3^-]}{\partial t} = -(k_1^+ + k_2^+[OH^-])[CO_2]$$
$$+ (k_a[H^+] + k_2^-)[HCO_3^-]\tag{7.7b}$$

$$-\frac{\partial [CO_3^{2-}]}{\partial t} + D\frac{\partial^2 [CO_3^{2-}]}{\partial z^2} = 0\tag{7.7c}$$

$$-\frac{\partial [Ca^{2+}]}{\partial t} + D\frac{\partial^2 [Ca^{2+}]}{\partial z^2} = 0.\tag{7.7d}$$

The brackets denote concentrations. The terms on the right side of Eqs. (7.7) result from CO_2 conversion and are dependent on the pH of the solution. Since we assume all the other species, i.e. CO_3^{2-}, HCO_3^-, H_2CO_3, H^+ and OH^-, to be in equilibrium, we can obtain two more equations from mass action laws:

$$[H^+][OH^-] = K_w/\gamma_H\gamma_{OH}; \quad [H^+][CO_3^{2-}] = \frac{K_2[HCO_3^-]\cdot\gamma_{HCO_3}}{\gamma_H\gamma_{CO_3}}.\tag{7.8}$$

The γ represent the activity coefficients and are calculated by the Debye-Hückel approximation. These six equations constitute a coupled system which has to be solved with the boundary conditions represented in Fig. 7.1.

These boundary conditions are:

$$-D\frac{[Ca^{2+}]}{\partial z/\delta} = F = k_1(H^+)_\delta + k_2(H_2CO_3^*)_\delta + k_3$$
$$- k_4(Ca^{2+})_\delta(HCO_3^-)_\delta\tag{7.9}$$

$$+D\frac{[HCO_3^-]}{\partial z/\delta} + D\frac{[CO_3^{2-}]}{\partial z/\delta} = -F\tag{7.10}$$

Equation (7.10) follows from the fact that the total flux of carbon from the surface has to be equal to the flux of Ca^{2+}.

Since CO_2 does not react at the surface we have furthermore

$$\frac{\partial [CO_2]}{\partial z/\delta} = 0 . \qquad (7.11)$$

At $z = 0$, the liquid-gas interface, we have:

$$[CO_2]_0 = K_H p_{CO_2} ; \qquad -D \frac{\partial [CO_2]}{\partial z/0} = F , \qquad (7.12)$$

where the second equation is due to the stoichiometric relation by which one molecule of CO_2 is consumed for each Ca^{2+} ion released. Under open system conditions the transfer of CO_2 to the solution in all cases is sufficiently fast and the CO_2 concentration in the solution remains constant. All the other species, however, cannot be transferred into the gas phase and therefore no flux of these species exists at $z = 0$:

$$\frac{\partial [HCO_3^-]}{\partial z/z = 0} = \frac{\partial [CO_3^{2-}]}{\partial z/z = 0} = \frac{\partial [Ca^{2+}]}{\partial z/z = 0} = 0 . \qquad (7.13)$$

As a first step to solve the system of coupled equations we realize that the equation for the calcium ion concentration can be solved if the flux F is known (cf. Eq. 7.9). This results from the fact that no chemical reactions of Ca^{2+} with other species are taken into account. Consideration of ion pairs, such as $CaCO_3^0$ and $CaHCO_3^+$, changes the results by only a few percent (Buhmann and Dreybrodt 1985a). Therefore, they can safely be neglected.

If we assume that dissolution proceeds slowly such that during a given time interval t the composition of the solution remains practically constant, we have the situation of mass transfer with prescribed constant flux F into the solution. In this case the solution of Eq. (7.7d) with boundary conditions (7.9) and (7.13) is given by Carslaw and Jaeger (1959) (cf. Eq. 3.23):

$$[Ca^{2+}](z, t) = \left\{ \frac{F \cdot t}{\delta} + \frac{F \cdot \delta}{D} \left(\frac{3z^2 - \delta^2}{6\delta^2} \right) - \frac{F \cdot \delta}{3D} + [Ca^{2+}]_\delta \right\}$$
$$- \frac{F\delta}{D} \cdot \sum \frac{(-1)^n \cdot 2}{n^2 \pi^2} \cos \frac{n\pi z}{\delta} \exp \left(-\frac{Dn^2 \pi^2}{\delta^2} t \right) . \qquad (7.14)$$

$[Ca^{2+}]_\delta$ is the concentration at $z = \delta$ at time $t = 0$. The exponentials in expression (7.14) decay with time constant $T_d = \delta^2/D\pi^2$. Therefore, after this time a stationary distribution $[Ca^{2+}](z)$ is quickly established and the concentration increases linearly with time. Thus, if the time t, in which the solution needs to achieve equilibrium, is large, i.e. $t > T_d$, we may assume that the concentration profile $[Ca^{2+}](z)$ changes very slowly and can practically considered to be constant:

$$[Ca^{2+}](z) = \frac{F \cdot t}{\delta} + \frac{F\delta}{D} \left(\frac{3z^2 - \delta^2}{6\delta^2} \right) - \frac{F\delta}{3D} + [Ca^{2+}]_\delta . \qquad (7.15)$$

As we will see later, the experiments show that during the dissolution process calcite equilibrium is achieved exponentially with a time constant in the order of $T \approx 10^4$ s. The decay time $T_d = \delta^2/D\pi^2$ is $T_d = 10^3$ s for $\delta = 0.3$ cm. Thus, for geologically relevant values of $\delta < 0.3$ cm we can treat the problem as quasi-stationary with solution (7.15) for Eq. (7.7).

To solve the coupled Eqs. (7.7a,b,c) we use a method which has been proposed by Quinn and Otto (1971).

Adding Eqs. (7.7a,b,c) leads to:

$$\frac{\partial[CO_2]}{\partial t} + \frac{\partial[HCO_3^-]}{\partial t} + \frac{\partial[CO_3^{2-}]}{\partial t}$$

$$= D_{CO_2}\frac{\partial^2[CO_2]}{\partial z^2} + D\frac{\partial^2[HCO_3^-]}{\partial z^2} + D\frac{\partial^2[CO_3^{2-}]}{\partial z^2}. \tag{7.16}$$

In the quasi-stationary approximation, we have the following condition. The change of carbonate concentration due to dissolution is approximated as:

$$D\frac{\partial\overline{[HCO_3^-]}}{\partial t} + D\frac{\partial\overline{[CO_3^{2-}]}}{\partial t} = (F_{HCO_3} + F_{CO_3} + F)\cdot\frac{1}{\delta} ; \quad \frac{\partial[CO_2]}{\partial t} = 0 , \tag{7.17}$$

where F_{CO_3} and F_{HCO_3} are the fluxes of the corresponding species from the surface into the solution and the bar denotes an average value. The term F/δ arises from the fact that for each Ca^{2+} ion one molecule of CO_2 is converted into HCO_3^-. Under open system conditions within our approximation $\partial[CO_2]/\partial t = 0$, since CO_2 remains constant. Furthermore, we have the boundary condition:

$$F_{HCO_3} + F_{CO_3} = F , \tag{7.18}$$

since each dissolved $CaCO_3$ releases one C-atom.

With these conditions, Eqs. (7.17) and (7.18), we obtain by inserting into Eq. (7.16)

$$D_{CO_2}\frac{\partial^2[CO_2]}{\partial z^2} + D\frac{\partial^2[HCO_3^-]}{\partial z^2} + D\frac{\partial^2[CO_3^{2-}]}{\partial z^2} = \frac{2F}{\delta} . \tag{7.19}$$

Integration yields:

$$\frac{2F}{\delta}z + C_1 = D_{CO_2}\frac{\partial[CO_2]}{\partial z} + D\frac{\partial[HCO_3^-]}{\partial z} + D\frac{\partial[CO_3^{2-}]}{\partial z} . \tag{7.20}$$

The integration constant C_1 is determined by the boundary conditions at $z = 0$. These are:

$$-D_{CO_2}\cdot\frac{\partial[CO_2]}{\partial z/0} = F_{CO_2} = F ; \quad \frac{\partial[HCO_3^-]}{\partial z/0} = \frac{\partial[CO_3^{2-}]}{\partial z/0} = 0 , \tag{7.21}$$

which yields $C_1 = -F$.

A second integration of Eq. (7.20) yields:

$$\frac{F}{\delta}z^2 - Fz + C_2 = D_{CO_2}[CO_2] + D[HCO_3^-] + D[CO_3^{2-}] . \tag{7.22}$$

The integration constant C_2 is determined from the concentrations $[\ ]_\delta$ at $z = \delta$. Finally, we obtain:

$$\frac{F \cdot \delta}{z^2} - Fz + D_{CO_2}[CO_2]_\delta + D([HCO_3^-]_\delta + [CO_3^{2-}]_\delta)$$

$$= D_{CO_2}[CO_2] + D([HCO_3^-] + [CO_3^{2-}]) . \tag{7.23}$$

To solve Eq. (7.7a), we have to express $[HCO_3^-]$ as a function of $[CO_2]$. This is achieved by using the equation of electroneutrality (neglecting ion pairs):

$$2[Ca^{2+}] + [H^+] = [HCO_3^-] + 2[CO_3^{2-}] + [OH^-] \tag{7.24}$$

and the mass action laws

$$[CO_3^{2-}] = \frac{K_2 \gamma_{HCO_3}[HCO_3^-]}{\gamma_H \gamma_{CO_3}[H^+]}; \qquad K_w = \gamma_H \gamma_{OH}[H^+][OH^-] . \tag{7.25}$$

Inserting Eq. (7.25) into (7.24) gives a quadratic equation for $[H^+]$ with the solution:

$$[H^+] = -\frac{1}{2}(2[Ca^{2+}] - [HCO_3^-])$$

$$+ \sqrt{\frac{(2[Ca^{2+}] - [HCO_3^-])^2}{4} + \frac{K_w}{\gamma_H \gamma_{OH}} + \frac{2K_2 \gamma_{HCO_3}}{\gamma_H \gamma_{CO_3}}[HCO_3^-]} . \tag{7.26}$$

From this the concentration $[CO_3^{2-}]$ is calculated as a function of $[HCO_3^-]$ by inserting Eq. (7.26) into Eq. (7.23). Inserting $[CO_3^{2-}]$ into Eq. (7.23) yields a cubic relation for $[HCO_3^-]$ as a function of $[Ca^{2+}]$ and $[CO_2]$:

$$L[HCO_3^-]^3 + M[HCO_3^-]^2 + N[HCO_3^-] + P = 0 \tag{7.27}$$

with

$$L = K_2 \frac{\gamma_{HCO_3}}{\gamma_H \gamma_{CO_3}};$$

$$M = \frac{K_w}{\gamma_H \gamma_{OH}} - L^2 + 3BL + 2L[Ca^{2+}];$$

$$N = 2B\left(\frac{K_w}{\gamma_H \gamma_{OH}} + L[Ca^{2+}] + BL\right);$$

$$P = B^2 \cdot (K_w/\gamma_H \gamma_{OH}); \tag{7.28}$$

and

$$B = \frac{1}{D}\left\{-\frac{F}{\delta}z^2 + Fz - D_{CO_2}([CO_2]_\delta - [CO_2]) - D[CO_3^{2-}]_\delta - D[HCO_3^-]_\delta\right\} . \tag{7.29}$$

Inserting the solution of Eq. (7.27) into Eq. (7.23) we obtain a relation which expresses $[HCO_3^-](z)$ in terms of the variable z and $[CO_2](z)$. Similarly, $[H^+](z)$ can be expressed as a function of z and $[CO_2](z)$. Introducing these functions for $[HCO_3^-](z)$ and $[H^+](z)$ into the transport Eq. (7.7a) for $[CO_2](z)$, the right-hand side of this equation becomes an expression depending only on z and $[CO_2](z)$ and also on the concentrations at the solid-liquid boundary. Remembering that in the open system $[CO_2]$ is independent of time and therefore the derivative with respect to time is zero, we have:

$$D\frac{\partial^2 [CO_2](z)}{\partial z^2} = g\{z, [CO_2](z), [Ca^{2+}]_\delta, [HCO_3^-]_\delta, [CO_2]_\delta\}\,. \qquad (7.30)$$

This is a second-order, decoupled differential equation for $[CO_2](z)$ with boundary conditions (7.11) and (7.12).

The procedure described above cannot be written down analytically, but has to be carried out numerically by computer. To do so, we choose as starting values $[Ca^{2+}]_\delta$ and $[HCO_3^-]_\delta$, and $[CO_2]_\delta$. From these we calculate $F([Ca^{2+}]_\delta, [HCO_3^-]_\delta, [H_2CO_3^*]_\delta)$ by employing the PWP equation and calculating $[H_2CO_3^*]_\delta$ in equilibrium with $[HCO_3^-]_\delta$.

With the knowledge of F the calcium ion concentration $[Ca^{2+}(z)]$, Eq. (7.15) is computed. $[H^+]_\delta$ is obtained from Eq. (7.26).

With the boundary conditions chosen for $[CO_2]$ at $z = \delta$ a Runge-Kutta procedure is started for Eq. (7.30) by employing for each step a procedure for the solution of the cubic equation to obtain the function g in Eq. (7.30).

From the solution of this procedure the flux of CO_2 at $z = 0$ is computed and compared to that of Ca^{2+} at the liquid-solid boundary $z = \delta$. $[HCO_3^-]_\delta$ and $[CO_2]_\delta$ are then varied until the boundary conditions (7.12) are fulfilled.

The final result then provides the concentration profiles of $[Ca^{2+}](z)$, $[CO_3^{2-}](z)$, $[HCO_3^-](z)$ and $[H^+](z)$. Since $[HCO_3^-]$ and $[CO_3^{2-}]$ are in equilibrium, $[CO_3^{2-}](z)$ is calculated by mass action. The flux is obtained as a function of $[Ca^{2+}]_\delta$ and p_{CO_2}. The average calcium concentration in the liquid can be calculated from $[Ca^{2+}](z)$. Thus, one finally obtains the flux F, i.e. the dissolution rate of calcite as a function of p_{CO_2} and the average calcium ion concentration. Thus, these two parameters under given conditions, such as temperature, layer thickness and hydrodynamic conditions, i.e. coefficient of diffusion, determine the dissolution rates entirely.

It should be noted that by employing this procedure it is also possible to calculate precipitation rates. This results from the fact that the PWP rate equation can also be applied for supersaturated solutions. In this case the sign of F changes, since now mass transport is directed from the bulk to the solid. From this, according to the boundary conditions employed, the flux of CO_2 also changes sign and there is now transfer of CO_2 from the solution to the atmosphere.

7.3.2 Results for the Open System

7.3.2.1 Dissolution Rates at Various Conditions

In the following we will discuss the results of the calculations described above for a variety of conditions relevant to karst environments. Further details can be taken from the original work of Buhmann and Dreybrodt (1985a).

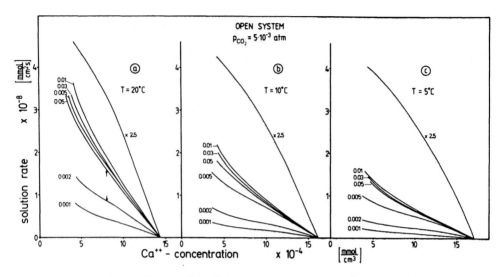

Fig. 7.5. Open system: Theoretical dissolution rates at $p_{CO_2} = 5 \times 10^{-3}$ atm for different temperatures and layer thicknesses δ. The *numbers* at the *solid lines* give the thickness δ in cm. The *uppermost curves* have to be multiplied by a factor of 2.5 as indicated in the figure. The *arrows* in **a** indicate the situation for which the concentration profiles in Fig. 7.6 are calculated

Figure 7.5 shows the theoretical dissolution rates calculated for a system open to an atmosphere containing CO_2 at a partial pressure of $p_{CO_2} = 5 \times 10^{-3}$ atm. The rates have been calculated for a region of temperatures as they are common in karst. Flow is assumed to be laminar with the exception of the uppermost curves where flow is assumed to be turbulent.

The rates are given as a function of average $[Ca^{2+}]$ already dissolved and for various thicknesses δ of the water layer.

Figure 7.5 exhibits three regimes of dissolution depending on the thickness of the layer:

Regime 1. CO_2 conversion: For $\delta < 0.005$ cm the dissolution rates increase linearly with δ. In this region CO_2 conversion into H_2CO_3 is the rate-limiting step. The amount of CO_2 converted into H_2CO_3 given by:

$$\frac{dCO_2}{dt} = V \cdot \frac{d[CO_2]}{dt}, \tag{7.31}$$

which is proportional to the volume of the solution. From stoichiometry the amount of Ca^{2+} dissolved has to be equal to the amount of CO_2 converted, thus:

$$F = \frac{V}{A} \frac{d[CO_2]}{dt} = \delta \frac{d[CO_2]}{dt} = \delta(k^+[CO_2] - k^-[HCO_3^-]), \tag{7.32}$$

where A is the area of dissolving material and F the Ca^{2+} dissolution rate. The rate constants k^- and k^+ depend on the pH of the solution (cf. Sect. 4.3.1). The local composition of the solution depends only slightly on δ since diffusion is fast at such small values of δ. Therefore, the conversion rate is practically constant for given values of $[Ca^{2+}]$ and $[CO_2]$. Thus, at sufficiently small δ the total amount of CO_2 converted which is proportional to δ determines the dissolution rates.

This can also be seen by calculating the dissolution rates after multiplying all the rate constants k of CO_2 conversion in Eq. (7.9) by a common factor of 10^6. This simulates a drastic increase in CO_2 conversion, which therefore is no longer rate-limiting. The results of such calculations show dissolution rates independent of δ for $\delta \le 0.005$ cm, since now they are neither limited by CO_2 conversion nor by diffusional mass transport due to the small layer thickness. The calculation shows that in this hypothetical case rates are determined entirely by the PWP equation, i.e. heterogeneous reactions at the surface of the calcite.

Regime 2. Mixed kinetics: For the range of 0.005 cm $< \delta <$ 0.05 cm the dissolution rates all crowd into a narrow region and are practically independent of δ. This is the regime of mixed kinetics, where the rates are controlled by molecular diffusion and the homogeneous reaction of CO_2 conversion. This regime is characterized by the diffusion length λ defined by Eq. (4.59):

$$\lambda = \sqrt{\frac{D}{k}}, \tag{7.33}$$

where k is a representative conversion constant. To understand this mechanism of mixed kinetics in the complex case of calcite dissolution, which is also controlled by heterogeneous surface reactions, it is useful to consider the distribution of species concentrations in the water layer. This is shown by Fig. 7.6 for the situations indicated by arrows in Fig. 7.5a. For small $\delta \ll \lambda$ the concentrations are almost constant across the layer. If $\delta > \lambda$ we find steep changes in the concentrations of H^+ and CO_3^{2-}. These changes define a region adjacent to the solid surface. The extension of this reaction zone is given by λ. CO_2 diffuses from the surface towards this zone, where CO_2 concentration is almost constant, showing that only small amounts of CO_2 are converted to H^+ and HCO_3^-. In the reaction layer the CO_3^{2-} curve bends over, since in this region the conversion of CO_2 takes place and delivers the hydrogen ions for the reaction with CO_3^{2-}. Therefore, the total amount of CO_2 conversion and thus calcite dissolution depends only on the depth λ of the reaction zone for $\delta > \lambda$ and thus becomes independent of δ. The rate of conversion is therefore proportional to $\lambda \cdot k = \sqrt{k \cdot D}$. This is also reflected by the temperature dependence of the rates in both regions. In regime 1 at fixed δ and $[Ca^{2+}]$ the rates increase with k, i.e. a factor of two, when the temperature is raised from 10° to 20°C. In regime 2 accordingly for $\delta = 0.01$ cm one finds a change of $\sqrt{2}$ corresponding to a dependence on $\sqrt{k}$, since the coefficient of diffusion depends only slightly on temperature. The heterogeneous surface reactions have no significant influence on the dissolution rates in regime 2.

Changing in a numerical simulation all the rate constants in the PWP equation simultaneously by a factor from 0.5 up to 2, i.e. increasing or decreasing the surface reaction rates, yields changes of less than 10% in these calculations. For values of $\delta \gg \lambda$, i.e. $\delta > 0.05$ cm, the rates start to decrease. With increasing δ the rates become

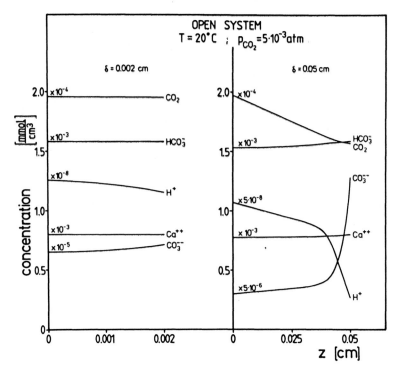

Fig. 7.6. Open system: Concentration profiles of CO_2 and the four most important ions at $[Ca^{2+}]_\delta = 8 \times 10^{-4}$ mmol cm^{-3} across a water film of 0.002 and 0.05 cm thickness respectively. The concentration scale for the individual species has to be multiplied by the factor given at the corresponding line

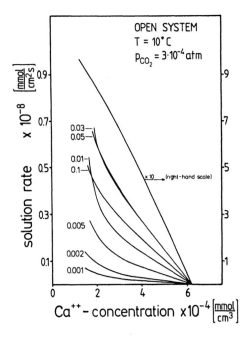

Fig. 7.7. Theoretical dissolution rates at $p_{CO_2} = 3 \times 10^{-4}$ atm and T = 10°C at various film-thicknesses δ. *Numbers* on curves give film thickness δ in cm. The drop of the dissolution rates for $\delta = 0.1$ results from increasing rate control by diffusion of Ca^{2+} ions into the bulk. The *uppermost curve* is for turbulent flow and $\delta = 1$ cm (see text) (Buhmann and Dreybrodt 1985a)

lower since now diffusion of $[Ca^{2+}]$ into the bulk becomes rate-limiting. This region is not shown in Fig. 7.5 since it would obscure the other curves.

Regime 3. Diffusion control: Figure 7.7 shows the dissolution rates for $p_{CO_2} = 3 \times 10^{-4}$ atm, i.e. atmospheric pressure of CO_2 for δ up to 0.1 cm. The upper limit of rates is reached at $\delta = 0.05$ cm. There is a significant drop at $\delta = 0.1$ cm from this upper limit.

   This drop results from the fact that with increasing thickness of the solution layer diffusion becomes more and more rate-controlling. One can understand this from the finding that the reaction zone of extension λ essentially is unaffected upon increase of δ. Therefore, fixed values of the surface concentration $[Ca^{2+}]_\delta$ are established at the surface of the dissolving calcite. The results of the calculations show that one obtains a limiting dissolution rate F, which can be approximated by:

$$F = \alpha_{lim}([Ca^{2+}]_{eq} - [Ca^{2+}]_\delta).\tag{7.34}$$

This approximation can be visualized from the upper limiting curve at $\delta_{lim} = 0.01$ for $p_{CO_2} = 5 \times 10^{-3}$ atm in Fig. 7.5 or $\delta_{lim} = 0.05$ for $p_{CO_2} = 3 \times 10^{-4}$ atm, (Fig. 7.7).

   Equation (7.34) expresses the fact that the driving force of dissolution is due to the difference of the concentration at saturation to the actual concentration $[Ca^{2+}]_\delta$ at the surface. In the case where F and δ are small, practically no concentration gradients for Ca^{2+} exist, cf. Fig. 7.6, and the average concentration of Ca^{2+} equals that at the surface. With increasing δ, $\delta > \delta_{lim}$ this is no longer valid. To express the dissolution rates as a function of the average concentration, a relation between this value and $[Ca^{2+}]_\delta$ is needed.

   By integrating the concentration profile given by Eq. (7.15) one obtains this relation between $[Ca^{2+}]_\delta$ and the average $[Ca^{2+}]$ value:

$$[Ca^{2+}] = \frac{1}{\delta}\int_0^\delta [Ca^{2+}](z)dz = [Ca^{2+}]_\delta - \frac{F\delta}{3D}.\tag{7.35}$$

Introducing this into Eq. (7.34) and solving for F one finds:

$$F = \frac{\alpha_{lim}}{1 + \dfrac{\alpha_{lim}\delta}{3D}} \cdot ([Ca^{2+}]_{eq} - [Ca^{2+}]).\tag{7.36}$$

For $\delta = \delta_{lim}$ the difference between $[Ca^{2+}]$ and $[Ca^{2+}]_\delta$ is small for all relevant values of F and furthermore $\alpha_{lim}\delta_{lim} \ll 3 \cdot D$.

   With increasing δ, however, the correction in the denominator contributes and the dissolution rates drop. In the limit of very large δ one obtains:

$$F = \frac{3D}{\delta}([Ca^{2+}]_{eq} - [Ca^{2+}]).\tag{7.37}$$

This is the region where transport is controlled entirely by diffusion, and dissolution rates become very small compared to those resulting from Eq. (7.33).

   All the curves discussed so far have been calculated for laminar flow, where diffusion is controlled by the coefficient of molecular diffusion.

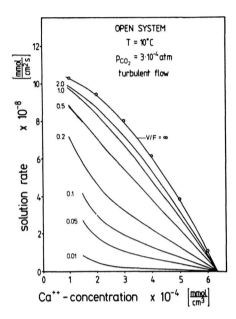

Fig. 7.8. Theoretical dissolution rates under turbulent flow conditions for various film thicknesses δ. The *number* at the solid curves gives δ = V/F in cm. The *uppermost curve* indicates the limit of infinite film thickness and is calculated for δ = 10 cm. The *open circles* are calculated according to the equilibrium theory as described in the text (Buhmann and Dreybrodt 1985a)

The uppermost curves in Figs. 7.5 and 7.7, however, are calculated for turbulent flow. In this case diffusion is controlled by turbulent eddies and the effective coefficient of diffusion is assumed to be by a factor of 10^4 larger than molecular diffusion coefficient everywhere in the fluid. In this first approximation the existence of a diffusion boundary layer is neglected. Furthermore, we have assumed δ = 1 cm. This is realistic in karst, since in conduits of this dimension flow is generally turbulent (Dreybrodt 1981a). Note that in Fig. 7.5 the scale has been changed by a factor of 2.5 and in Fig. 7.7 by a factor of 10. Thus, one observes a drastic increase of dissolution rates. This increase is of utmost geological importance as a threshold in karst evolution. The first conduit tubes reaching dimensions sufficiently large for the onset of turbulent flow will grow faster than those where laminar flow still exists. Therefore, they will draw flow away from them, and the pattern of the mature cave system will develop along these most effective drainage pathways.

To understand the mechanism of dissolution in turbulent flow, Fig. 7.8 shows results for dissolution at p_{CO_2} = 3 × 10^{-4} atm calculated with $D_e = 10^4 \cdot D_m$ for various values of δ. For very small values of $\delta \le 0.005$ the results are identical to those calculated for laminar flow (not shown in Fig. 7.8). This is easily explained from the fact that at small values of δ no significant concentration gradients exist (cf. Fig. 7.6) and diffusion is not rate-limiting. Therefore, the results are by no means sensitive to the value of the diffusion constant employed in the calculation.

In both cases, molecular or eddy diffusion, at small δ, rates are entirely controlled by CO_2 conversion and are therefore equal. With increasing δ, CO_2 conversion increases (Eq. 7.31) and therefore the rates in Fig. 7.8 increase accordingly. They do not reach the limit of mixed kinetics as in the case of laminar flow, since because of the large diffusion coefficient, diffusion is no longer rate controlling.

With increasing δ the rates converge to a new uppermost limit, where the heterogeneous surface reactions control rates exclusively. This results from the fact that due to the effective transport by eddies, the bulk is well mixed and no concentration gradients exist. Therefore, in this approximation the concentration of all species in the bulk is equal to that at the calcite surface. Furthermore, because of the large δ only a small deviation of $[CO_2]$ and $[HCO_3^-]$ from equilibrium is necessary to deliver sufficient amounts of hydrogen ions and carbonic acid. This situation is entirely equivalent to a well-mixed solution with all species in equilibrium with each other. The rates can therefore also be calculated by using the equilibrium model of Chapter 2 to calculate all species from $[Ca^{2+}]$ and p_{CO_2}. Employing these results and the PWP rate equation, the dissolution rates are obtained as shown in section 6.2.2. The open circles show the results in Fig. 7.8 to be in perfect agreement to the curve calculated from the complete theory.

To condense the findings of all calculations performed for various temperatures and p_{CO_2} Buhmann and Dreybrodt (1985a) have approximated their results by a linear expression:

$$F = \alpha([Ca^{2+}]_{eq} - [Ca^{2+}]); \qquad [Ca^{2+}] \geq 0.2[Ca^{2+}]_{eq} . \qquad (7.38)$$

The kinetic constant α depends on temperature, δ, p_{CO_2} and hydrodynamic flow conditions. Table 7.1 lists α and $[Ca^{2+}]_{eq}$ for various conditions. It can be used by interpolation to estimate dissolution rates for the open system in the range of 3×10^{-4} atm $\leq p_{CO_2} \leq 5 \times 10^{-3}$ atm and $5°C \leq T \leq 20°C$ for laminar and turbulent flow. It thus is possible to estimate dissolution rates for natural environments from the knowledge of the chemical composition of the solution and flow conditions.

Table 7.1. Numerical values of $\alpha = \alpha(T, \delta)$ in 10^{-5} cm s^{-1} at various partial pressures of CO_2 (in atm%): (a) for laminar flow; (b) for turbulent flow ($D_e = 10^4 D_{CO_2}$); (c) gives the calculated Ca^{2+} concentrations at saturation, $[Ca^{2+}]_{eq}$ in 10^{-4} mmol cm^{-3}. Open system

δ[cm]	$p_{CO_2} = 3 \times 10^{-4}$			$p_{CO_2} = 1 \times 10^{-3}$ atm			$p_{CO_2} = 5 \times 10^{-3}$ atm		
	5°C	10°C	20°C	5°C	10°C	20°C	5°C	10°C	20°C
(a) 0.001	0.041	0.0714	0.232	0.0619	0.11	0.358	0.153	0.25	0.679
0.002	0.076	0.135	0.407	0.134	0.233	0.75	0.294	0.467	1.36
0.005	0.197	0.341	1.04	0.33	0.583	1.69	0.647	1.11	2.74
0.01	0.37	0.635	1.84	0.619	1.08	2.76	1.00	1.54	3.13
0.03	0.803	1.29	2.91	0.979	1.61	3.29	0.969	1.47	2.87
0.05	0.915	1.29	2.57	0.907	1.40	2.64	0.926	1.35	2.63
0.1	0.636	0.919	1.60	0.68	1.0	1.63	0.85	1.21	2.0
(b) 0.1	3.3	5.7	13.5	5.2	8.0	14.7	7.25	8.3	11.0
0.2	5.9	9.5	20.0	8.4	11.5	17.5	7.9	8.8	12.0
0.5	11.5	16.5	26.0	12.0	14.5	18.5	8.0	9.0	12.0
1.0	16.0	20.6	28.5	13.5	15.0	19.0	8.0	9.0	12.0
2.0	20.0	24.0	30.0	13.5	15.0	19.0	8.0	9.0	12.0
(c) $[Ca^{2+}]_{eq}$	6.75	6.3	5.6	10.0	9.3	8.3	17.0	16.2	14.3

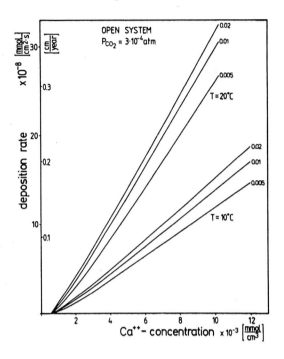

Fig. 7.9. Theoretical deposition rates at T = 10°C and T = 20°C respectively. The CO_2 pressure of the surrounding atmosphere is 3×10^{-4} atm. The *number* at the solid curves gives the film thickness δ in cm (Buhmann and Dreybrodt 1985a)

7.3.2.2 Precipitation Rates Under Various Conditions

As already pointed out precipitation rates can also be calculated from the formalism developed in the previous sections. Precipitation of calcite occurs on speleothems where $CaCO_3$ is deposited from a thin layer of supersaturated solution. In the formation of regular stalagmites this layer is stagnant and about 0.01 cm thick (Dreybrodt 1980). Precipitation can also occur from turbulent flowing water layers with largely varying thickness.

To estimate rates occurring under these conditions Fig. 7.9 shows theoretical deposition results for stagnant water layers at 10° and 20°C for supersaturated solutions at $p_{CO_2} = 3 \times 10^{-4}$ atm. The deposition rates are also scaled in growth rates of cm year^{-1} by the relation (Dreybrodt 1980):

$$\frac{dl}{dt} \cdot [\text{cm year}^{-1}] \cong 1.174 \times 10^6 [\text{mmol cm}^{-2}\text{s}^{-1}].$$

From this figure growth rates of stalagmites can be easily estimated. There is a significant increase of precipitation rates with temperature, which might be one reason why caves in warmer climates are heavily decorated by speleothems. Figure 7.10 shows growth rates for various p_{CO_2} which exist in cave environments. Growth rates between 10^{-2} cm year^{-1} up to 5×10^{-2} cm year^{-1} depending on supersaturation of the solution, as read from this figure, are in agreement with observations in the field (Geyh 1970, Pielsticker 1970, D.C. Ford and Drake 1980) (cf. Chap. 10).

Figure 7.11 plots maximal growth rates for the case of turbulent flow. The amount of calcite deposited depends heavily on the depth of the water streaming

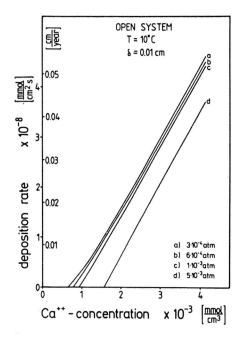

Fig. 7.10. Theoretical deposition rates of CaCO₃ at various CO₂ pressures of the surrounding atmosphere for geologically relevant initial Ca²⁺ concentrations (Buhmann and Dreybrodt 1985a)

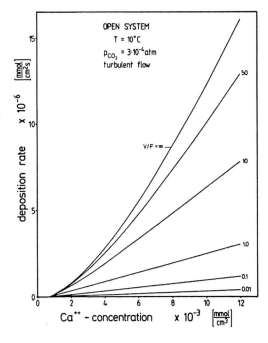

Fig. 7.11. Theoretical deposition rates under turbulent flow conditions for various film thicknesses δ (*numbers* at the solid curves, in cm). $\delta = V/F$. The diffusion constants D are assumed to be so high that no concentration gradients build up. The *uppermost curve* indicates the limit of infinite film thickness which yields the largest possible deposition

across the sinter. For $[Ca^{2+}]$ values as they are realistic in cave environments growth rates in the order of up to 1 cm year^{-1} are predicted. These values constitute an upper limit since we have not considered diffusion boundary layers in our turbulent model. We will discuss this later in this chapter.

7.3.2.3   Experimental Verification of Predicted Rates

The theoretical predictions have been verified by experiments under both laminar and turbulent flow conditions by Buhmann and Dreybrodt (1985a). Recalling Section 7.2, the time evolution of the chemical composition of a stagnant layer of solution on a calcite surface is directly related to the spatial evolution of chemical composition of a fluid flowing with average velocity v_y along the calcite surface. Thus, experimentally one can simulate geological conditions with a stagnant water layer on top of a calcite surface in the case of laminar flow, or turbulently stirred in the case of turbulent flow.

From the rate equations approximated by:

$$F = \alpha\{[Ca^{2+}]_{eq} - [Ca^{2+}](t)\} , \tag{7.39}$$

and using conservation of mass we find a differential equation for the Ca^{2+} concentration:

$$V \cdot \frac{d[Ca^{2+}](t)}{dt} = A \cdot \alpha\{[Ca^{2+}]_{eq} - [Ca^{2+}](t)\} \tag{7.40}$$

where V is the volume of the solution and A the area of calcite in contact to it.

The solution of this differential equation is:

$$[Ca^{2+}](t) = \left[1 - \exp\left(-\frac{\alpha A}{V}t\right)\right][Ca^{2+}]_{eq} . \tag{7.41}$$

Thus, equilibrium is approached exponentially with a mean time T, after which 63.2% of saturation is achieved:

$$T = \frac{V}{\alpha \cdot A} , \tag{7.42}$$

In case of a stagnant water layer of thickness δ on a plane calcite surface, the volume-surface ratio is equal to δ. Buhmann und Dreybrodt (1985a) performed measurements on the time evolution of Ca^{2+} ion concentration for thin stagnant films of thickness 0.03 cm $\leq \delta \leq$ 0.15 cm under controlled conditions of temperature and p_{CO_2} on various calcite samples. In these experiments a layer of distilled water was evenly distributed on a plane calcite surface and the whole setup was brought into a CO_2-containing atmosphere of known p_{CO_2} and temperature. Equilibration of aqueous CO_2 with the atmosphere occurs by diffusion (cf. Sect. 3.2.1.3). The mean time needed for equilibration is in the order of 10 min at its maximum. The solution was kept on the calcite for a given time and then analyzed for calcium

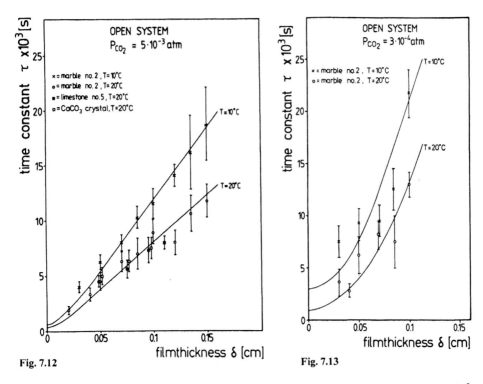

Fig. 7.12

Fig. 7.13

Fig. 7.12. Time constants of $CaCO_3$ solution measured for various film thicknesses at $p_{CO_2} = 5 \times 10^{-3}$ atm and $T = 10°$ and $20°C$ respectively. For $T = 20°C$ measurements on limestone and a pure calcite crystal are included. The *solid lines* are the theoretical curves fitted to the experiments by a factor f_c (see text) (Buhmann and Dreybrodt 1985a)

Fig. 7.13. Time constants of a $CaCO_3$ solution measured for various film thicknesses at $p_{CO_2} = 3 \times 10^{-4}$ atm and $T = 10°$ and $20°C$ respectively. The *solid lines* are the theoretical curves fitted to the experiments by a factor f_c (see text) (Buhmann and Dreybrodt 1985a)

concentration. From the exponential time dependence of $[Ca^{2+}]$ the mean time T was determined and compared to the theoretical value, derived from the calculation of α.

Figures 7.12 and 7.13 show the results for different p_{CO_2} values. The agreement of the experimental data to theory is remarkable, provided all the theoretical data are multiplied by a common factor, $f_c = 2.0$. The reason for this discrepancy is not clear. Since the experimental dissolution rates are slower by a factor of two, consistently in all open system experiments, one might speculate that there is some additional diffusion resistance to CO_2 diffusion built into the gas-liquid boundary (Stumm and Morgan 1981). This supported by the findings of analogous experiments under closed system conditions where one finds agreement to the theoretical data (cf. Sect. 7.4).

One further important point should be noted. In Fig. 7.12 data are shown for a variety of calcite samples, e.g. single crystal Iceland spar, Carrara marble and natural pure limestone (Sollnhofen, West Germany). They all agree well with the theory and there is no influence of lithology. This is what one expects, since all the experiments are conducted in regime 2, where only diffusion and CO_2 conversion control dissolution rates.

From this an important conclusion related to the initial state of karstification can be drawn. In this state, where the solution is confined in narrow partings of less than 0.01 cm, lithology does not control dissolution rates. Although this statement relates to closed system conditions, the experimental data on open system dissolution can be used to derive it. As we will see in the next section, there are analogous regimes 1 to 3 also under closed system conditions.

To simulate turbulent flow marble samples were cut into discs of 5-cm diameter. These samples were surrounded by a rim of Teflon and water of known volume was put into the vessel so formed. The depth of the water layers was between 0.5 to 1 cm. The solution was stirred with a glas propeller, such that an eddy diffusion constant of $D_e = 10^3 D_m$ resulted.

The time constants found from these experiments are shown in Fig. 7.14. These data show an increase of the kinetic constant α by about one order of magnitude compared to laminar flow situations. As in the experiments of laminar flow a factor of $f_c = 2.5$ is necessary to scale consistently the theoretical data to the experimental ones. In the regime of turbulent flow surface reactions are rate-controlling. The PWP coefficients are accurate within a factor of two (cf. Chap. 6). Furthermore, we have not considered diffusion boundary layers. As will be discussed later in this chapter, neglecting this extra resistance to diffusion, yields rates which are higher by about a factor of two in this type of experiment. One further remark to these experiments is important. All the data have been obtained at a composition of the solution far from equilibrium. They cannot be extrapolated uncritically close to equilibrium (cf. Chap. 6). Therefore, they can be applied only to situations in karst, where one has conduit flow and mineralization of water is sufficiently low. We will discuss this topic in later chapters in more detail.

Buhmann and Dreybrodt (1985a) have also performed experiments on calcite precipitation. They simulated conditions existing in the growth of regular stalagmites (Dreybrodt 1980, 1981a). A saturated solution of $CaCO_3$ equilibrated at a p_{CO_2} of 1, 0.3 or 0.05 atm respectively, was dripped at a rate of three drops per minute to a $CaCO_3$ sample cut from a speleothem. The surface area of the sample embedded into a plastic casting was 1×1 cm^2. The plastic casting was ground and polished together with the calcite sample. Thus, one even surface was prepared onto which the drops evenly distributed to a thin layer of 0.01 cm depth. This layer was in contact with an atmosphere of $p_{CO_2} = 3 \times 10^{-4}$ atm. Since equilibration of CO_2 to this atmosphere is established within a few seconds in average, the CO_2 in the layer is close to equilibrium with the surrounding atmosphere.

The $CaCO_3$ samples were left in place for a few days. The precipitation rates were measured by weighing the samples to determine an increase in weight. Figure 7.15 shows the experimental results in comparison to the theoretically calculated data (full lines). Agreement is remarkable and no scaling factor is necessary in the case of precipitation.

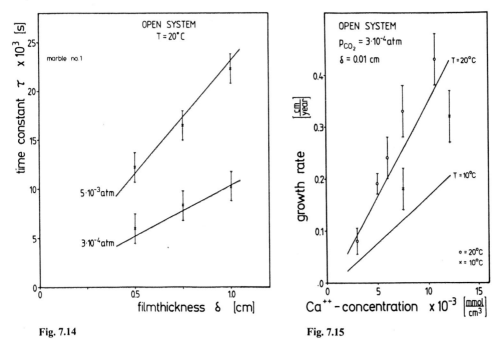

Fig. 7.14   Fig. 7.15

Fig. 7.14. Time constants of $CaCO_3$ solution measured under turbulent conditions for various film thicknesses and CO_2 pressures at T = 20°C. The speed of rotation of the glass stirrer is 200 rpm, which gives an effective diffusion constant $D_e = 1000\ D_m$. The *solid lines* are calculated with this value of D_e and fitted to the experiments by a factor $f_e = 2.5$ (Buhmann and Dreybrodt 1985a)

Fig. 7.15. Experimental growth rates of $CaCO_3$ at T = 10°C and T = 20°C respectively. The initial Ca^{2+} concentrations are given on the horizontal axis. The CO_2 pressure of the surrounding atmosphere is 3×10^{-4} atm. The *solid lines* are calculated with $\delta = 0.01$ cm and $p_{CO_2} = 3 \times 10^{-4}$ atm (Buhmann and Dreybrodt 1985a)

7.4   Dissolution and Precipitation Kinetics in the Closed System

7.4.1   Calculation of the Rates

The theory to calculate rates for closed system conditions in principle is analogous to that of the open system. The only difference stems from the boundary conditions, shown by Fig. 7.2. They are different to those of the open system (cf. Eqs. 7.9 to 7.13).

Since the geometry is mirror symmetrical with respect to the plane at z = 0, one finds a solution to the problem by treating only one-half of the region, i.e. $0 \le z \le \delta$. The complete solution is then found by symmetry. This implies, since concentrations have to be continuous and differentiable functions, that their derivatives with respect to z are zero at z = 0 for all species. This is in contrast to the open system, where there is a flux F_{CO_2} of CO_2 across z = 0. This is the only change with respect to the geometry from the open to the closed system. A second change comes from the chemical conditions. Since there is no flux of CO_2 into the system, the CO_2

concentration in the solution is lowered because of the stoichiometry of the reaction, which yields:

$$F([Ca^{2+}]_\delta, [HCO_3^-]_\delta) = - \int_0^\delta \frac{d[CO_2](z)}{dt} dz \, , \qquad (7.43)$$

where $\frac{d[CO_2]}{dt}$ is the conversion rate of CO_2 given by the right-hand side of Eq. (7.7a). Equation 7.43 simply states that at each time during dissolution the total amount of CO_2 converted in the solution has to equal the amount of Ca^{2+} released from the solid.

With these two changes the problem is treated analogously as in section 7.3. The details of this calculation can be taken from the work of Buhmann and Dreybrodt (1985b). The dissolution and precipitation rates are found as functions of the average Ca^{2+} concentration in the solution and of the initial CO_2 pressure $p_{CO_2}^i$. This is the pressure to which the pure H_2O–CO_2 solution has equilibrated prior to any dissolution of calcite.

The relation between the concentration $[H_2CO_3^*]$ prior to dissolution and that after the calcium concentration $[Ca^{2+}]$ has been attained, has been discussed in Chapter 2, (cf. Eq. 2.31).

7.4.2 Results for the Closed System

7.4.2.1 Dissolution Rates Under Various Conditions

Figure 7.16 shows dissolution rates calculated for the closed system with an initial $p_{CO_2}^i = 5 \times 10^{-2}$ atm as a function of the calcium ion concentration already achieved by dissolution. The curves are similar to those discussed in the open system.

For small values of δ the dissolution rates at a given value of $[Ca^{2+}]$ depend linearly on δ, which characterizes regime 1, where CO_2 conversion is the rate-limiting step. Note that the distance between the two calcite-dissolving planes is 2δ (cf. Fig. 7.2).

In the region 0.005 cm $< \delta <$ 0.1 cm the rates become independent of δ. This is regime 2, where diffusion and CO_2 conversion have comparable influence. If one increases the rate constants of conversion in this region of δ and also the coefficient of molecular diffusion by a common factor of 10^6, one simulates a situation, where neither CO_2 conversion nor diffusion are of rate-limiting influence. From these calculations one obtains the uppermost curves in Fig. 7.16, where the rates are determined entirely by the PWP rate equation of the heterogeneous surface reaction.

These curves are also obtained by leaving the rate constants of CO_2 conversion unchanged and using instead $\delta >$ 0.1 cm, which also increases CO_2 conversion accordingly due to the increase in volume [cf. Eq. (7.31)]. The diffusion coefficients are still higher by a factor 10^4 compared to molecular diffusion. This simulates

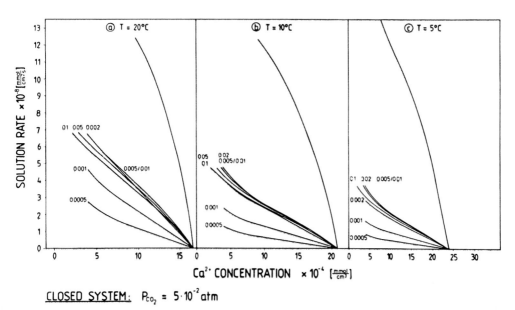

CLOSED SYSTEM: P_{CO_2} = 5·10⁻² atm

Fig. 7.16. Closed system: Calculated dissolution rates at p_{CO_2} = 5 × 10⁻² atm for various film thicknesses and temperatures. The *numbers* on the solid lines give the film thicknesses δ in cm. The *uppermost curves* represent turbulent flow conditions and are calculated for δ = 1 cm (Buhmann and Dreybrodt 1985b)

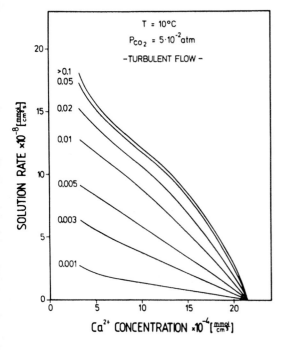

Fig. 7.17. Calculated dissolution rates under turbulent flow for various film thicknesses δ. The *numbers* on the solid lines give δ in cm. The *uppermost curve* indicates the limit of infinite film thickness (Buhmann and Dreybrodt 1985b)

turbulent flow, where, as in the open system, the rates are drastically increased in comparison to laminar flow.

Figure 7.17 shows calculations simulating turbulent flow by increasing the coefficient of diffusion for various $\delta \leq 0.1$ cm. Since diffusion now is extremely fast, only CO_2 conversion and heterogeneous reactions determine the rates. For small δ, therefore, CO_2 conversion is rate-limiting. Since the amount of CO_2 converted per time unit is proportional to the concentration of CO_2 and also to the thickness of the layer, the product $p_{CO_2} \cdot \delta$ determines the limit, where CO_2 conversion loses control. CO_2 pressure in the system ranges from 5×10^{-2} atm prior to dissolution to 2×10^{-2} atm at equilibrium. The limit of δ, where CO_2 loses control, is at $\delta = 0.1$ cm. In Fig. 7.8 the analogous situation for open system conditions is shown. There p_{CO_2} is 3×10^{-4} atm and correspondingly $\delta > 2$ cm had to be employed to reach the limit of surface control.

The equivalence of regimes 1 and 2 for open and closed systems correspondingly can also be seen from Fig. 7.18, which plots the concentration profiles for dissolution at $p_{CO_2} = 0.01$ atm for two values of δ representative to these regimes in the closed system. For small δ no significant changes in concentration develop. This is different for $\delta > \lambda$, where the reaction zone shows steep gradients in $[H^+]$ and $[CO_3^{2-}]$. Also, at still higher δ the influence of diffusion as a rate-controlling process is analogous to that discussed in section 7.3.

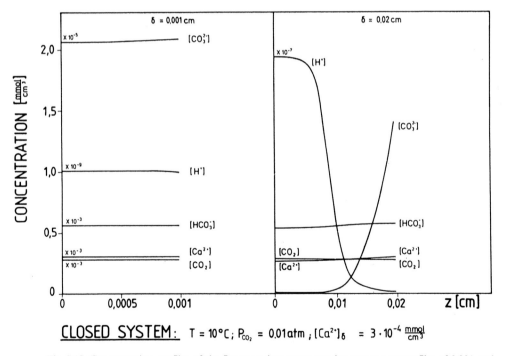

CLOSED SYSTEM: $T = 10°C$; $P_{CO_2} = 0.01$ atm; $[Ca^{2+}]_\delta = 3 \cdot 10^{-4} \frac{mmol}{cm^3}$

Fig. 7.18. Concentration profiles of the five most important species across a water film of 0.001 and 0.02 cm thickness respectively. The concentration scale for the individual species has to be multiplied by the factor given on the corresponding curve (Buhmann and Dreybrodt 1985b)

Table 7.2 Calculated values of α in 10^{-5} cm s^{-1} for various pressures of CO_2, temperatures and film thicknesses δ: (a) laminar flow; (b) turbulent flow; (c) gives the calculated Ca^{2+} concentrations at saturation, $[Ca^{2+}]_{eq}$ in 10^{-4} mmol cm^{-3}. Closed system

$\delta[m]$	$p_{CO_2} = 1 \times 10^{-2}$ atm			3×10^{-2} atm			5×10^{-2} atm			1×10^{-1} atm		
	5°C	10°C	20°C	5°C	10°C	20°C	5°C	10°C	20°C	5°C	10°C	20°C
(a) 0.001	0.714	1.6	7.14	0.606	1.1	3.33	0.666	1.19	3.33	0.833	1.39	3.23
0.002	1.3	2.8	13.3	1.18	2.1	6.25	1.25	2.25	5.26	1.05	1.73	3.33
0.005	3.13	6.4	21.7	2.13	3.5	8.93	1.49	2.5	5.0	1.06	1.73	3.33
0.01	5.26	9.6	21.7	2.04	3.5	8.7	1.49	2.47	4.76	1.06	1.73	3.23
0.02	5.13	9.5	18.2	2.0	3.46	6.9	1.46	2.44	4.76	1.05	1.70	3.17
0.05	4.76	8.4	16.1	2.0	3.3	6.5	1.43	2.43	4.67	1.04	1.68	3.07
0.1	4.08	7.7	12.7	1.89	3.25	5.9	1.4	2.29	4.46	1.01	1.61	2.95
(b) 0.001	1.08	3.3	9.75	0.63	1.08	4.1	0.655	1.17	3.37	0.82	1.37	3.5
0.003	2.9	7.3	22.0	1.8	3.3	9.5	1.97	3.32	8.14	2.28	3.5	6.0
0.005	5.08	10.2	27.5	2.9	5.1	13.0	3.13	5.09	10.4	3.3	4.7	7.8
0.01	8.6	15.1	34.0	5.4	8.75	17.0	5.2	7.3	13.0	4.54	6.05	9.0
0.02	13.2	20.5	45.0	8.2	11.3	19.0	6.9	8.9	13.9	5.5	7.0	10.2
(c) $[Ca^{2+}]_{eq}$	6.4	5.5	4.1	16.6	14.4	11.0	24.3	21.4	16.9	38.8	35.1	28.0

The results in the closed system have been summarized in a linear relationship, similarly to the open system as (Eq. 7.38):

$$F = \alpha([Ca^{2+}]_{eq} - [Ca^{2+}]).$$

Table 7.2 gives the values of α and $[Ca^{2+}]_{eq}$ for various $p_{CO_2}^i$ and δ for laminar and turbulent flow. Although, turbulent flow is unlikely to occur at $\delta < 0.1$ cm, the values for turbulent flow are given for completeness.

The rates under turbulent flow conditions for $\delta > 1.0$ cm are shown in Fig. 7.19. These curves can no longer be approximated by one linear relation. Two regions of linear approximation have to be used as shown by Fig. 7.20.

For $[Ca^{2+}] < [Ca^{2+}]_{eq}$ one has:

$$F = \alpha_1([Ca^{2+}]_{eq} - [Ca^{2+}]); \qquad [Ca^{2+}] \geq 0.7[Ca^{2+}]_{eq},$$

whereas for $[Ca^{2+}] < 0.7[Ca^{2+}]_{eq}$

$$F = \alpha_2([Ca^{2+}]_{as} - [Ca^{2+}]).$$

The definition of the slopes α_1 and α_2 and $[Ca^{2+}]_{as}$ can be visualized from Fig. 7.20. Table 7.3 gives values of $\alpha_1, \alpha_2, [Ca^{2+}]_{as}$ for turbulent flow with $\delta \geq 0.1$ cm. These data can be used to estimate dissolution rates in real karst conduits.

7.4.2.2 Precipitation Rates Under Various Conditions

In general, calcite-aggressive solutions entering from the soil zone, where p_{CO_2} is much larger than atmospheric p_{CO_2}, into fissures of limestone rock come quickly to

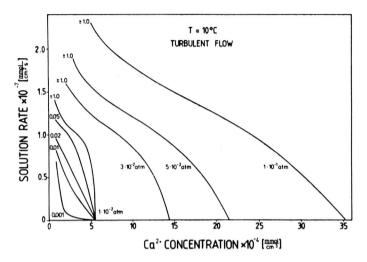

Fig. 7.19. Calculated dissolution rates under turbulent flow for various CO_2 pressures and various film thicknesses δ in the case of $p_{CO_2} = 1 \times 10^{-2}$ atm. The *numbers* on the solid lines give δ in cm. (Buhmann and Dreybrodt 1985b)

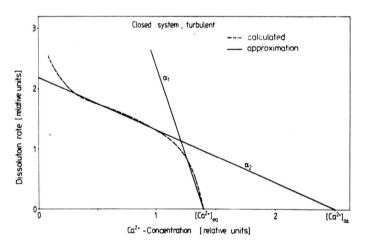

Fig. 7.20. Approximation of calculated dissolution rates at turbulent flow in the closed system by two linear relationships with different slopes α_1, α_2 and different apparent saturation concentrations $[Ca^{2+}]_{as}$, $[Ca^{2+}]_{eq}$ respectively (Buhmann 1984)

Table 7.3. Calculated values of α_i for various pressures of CO_2 and temperatures at turbulent flow and film thicknesses > 0.02 cm. Closed system

δ[cm][a]	$p_{CO_2} = 1 \times 10^{-2}$ atm			3×10^{-2} atm			5×10^{-2} atm			1×10^{-1} atm		
	5°C	10°C	20°C	5°C	10°C	20°C	5°C	10°C	20°C	5°C	10°C	20°C
(a) α_1												
0.05	2.4	4.2	11.8	1.4	2.1	2.6	1.0	1.4	2.3	0.76	0.93	1.8
0.1	4.0	6.7	17.3	1.7	2.5	2.8	1.2	1.5	2.4	0.8	1.0	1.8
0.3	9.8	14.3	29.2	1.8	2.5	2.8	1.3	1.5	2.4	0.8	1.0	1.8
$[Ca^{2+}]_{eq}$	6.4	5.5	4.1	16.6	14.4	11.0	24.3	21.4	16.9	38.8	35.1	28.0
(b) α_2												
0.05	1.5[b]	1.3[b]	1.6	0.6	0.78	1.3	0.54	0.67	1.0	0.44	0.56	0.7
0.1	0.85	1.0	1.7	0.63	0.82	1.35	0.56	0.69	1.02	0.45	0.57	0.7
0.3	0.9	1.0	1.7	0.65	0.82	1.35	0.56	0.69	1.02	0.45	0.57	0.7
$[Ca^{2+}]_{as}$	8.0[b]	9.7[b]										
	14.2	13.3	9.3	24.0	20.5	16.0	31.0	27.5	22.0	47.5	41.5	34.0

[a] Values of α_i in 10^{-4} cm s^{-1}; $[Ca^{2+}]_{eq}$ and $[Ca^{2+}]_{as}$ in 10^{-4} mmol cm^{-3}. $[Ca^{2+}]_{as}$ in an apparent saturation concentration of Ca^{2+} as discussed in the text.
[b] Corresponding values for T $= 5°C$ and T $= 10°C$ in this special case.

an equilibrium with calcite. The equilibrium p_{CO_2} mostly still exceeds atmospheric p_{CO_2}. If such a solution enters into vadose parts of a cave, it becomes quickly supersaturated by outgassing of CO_2. These solutions eventually penetrate into joints, where under closed system conditions calcite precipitates from the solution. Joints may thus be closed to further flow of water and percolation routes are changed accordingly.

Figure 7.21 shows an example of deposition rates of a supersaturated solution which has equilibrated its CO_2 concentration to an atmosphere with $p_{CO_2} = 3 \times 10^{-4}$ atm. Flow of this solution is assumed to be turbulent and the aperture of the fissures is given in centimetres at the particular curves.

Average growth rates of calcite deposition are in the order of 0.1 cm year^{-1}. In laminar flow through narrow partings deposition rates are lower by a factor of ten as shown in Fig. 7.22, where two supersaturated solutions with different initial Ca^{2+} concentrations $[Ca^{2+}]_S$ upon entering into the closed system are shown. Even in the case that outgassing of CO_2 is not complete, remarkable deposition rates are possible. Figure 7.23 shows precipitation rates for supersaturated solutions with different p_{CO_2}, achieved by outgassing CO_2 when entering into a large fissure with $\delta = 1$ cm. The numbers on the three sets of curves refer to different p_{CO_2}. Each set of curves furthermore refers to different $[Ca^{2+}]_S$.

Figures 7.21 to 7.23 can be used to estimate times which may be necessary to close joints for further flow of water. Under favourable conditions these times may be very short, e.g. in the order of 10 years.

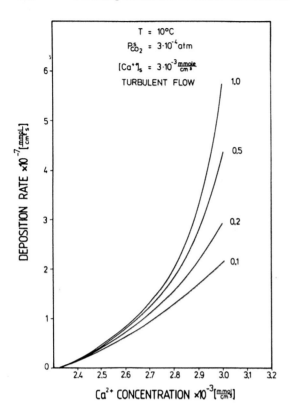

Fig. 7.21. Calculated deposition rates under turbulent flow conditions at various film thicknesses δ. The solution enters the closed system with $p_{CO_2} = 3 \times 10^{-4}$ atm and a Ca^{2+} concentration of 3×10^{-3} mmol cm^{-3}. The *numbers* on the solid lines give δ in cm (Buhmann and Dreybrodt 1985b)

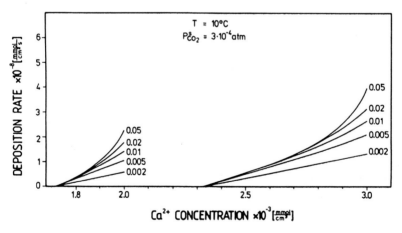

Fig. 7.22. Calculated deposition rates at various film-thicknesses δ and two different initial Ca^{2+} concentrations of 2×10^{-3} and 3×10^{-3} mmol cm^{-3} respectively. The *numbers* on the curves give the values of δ in cm. The solution enters the closed system with $p_{CO_2} = 3 \times 10^{-4}$ atm (Buhmann and Dreybrodt 1985b)

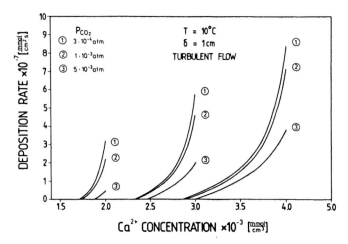

Fig. 7.23. Calculated deposition rates under turbulent flow conditions at various initial CO_2 pressures and initial Ca^{2+} concentrations (Buhmann and Dreybrodt 1985b)

7.4.2.3 Experimental Verification for Predicted Rates in the Closed System

To prove the theoretical results one proceeds in a similar way as in the experiments for open system conditions. One measures the time dependence of $[Ca^{2+}]$ during dissolution in a stagnant H_2O film enclosed by two parallel calcite surfaces with distance 2δ. As long as the dissolution rates are given by a linear relation according to Eq. (7.41) the time dependence of the average $[Ca^{2+}]$ concentration is:

$$[Ca^{2+}](t) = [Ca^{2+}]_{eq}\left[1 - \exp\left(-\frac{t}{T}\right)\right], \qquad T = \frac{\delta}{\alpha}. \qquad (7.44)$$

Thus, from the exponential decay of the quantity $[Ca^{2+}]_{eq} - [Ca^{2+}](t)$ towards zero, the time constant T can be measured and compared to the theoretical value.

The calcite specimens are flat plates with dimensions $5 \times 10 \times 1$ cm, cut from white marble (Carrara). To simulate the conditions of laminar flow two of these plates were mounted with a distance $d = 2\delta$ towards each other. The distance was achieved by placing distance pieces out of Teflon at the rims. Then, the volume between the plates was filled with bidistilled water of known CO_2 content and the system was closed with respect to the surroundings to prevent CO_2 exchange. After a time t the solution was extracted with a syringe, which was mounted into one of the calcite plates. The solution was then analyzed for its Ca^{2+} concentration by atomic absorption spectroscopy.

All the experiments were performed in a temperature regulated thermostatic chamber (T $\pm$ 0.5°C). The solutions with fixed CO_2 content were prepared by bubbling an artificial, commercially available atmosphere with known CO_2 content through bidistilled water and monitoring pH to control equilibrium.

To simulate the conditions of turbulent flow, a calcite specimen of 6-cm diameter and 1-cm thickness was mounted from below, by using O-ring seals of Teflon, to a cylindrical Teflon beaker with an inner diameter of 3.8 cm and of

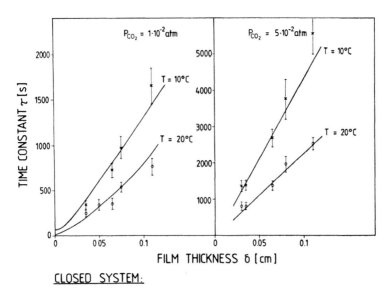

CLOSED SYSTEM:

Fig. 7.24. Time constants of $CaCO_3$ dissolution in the closed system measured for various film thicknesses, temperatures and CO_2 pressures. The *solid lines* are the theoretical curves; the measurements were performed on pure marble (Buhmann and Dreybrodt 1985b)

variable inner height. Through the upper side also a Teflon plate a glass stirrer was introduced, which was tightened by a Simmering® seal. The CO_2-containing solution is introduced through one inlet until the solution flows out at an outlet. Thus, one could ascertain that no air bubbles remained in the chamber. The inlet and the outlet were then sealed against the outer atmosphere. By stirring the solution the effective diffusion coefficient was enhanced by a factor of 10^3. The thickness of the solution in these experiments was 1, 1.5 and 2 cm. The solution was extracted from the container after a time t and analyzed for $[Ca^{2+}]$. From the time dependence the mean time T is determined.

Figure 7.24 shows the results for laminar flow in comparison to the theoretical values at temperatures of 10° and 20°C for two initial $p_{CO_2}^i$. The agreement to the experimental data is remarkable. Note that in the case of laminar flow in the closed system no scaling factor is needed. This corroborates the assumption that in the open system, where this factor is needed, some additional resistance to CO_2 diffusion is present in the gas-liquid boundary, which is missing in the closed system. In turbulent flow the experimental data show a significant increase of dissolution rates compared to laminar flow, similarly as in the open system.

7.5   The Influence of the Diffusion Boundary Layer to Dissolution and Precipitation Rates in Turbulent Flow

In the treatment of the kinetics of calcite dissolution and precipitation the transition from laminar to turbulent flow so far has been approximated by simply replacing the molecular coefficient of diffusion by that of eddies everywhere in the solution. This simple first approach is very convenient since both cases can then be dealt with by employing the same mathematical formulations. It is assumed in this approach, however, that a high eddy diffusion constant is also present close to the calcite surface. This is not the case. As we have discussed in Chapter 5 (cf. Eq. 5.30), there is a laminar layer of thickness ε_n adjacent to the wall of the flow conduit. In this layer transport of momentum is determined by molecular properties of the fluid, i.e. viscosity. Similarly, related to this hydrodynamic boundary layer, a diffusion boundary layer exists, where mass transport is controlled by molecular diffusion. The thickness ε of this layer is related to that of the hydrodynamic layer by the Schmidt number (Beek and Mutzall 1975, Kay and Nedderman 1985; cf. Chap. 6):

$$\varepsilon = \varepsilon_h N_{Sc}^{-1/3}, \qquad N_{Sc} = \frac{\nu}{D_m} . \tag{7.45}$$

When mass transport in this layer is entirely by molecular diffusion, a Nernst-layer model is employed. The flux effecting mass transfer is then simply (cf. Eq. 3.27):

$$F = \frac{D_m}{\varepsilon}(c_{eq} - c_B) , \tag{7.46}$$

where c_B is the concentration in the turbulent core.

If the detailed mechanism of mass transport is not clear, an empirical mass transfer coefficient h is used, defined by:

$$F = h(c_{eq} - c_B) , \qquad h = \frac{D_m}{\varepsilon} . \tag{7.47}$$

To describe mass transport one often uses the dimensionless Sherwood number:

$$N_{Sh} = \frac{hL}{D_m} , \tag{7.48}$$

where L is a characteristic dimension of the flow conduit. This number can be expressed as a function of the dimensionless Reynolds and Schmidt numbers. By dimensional analysis one finds a correlation:

$$N_{Sh} = C \cdot N_{Re}^n \cdot N_{Sc}^m + B . \tag{7.49}$$

There are a variety of semi-empirical correlations giving B, C and the exponents n and m for different geometrical conditions (Beek and Mutzall 1975, Bird et al. 1960, Skelland 1974). Table 7.4 lists some of these correlations from the current literature.

From the knowledge of the Sherwood number it is possible to estimate ε, by inserting Eq. (7.47) into Eq. (7.48):

$$N_{Sh} = \varepsilon/L . \tag{7.50}$$

Thus, from the knowledge of N_{Sh} and the characteristic dimension L of the conduit (L = diameter, in the case of a circular tube) one can give an estimation for ε to define the problem.

Table 7.4. Experimental correlations of the Sherwood number (N_{Sh}) to Reynolds number (N_{Re}) and the Schmidt number (N_{Sc}) (Skelland, 1974)[a]

Flow situation	Sherwood number $N = L/\varepsilon$
Flow through a straight circular pipe with diameter d_1, $$N = \frac{v\,d_1}{v}$$	$N_{Sh} = 0.023\,N_{Re}^{0.83}\,N_{Sc}^{1/3}$ $0.6 \leq N_{Sc} \leq 2500$ $2000 \leq N_{Re} \leq 35000$
Flow along a flat plate with length L along one side $$N = \frac{v \cdot L}{v}$$	$N_{Sh} = 0.037\,N_{Re}^{0.8}\,N_{Sc}^{1/3}$ (turbulent flow) $N_{Sh} = 0.664\,N_{Re}^{0.5}\,N_{Sc}^{1/3}$ (laminar flow)
Flow around sphere of diameter d_r $$N = \frac{v\,d_r}{v}$$	$N = 0.58\,N_{Re}^{0.5}\,N_{Sc}^{1/3}$ $300 \leq N_{Re} \leq 7600$ $N_{Sc} = 1210$

[a] N_{Sh} is related to the diffusion boundary thickness by $N_{Sh} = \varepsilon/L$, where L is the characteristic dimension of the geometrical configuration.

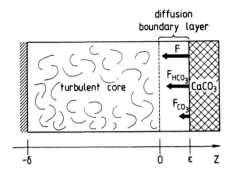

Fig. 7.25. Model of $CaCO_3$ dissolution for a system with a turbulent core and a diffusion boundary layer ($\delta \gg \varepsilon$)

The boundary conditions for calcite dissolutions can now be defined, once ε is known. Figure 7.25 gives the geometrical model. Adjacent to the calcite surface we have a diffusion boundary layer of thickness ε. The turbulent core extends to a thickness δ.

If δ is sufficiently large (i.e. $\delta \geq 1$ cm) and the CO_2 concentration of the solution is known, the dissolution rates can be calculated according to this model for both open and closed systems. This is due to the fact that at a sufficiently large volume of the solution the change in concentration of CO_2 is small during dissolution of $CaCO_3$. Provided a sufficiently short time is observed, the composition of the solution in the turbulent core stays practically constant. Furthermore, the boundary conditions at $z = 0$ due to the large eddy coefficient of diffusion are practically the same for open and closed systems, if the chemical compositions of the solutions are identical. Therefore, to calculate dissolution rates for the closed system, one has to know the initial $p_{CO_2}^i$ and thus $[H_2CO_3^*]_i$. The dissolution rates, after the system has evolved under a closed system reaction path to a calcium concentration $[Ca^{2+}]$, can then be obtained for the closed system by calculating the rates with the open system formulation by using $[H_2CO_3^*]$ now attained in the solution, which is given (cf. Chap. 2) by:

$$[H_2CO_3^*] = [H_2CO_3^*]_i - [Ca^{2+}] . \tag{7.51}$$

7.5.1 Calculation of Dissolution and Precipitation Rates

Figure 7.25 shows the geometry and boundary conditions. In the turbulent bulk we have a well-mixed core where all concentrations are spatially constant. Thus, at $z = 0$, the boundary between the turbulent core and the diffusion boundary layer, all concentrations are equal to those in the bulk. At the solid surface there is flux of Ca^{2+} and carbonate species across the boundary layer. This flux is related to the composition of the solution at $z = \varepsilon$ according to the PWP equation. In any case stoichiometry requires that at any time the flux of Ca^{2+} released is equal to the conversion of CO_2:

$$\int_{-\delta}^{\varepsilon} R_{CO_2} \cdot dz = F . \tag{7.52}$$

Using this model Dreybrodt and Buhmann (1988) have calculated dissolution and precipitation rates by similar methods as discussed in the previous sections. Figure 7.26 shows dissolution rates for the open system at various pressures of CO_2 in equilibrium with the solution. Curves from 0 to 8 are for increasing ε. The uppermost curves with $\varepsilon = 0$ are identical to the approximation of turbulent flow by neglecting the diffusion boundary layer. With increasing ε the rates decrease until finally they converge to a lower limit at $\varepsilon \approx 0.02$ cm. This lower limit corresponds to regime 2, discussed in section 7.3 and 7.4, where both diffusion and CO_2 conversion in the diffusion boundary layer determine the rates. Figure 7.27 illustrates the concentration profiles across the boundary layer for the situation indicated in Fig. 7.26c by an arrow. As in the case of laminar flow we find a reaction zone with extension λ smaller than ε, where CO_2 conversion is effected. If flow is highly turbulent and $\varepsilon \ll \lambda$ we have in the limit $\varepsilon = 0$. With increasing ε, due to diffusional resistance, the rates drop until $\varepsilon \approx \lambda$. For still higher ε, as in the case of laminar flow, a limiting rate is reached, which is determined by λ.

Precipitation rates for the same parameters as in Fig. 7.26 are shown in Fig. 7.28. They show similar behaviour with respect to ε. The large variation of the rates with ε shows that a sufficiently correct estimation of rates requires knowledge of the hydrodynamic conditions as well in the experiment as in nature. Since the extension of boundary layers in all batch experiments, as discussed in Chapter 6, are at least 10^{-3} cm, uncertainties in the reproducibility of the rates by a factor of two can now be also understood by inspection of Figs. 7.26 and 7.28.

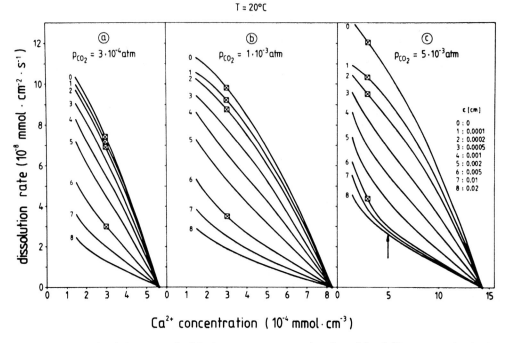

Fig. 7.26. Dissolution rates of calcite in open systems as a function of the Ca^{2+} concentration in the solution at three different CO_2 pressures. The *arrow* in **c** indicates the situation for which the concentration profiles of Fig. 7.27 are calculated. The *numbers* on the curves are related to the thickness ε of the boundary layer as listed in **c**. Thickness of the turbulent core: $\delta = 1$ cm

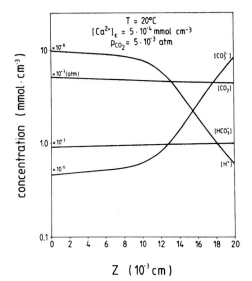

Fig. 7.27. Concentration profiles of the main species across a laminar boundary layer of thickness $\varepsilon = 0.02$ cm (cf. Fig. 7.26)

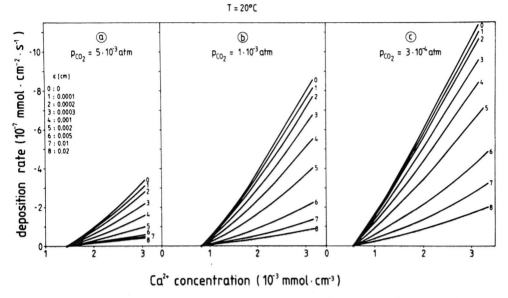

Fig. 7.28. Deposition rates of calcite in open systems as a function of the Ca^{2+} concentration in the solution at different atmospheric CO_2 pressures. The *numbers* on the curves are related to the thickness ε of the boundary layer as listed in **a**. Thickness of the turbulent core: $\delta = 1$ cm

Furthermore, the scaling factor, which was necessary to scale the theoretical data to the experimental finding in the experiments simulating turbulent flow of Buhmann and Dreybrodt (1985a,b) can now be explained. In these experiments ε can be estimated (Beek and Mutzall 1975) to be also in the order of 10^{-3} cm.

7.5.2 Experimental Verification

There is only one systematic study on the influence of ε on dissolution rates performed by Herman (1982) using rotating disc techniques (cf. Chap. 6). By varying rotational speeds from 7 rpm to 900 rpm a range of ε from 0.018 cm to 0.0016 cm was covered.

The experiments were performed far from equilibrium under open system conditions with high $p_{CO_2} = 1$ atm. The rates were measured as initial rates, where the calcium ion concentration in the solution was less than 1×10^{-4} mol l^{-1}, and also at concentrations of 2.2×10^{-3} mol l^{-1} and 5×10^{-3} mol l^{-1}. Figure 7.29 shows the comparison of calculated curves with the experimental data.

Curves 1 and 2 represent the lower and upper limit of $[Ca^{2+}]$ for which initial rates were measured. Curve 1 has been calculated for $[Ca^{2+}] = 10^{-5}$ mol l^{-1} and curve 2 for $[Ca^{2+}] = 10^{-4}$ mol l^{-1}. Although the experimental points are higher by a factor of two, they show the same behaviour with decreasing ε. For $\varepsilon < 0.004$ cm there is a sharp increase in the dissolution rates. In the case of higher calcium

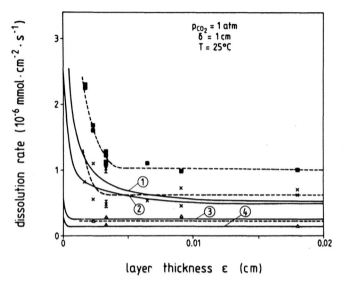

Fig. 7.29. Dissolution rates of calcite as a function of the layer thickness ε at different Ca^{2+} concentrations of the solution. *Dashed curves:* experimental values from Herman (1982), obtained by rotating disc experiments. ■: Initial rates $[Ca^{2+}] \leq 1 \times 10^{-4}$ mmol cm^{-3}; x: $[Ca^{2+}] = 2.2 \times 10^{-3}$ mmol cm^{-3}; ▲: 5×10^{-3} mmol cm^{-3}. *Solid curves:* calculated rates; the *numbers* on the curves refer to the Ca^{2+} concentration in the solution (mmol cm^{-3}) *1:* 1×10^{-5}; *2:* 1×10^{-4}; *3:* 2×10^{-3}; *4:* 5×10^{-3}

concentrations this increase is missing in both the theoretical and experimental data. The agreement in the behaviour of the calculated and experimental data provides evidence for the importance of diffusion boundary layers to dissolution rates.

7.6   Dissolution in Natural Systems

7.6.1   Influence of Foreign Ions on Dissolution Rates Far from Equilibrium

Natural waters in karst do not contain calcium and carbonate species only. One also encounters foreign ions such as Mg^{2+}, Na^+, SO_4^{2-} and Cl^-. The concentrations of these ions can be quite considerable, and therefore the question arises as to which extent dissolution kinetics is influenced by their presence. Sjöberg (1978) reported on the reduction of calcite dissolution rates in 0.7 KCl solutions in the presence of Mg^{2+} and SO_4^{2-}. These investigations were directed to dissolution kinetics in seawater.

To eludicate the influence of foreign ions in karst water, Buhmann and Dreybrodt (1987) performed a systematic investigation, both experimentally and theoretically, on the influence of Mg^{2+}, SO_4^{2-}, Cl^- and Na^+ to calcite dissolution rates. They added various concentrations of either NaCl, $CaCl_2$, $CaSO_4$, $MgCO_3$, $NaSO_4$ or $MgSO_4$ to the pure $H_2O-CO_2-CaCO_3$ system.

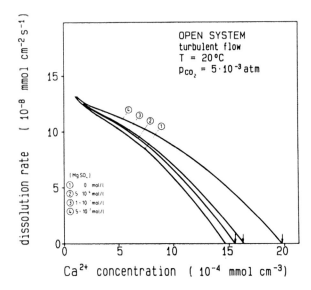

Fig. 7.30. Calculated rates of calcite dissolution upon addition of $MgSO_4$. Layer thickness $\delta = 1$ cm. The *arrows* indicate the calculated concentrations of Ca^{2+} using a simple equilibrium model (Buhmann and Dreybrodt 1987)

Theoretical predictions on the influence of these materials to the rates are obtained by extending the theory to the presence of the additional species. Using the methods discussed in section 7.3 dissolution rates were calculated for turbulent flow in the open system.

Figure 7.30 shows as a representative example dissolution rates calculated for various concentrations of $MgSO_4$.

With increasing concentration of $MgSO_4$ the calcite saturation concentration is shifted to higher values. This is mainly due to the ion-pair effect discussed in Chapter 2. The slopes of the rate curves, however, are not significantly changed upon addition of $MgSO_4$. Similar findings have been obtained for all the other substances investigated. Note that these calculations have been performed by extending the theory in its chemical part only, i.e. considering the influence of the foreign substances to the equilibria in the solution and also taking into account the presence of ion pairs. The heterogeneous PWP rate equations have been left unchanged. The rates shown in Fig. 7.30 are entirely determined by the surface reactions, i.e. by the PWP rate equation, because with $\delta = 1$ cm and $D_e = 10^4 \, D_m$, CO_2 conversion and diffusion are fast and not rate-limiting. Since changes in the kinetics of CO_2 conversion or the magnitude of diffusion constant are not expected by the presence of foreign ions, as specified above, changes of dissolution rates should occur due to the influence to the heterogeneous reactions at the calcite surface. Thus, by conducting experiments, in which dissolution rates are determined by this heterogeneous reaction, one can measure the effect of foreign ions by comparing these experimental results to the theoretical ones.

Using the corresponding experimental setup to simulate turbulent flow conditions, Buhmann and Dreybrodt (1987) measured the mean time T of approach to apparent equilibrium. They found good agreement to the theoretically predicted values in a concentration range of the employed foreign ions up to 1.5×10^{-3}

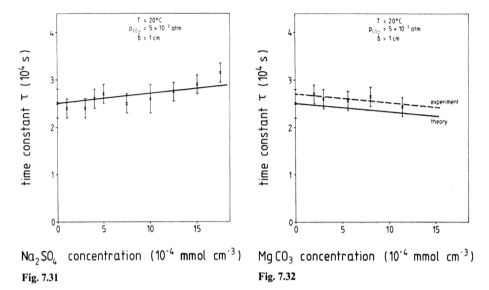

Na_2SO_4 concentration $(10^{-4}\ mmol\ cm^{-3})$ $MgCO_3$ concentration $(10^{-4}\ mmol\ cm^{-3})$

Fig. 7.31   **Fig. 7.32**

Fig. 7.31. Measured time constants of calcite dissolution upon addition of Na_2SO_4. *Solid line*: calculated variation of the time constants fitted to the experiments by a factor of 2.6 (Buhmann and Dreybrodt 1987)

Fig. 7.32. Measured time constants of calcite dissolution upon addition of $MgCO_3$. *Solid line*: calculated variation of the time constants; *dashed line*: fit through the experimental points (Buhmann and Dreybrodt 1987)

mol l^{-1}. Figures 7.31 and 7.32 show the comparison between theoretical and experimental data for Na_2SO_4 and $MgCO_3$ respectively, as foreign substances. From this good agreement one concludes that at least sufficiently far from equilibrium dissolution rates in the presence of foreign ions are changed only due to the differing chemical composition of the solution with respect to the pure system. There is no effect of these ions to the surface reactions. Therefore, one can estimate the rates in the presence of foreign ions as:

$$F = \alpha \cdot ([Ca^{2+}]^f_{eq} - [Ca^{2+}]) . \qquad (7.53)$$

For all cases of laminar and turbulent flow, as well as for open and closed systems, it is possible to derive dissolution and precipitation rates by using in Eq. (7.53) the kinetic constant α of the pure $H_2O-CO_2-CaCO_3$ system as tabulated in Tables 7.1 and 7.2. The equilibrium value of the calcium concentration $[Ca^{2+}]^f_{eq}$ used in Eq. (7.53) is that of the system with added foreign ions. It can be calculated by using the methods in Chapter 2.

7.6.2 Influence of Lithology on Dissolution Rates

Applying the results on dissolution kinetics, which were derived from experiments with pure $CaCO_3$, to real limestone one might question as to how the surface

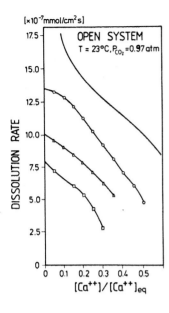

Fig. 7.33. Experimental dissolution rates on limestone samples as measured by Rauch and White (1977) in comparison to the rates calculated theoretically (see text). o Pure limestone; ∆ very pure limestone, □ impure limestone

kinetics might be changed from that of pure $CaCO_3$. Some information on that problem has been reported by Rauch and White (1977). They measured dissolution rates on a variety of limestones of different compositions, such as very pure limestone, pure limestone with silty streaks and impure limestone.

The experiments were performed by circulating a fixed volume of water with $p_{CO_2} = 1$ atm, kept constant during the entire run, through circular holes of 0.7-cm diameter and 30-cm length, drilled into specimens of limestone. The Reynolds number of flow was 1840. Therefore, one expects turbulent flow. The width of the diffusion boundary layer is 6×10^{-4} cm. Under these experimental conditions dissolution rates are controlled by surface reaction. (cf. Fig. 7.26).

Figure 7.33 shows representative rates for different limestones and the theoretically calculated data, assuming $\varepsilon = 0$. The rates are represented as functions of percentage of equilibrium concentration. Although the samples differ also in petrographic composition, the dissolution rates all behave similarly in agreement to that predicted by theory.

Buhmann and Dreybrodt (1987) have also performed experiments at $p_{CO_2} = 5 \times 10^{-3}$ atm with turbulently stirred water layers of 1-cm depth on top of natural limestone samples of high purity, but of differing petrography. Comparison of these data to pure marble samples showed only a small decrease in the kinetic constants by no more than 30% for four different specimens. There is a second important fact relating to these data. If one extrapolates the curves in Fig. 7.33 linearly, the points of intersection with the abscissa are in the region of 50 to 60% of saturation. This is similar to findings discussed in Section 6.4, confirming that in natural systems the surface kinetics are inhibited at concentrations of calcium above about 50% of the saturation value.

7.6.3   Dissolution in Porous Media

Partings in limestone rock, e.g. occurring between beddings or in joints, can be visualized as very thin porous media. They constitute the primary porosity of the rock. Their properties, with regards to hydrology and dissolution kinetics, play an overwhelming role in the initiation of first water routes. It is therefore of consider-able importance to understand calcite dissolution under these circumstances.

Partings in limestone are to be described as two-dimensional media, due to the small distance between the confining rocks. Nevertheless, we first solve the problem for a three-dimensional, homogeneous porous medium. As has been pointed out in Section 7.1 (Fig. 7.3), one can conclude that dissolution on the grain surfaces of a porous medium can be treated like that in a closed system confined by two parallel planes with distance $2\bar{\delta}$, where $2\bar{\delta}$ is the average diameter of the pores.

The amount of calcite dissolved in a porous medium per volume and per time is related to the flux F from the calcite surface into this volume by:

$$R_S = \alpha \cdot \frac{A}{V_P}(c_{eq} - \bar{c}) = k(c_{eq} - \bar{c}), \tag{7.54}$$

where A is the surface per volume of the porous medium and V_P the corresponding pore volume; $\bar{c}$ is the average concentration of Ca^{2+}. α is calculated for the closed system with spacing $2\bar{\delta}$.

The transport equation for the calcite concentration $\bar{c}$ through the porous medium reads:

$$\bar{D}\frac{\partial \bar{c}^2}{\partial x^2} - \bar{v}\frac{\partial \bar{c}}{\partial x} = -k(c_{eq} - \bar{c}), \tag{7.55}$$

where $\bar{D}$ is the coefficient of diffusion comprising also mechanical dispersion (cf. Chap. 3). It is given by:

$$\bar{D} = D_m + \bar{v}\bar{\delta} = D_m(1 + P_e), \tag{7.56}$$

where $\bar{v}$ is the average velocity of the liquid.

To determine the value of k experimentally Baumann et al. (1985) circulated a fixed volume of CO_2-containing distilled water through a column of marble grains with known average diameter. During the experiment, CO_2 pressure was kept constant by bubbling a CO_2 atmosphere through the solution in the reservoir vessel. The velocity of flow was fixed by a peristaltic pump. k was determined from the surface of the calcite sample employed, the volume of the circulating solution and by measuring the time dependence of the calcium concentration. Figure 7.34 shows the dissolution rates measured in comparison to the calculated ones. The upper curve results from an experiment, in which velocity is high and the Peclet number is 6800. The lower curve results from a run with $P_e = 110$. This shows the significant influence on dissolution rates by mass transport resulting from mechanical dis-persion. The full lines have been calculated according to theory and show excellent agreement to experimental data.

It should be noted here that the first systematic experiments on calcite dissolu-tion were performed on porous calcite packings by Weyl (1958). He passed water

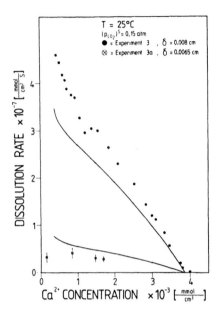

Fig. 7.34. Measured dissolution rates of a porous calcite sample at $p^s_{CO_2} = 0.15$ atm and two different flow velocities. *Upper curve*: $\bar{v} = 1.51$ cm s^{-1}; *lower curve*: $\bar{v} = 0.031$ cm s^{-1}. The *solid lines* represent the theoretically calculated curves (Baumann et al. 1985)

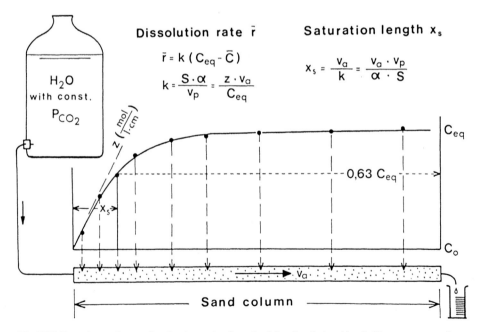

Fig. 7.35. Experimental setup for the determination of calcite dissolution kinetics in a porous medium. $\bar{r}$ is the dissolution rate $\bar{r} = R_s$; k kinetic constant (cf. Eq. 7.54); S surface area $S = A$; V_p volume of pores; $v_a = \bar{v}$ flow velocity of water (see text) (Schulz and Baumann 1985)

equilibrated with CO_2 at 1 atm into calcite packings of 20-cm length and measured the concentration of Ca^{2+} at the outlet of the column as a function of velocity.

Palciauskas and Domenico (1976) have given a general solution to Eq. (7.55). For a solution having passed a distance x with velocity v through the medium, it is:

$$\bar{c}(x) = c_{eq} - (c_{in} - c_{eq})\exp\left(-\frac{x}{x_s}\right) ; \qquad x_s = \frac{\bar{v}}{k} \qquad\qquad (7.57)$$

where c_{in} is the concentration at the entrance of the calcite pakking. Using this equation one can compute k from Weyl's experiments. They show good agreement to those calculated theoretically by Baumann et al. for the specific conditions of Weyl's experiments. Schulz and Baumann (1985) and Dahmke et al. (1986) extend Weyl's method. They used long pipes of 2 m filled with calcite-containing porous medium, i.e. inert quartz grains, 5% calcite grains. Water with fixed p_{CO_2} was pumped with known velocity through these "solution pipes". The conductivity at the outlet was monitored. After it had attained a constant value, indicating steady state, water samples were drawn from the tube at ten different sites. By analyzing them for their chemical composition, the concentration profile along the tube was obtained. Figure 7.35 shows the results. Clearly, an exponential approach to saturation according to Eq. (7.57) is observed from which x_s and thus k can be computed.

Values of k can also be calculated for these mixtures of calcite and inert quartz grains. One merely has to change the value of $\bar{\delta}$ in the calculation of α by a factor between one and two, depending on the percentage of calcite. This accounts for the different boundary conditions. At a low percentage of calcite grains, each calcite particle is surrounded by inert mineral grains. This reduces the active surface by a factor of 0.5 and accounts for a change of δ by a factor of two.

Part III
Conceptual Models of Karst Processes

8 Karst Systems

Karst systems, in general, can be observed only in their mature state of karstification. From their morphology at the surface and underground (caves) one has to infer the history of evolution of those systems. From the relation of tectonic structures in the rock, such as fracture systems, to the orientation of cave passages one agrees nowadays that secondary permeability for groundwater flow is established along those fracture planes, which represent the least resistance to water flow, and are therefore most effectively enlarged by dissolution of the confining limestone rock. Thus, karst systems evolve due to increasing secondary porosity and the corresponding secondary permeability by the dissolving action of calcite-aggressive water, penetrating into the rock along joints, bedding plane partings, faults or intersections of those structures.

The primary porosity of karstifiable limestone and accordingly the primary permeability is extremely low, since initial joint apertures are in the order of 5×10^{-3} cm. Smith et al. (1976) have summarized this situation, as shown in Fig. 8.1. Here, the porosities and permeabilities of the principal limestone rock types are plotted as a function of pore size. The full lines at the left show primary values; the dotted lines the secondary ones. The permeabilities, represented by the straight lines, are calculated by the assumption that the rock behaves as a bundle of straight, capillary tubes with diameters of the pore size and a density such that a given value of porosity results. This is a coarse approximation. The calculated values, however, show reasonable agreement to field and laboratory data, which are given by the numbers in the diagram. These values of the permeability K are given in m day^{-1}.

The primary values of massive limestone are determined by a pore size of about 1×10^{-3} cm and porosity less than 1%. By solutional widening of closed partings the permeability of this rock increases by a factor of 10^4. Moreover, the development of conduit tubes leading to caverns increases values of the permeability up to 10^5.

Bocker (1969) claimed that a threshold diameter of 10^{-3} cm is necessary for dissolving water to modify the surrounding rock. Therefore, no karstification is expected in rocks of pore size below this limit, D.C. Ford (1980) has added a boundary line for conditions favouring the evolution of large conduit systems from those that favour the development of more diffuse flow aquifers. This line, also shown in Fig. 8.1, starts at the threshold pore size of 0.01 mm. It indicates that at a high porosity, i.e. a high density of fissures, water input is diffuse and thus solutional attack is evenly distributed throughout the rock, favouring diffuse flow aquifers. On the other hand, at low porosity, but large pore size, water input may concentrate to a few points, thus enhancing evolution of conduits. From these considerations it is clear that karstification is controlled by the density of fissures penetrable by water.

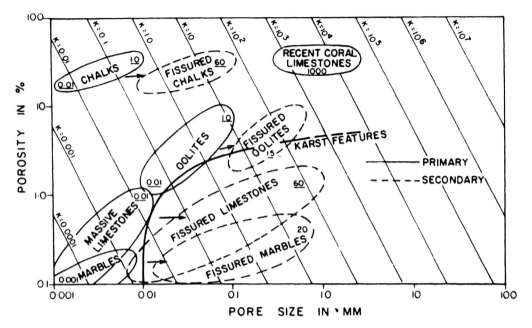

Fig. 8.1. Primary and secondary porosity, pore size and permeability of carbonate rocks. Values of K, the theoretical permeability, are in metres/day and derived from $Q/a = K(h/L)$, where Q ist the discharge, a the cross-section, h the head and L the length of system. The values are based on the assumption that the rock behaves as a bundle of straight, parallel capillary tubes. *Full lines indicate* primary permeability; *dashed lines* secondary permeability. The line designated as *"karst features"* separates the regions where development of enterable cave systems is favourable (*below this line*) from those where diffuse flow aquifers develop (D.C. Ford 1980)

One expects thickly bedded limestones which, as a consequence of thick bedding, have also low joint density (Price 1966), to exhibit more extensive karst features.

There is, however, no general validity of such statements. This is due to the extreme complexity of the processes:

1. In the initial state of karstification a hydraulic gradient for water flow along penetrable fractures has to be established by formation of a base level, e.g. incision of a river into a valley.
2. Calcite-aggressive water flowing down the pathways of a most favourable hydraulic gradient, due to solution, starts to enlarge fissures thus changing the pattern of hydraulic gradients and as a result, also the flow pattern.
3. There may be changes in the location of water inputs altering the evolution of the system.
4. Lithological and chemical composition of the rocks can contribute to varying dissolution rates, thus imposing additional complexity to the system.
5. Biological processes in the soil determine its CO_2 production. Thus rainwater seeping through that soil absorbs carbon dioxide and becomes highly aggressive with CO_2 pressures which can reach up to 0.1 atm. Thus the production of CO_2 in soils overlying karst areas determines the intensity of karstification.

By observing the geomorphology of karst areas, much can be inferred on the elements which determine karstification. In a series of papers D.C. Ford (1968, 1971) and D.C. Ford and Ewers (1978) condensed these findings into a general genetic model, which comprises the classical theories of vadose, deep phreatic and water-table caves. From the basis of these inferred elements a deeper understanding on karstification is possible by investigating these processes from the basis of physics and chemistry.

Therefore, first one has to describe the important structural features of karst. This will be done by presenting those topics, which to the author, seem to be of general importance, and which might represent also a key to physical interpretation, which will be given later.

8.1   Fractures and Discontinuities in Limestone Rocks

Since limestone, under the action of stress, behaves as a brittle material, once stress exerted by tectonic events exceeds a critical value, the rock shows failure resulting in fracture.

Fractures, which do not show any displacement of the rocks at either side of the fracture planes along them, are called joints. The scale of these fractures covers continuously a region from microscopic sizes to hundreds of metres. There are also planes of fracture which exhibit differential movement of the rock on either side of this fracture plane. These fractures are called faults. Faults are major-sized features, which due to the movement of one part of the rock mass, may bring beds of different lithology into contact with each other. A special type of fault is an overthrust. These faults can extend over tens of kilometres along planes dipping $10°$ or less. An illustration of fracturing by joints and faults is shown in Fig. 8.2. This block diagram shows orientations of these fractures in an unfolded, layered rock. F_1 to F_8 are different types of faults, whereas J_1 to J_4 represent sets of joints. Systematic classifications of these features is of no concern in this context and therefore omitted. The reader is referred to Price (1966).

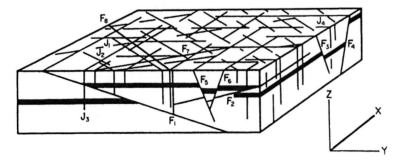

Fig. 8.2. Block diagram showing orientations of faults F and joints J in unfolded rocks (Price 1966)

8.1.1 Joints

In contrast to faults, which are single geological features, joints are encountered as systems of joint sets extending over large areas. These sets consist of planar and parallel fractures, which are mostly perpendicular to the bedding of the rock. Figure 8.3 shows a block diagram of joints of one bedding in an anticline. The joint system shows two perpendicular sets T_3, T_4 (T_1, T_2) of joints. The joints T_4 (T_2) are often called cross joints. They are oriented perpendicular to the axis b of the anticline, i.e. along the dip of the bedding. T_3 (T_1) represent orthogonal joints in relation to T_4 (T_2) and are called longitudinal. They are oriented in the strike direction. There are also sets of oblique joints, S_3, S_4 (S_1, S_2). Those joints are shear joints, whereas the cross and longitudinal joints result from tension (Price 1966).

    Many joints are restricted to only one bed of the rock, some may pass only a few beds. Those joints, which extend over many beds, are called master joints. T.D. Ford (1971) has given, as a minimum, a vertical extension of these joints of at least 20 m

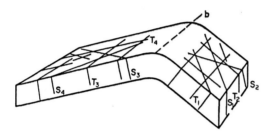

Fig. 8.3. Block diagram showing typical orientations of joints in an asymmetrical anticline of a folded rock (Price 1966)

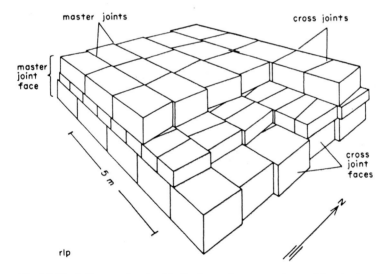

Fig. 8.4. Block diagram showing idealized sets of master and cross joints in three beds of rock (Powell 1977)

and generally much more. The lateral length of joints can amount to several hundred metres. Figure 8.4 shows schematically idealized sets of master and cross joints in three beds of rock, characteristic for joint patterns in Indiana, USA (Powell 1977). The faces of the cross joints show a horizontal extension which is larger than the vertical extension by about a factor of two. From Fig. 8.4 one can easily visualize that those joints passing several beds will be favourable sites of initial penetration of water into the rock, since they may show less resistance than routes along cross joints, where the flow route is alternately through joints and bedding plane partings.

The permeability of karstifiable rocks is determined by the frequency of joints and their average aperture. By idealizing a joint as a slit consisting of two parallel planes with constant distance (i.e. average aperture), it is possible to calculate the amount of flow transmitted through jointed rock.

Little is known on the distribution of joint frequencies and apertures, which can vary over a range of several orders of magnitude. There are some general relations between bed thickness and joint distance and bed thickness and aperture. The average distance Δx between two joints in a given set is proportional to the bed thickness. This can also be justified theoretically (Price 1966). The value of this ratio of bed thickness and joint distance, however, is different for different lithological types of rocks and can only be found empirically (Harris et al. 1960). The relation of the aperture of joints to bed thickness has been reviewed by Kiraly (1975). The data, reported by several workers, seem to support the relation:

$$\log d = A + B \log e ,\qquad (8.1)$$

where d is the aperture of the joint and e the bed thickness; A and B are constants. From the proportionality of joint distance Δx to the thickness of beds e, one also obtains:

$$\log d = A + B' \log(\Delta x) .\qquad (8.2)$$

For a set of small samples of limestone with Δx between 0.5 and 10 cm, a value of $A = -3.37$ and $B = 1.1$ is cited, where d and Δx are measured in metres. These numbers are to be considered with caution. From theoretical reasoning, for instance, one would expect joints originating from tension to be more open than those resulting from compression.

The relations above indicate that in thickly bedded rocks one would expect only few joints of comparatively large aperture. This seems to be favourable for the development of large cave systems, since there are only few but easily penetrable routes of initial flow. Thus, water is channelled into these pathways, which then are enlarged by dissolution.

8.1.2 Faults

Faults are fractures with a relative displacement of the two blocks of rock separated by the fracture. The displacement covers a wide range of scale, from a few centimetres to tenths of metres. The extension of the fracture zone ranges up to tenths of kilometres.

Faults are classified into three main groups:

1. Normal faults, which are generally inclined at angles greater than 45° from the horizontal, show a downward movement of the hanging wall, i.e. the block of limestone, which lies above the fault plane.
2. If the hanging wall moves upward, the fault is classified as a thrust fault, when the dip of the fault plane is less than 45° and a reverse fault for dip angles larger than 45°.
3. In contrast to normal and thrust faults, which are due to a dip-slip movement, wrench or tear faults show a lateral movement along the strike of the fault plane.

Since faults represent zones of geological weakness, they may be favourable sites of cave development. Their importance in controlling cave systems, however, is much less than that of joints and bedding planes (Waltham 1981, D.C. Ford and Ewers 1978). There are examples of caves, however, which have developed entirely in fault planes (Waltham 1971).

The influence of faults on cave development can be very different. If the fault contains significant amounts of primary porosity, water flow may be concentrated in this fracture zone. On the other hand, faults can act as impervious hydrological barriers due to mineralization or filling of their initial spacing. Thus, water flow is channelled against this barrier and cave development may occur in the minor fraction and joint sets associated with the fault.

One important influence of faults to the development of karst conduits was pointed out by T.D. Ford (1976). Figure 8.5a shows a situation, in which due to the movement of the hanging block, beds of favourable speleogenetic properties are displaced from each other, such that a seal is formed by other strata. In Fig. 8.5b, on the other hand, there is only a displacement of similar limestone beds, which can influence a possible cave.

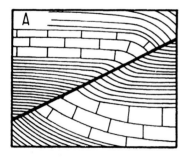 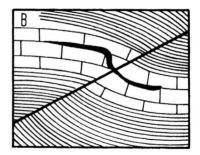

Fig. 8.5A,B. Diagrams showing thrust faulting and its effects on a limestone bed. **A** Limestone beds are separated by the fault and an impervious seal is formed between the limestones; **B** Limestone beds remain in contact due to small displacement of the beds. A possible cave profile is shown (T.D. Ford 1976)

8.1.3 Bedding Planes

Bedding planes are structures extending continuously over large areas to the boundaries of the limestone mass in all directions and over distances of several kilometres. They can also be exposed at the surface over large distances and therefore act as efficient inputs of water.

They result from a change or an interruption in steady limestone sedimentation. This can just be a complete break in sedimentation for a time, or washing in of non-carbonate sediments, depositing thin shale layers in the partings between two beds.

Bedding planes also can form from a change in the grain size of the sedimentary material. Under favourable circumstances these discontinuities show sufficiently high permeability to act as initial pathways for phreatic water flow.

In a first approximation one may regard a bedding plane parting as a two-dimensional porous medium with an average pore size comparable to the grain size of the confining rock. This approach has been taken by Ewers (1982) in simulating flow through these partings. Many caves are guided mainly by bedding planes, e.g. the large systems as Mammoth Cave in Kentucky, USA or the Hölloch in Switzerland. Which of the many bedding planes are finally selected cannot be generalized. There are reports that shale partings are preferred (Waltham 1971). Bögli (1969) has observed tectonic movements along bedding planes in various distant places in the Hölloch, giving rise to enhanced permeability.

8.2 Structural Segments and Tectonic Control of Caves

In the initial state of karstification cavern-bearing limestones develop as fracture aquifers by employing favourable flow routes along fracture planes or interceptions of them, which are then gradually enlarged by dissolution. Therefore, cave systems should be composed of many connected, individual segments, each of which is contained entirely in a distinct fracture structure, such as a bedding plane, a joint, a fault or an interception of those structures. These structural elements have guided the flow of groundwater during the early fracture conduit stage of the aquifer.

Deciphering these structural elements in cave systems is a valuable tool in identifying the initial pathways of flow, which are a key to the speleogenetic history of a distinct cave. Such a structural analysis has very carefully been performed by Jameson (1985) in the North Canyon of Snedegar Cave, Friars Hole, West Virginia, USA. This cave is particularly suited for this analysis, since after establishment of an early phreatic aquifer with small conduit tubes along the structural segments, rapid downward entrenchment by vadose flow has set in. Thus, the remnants of the early flow paths are well conserved in the roofs of the canyons, and can be analyzed by careful surveying.

Small enlargements of the early conduits appear as half tubes or fissures on the canyon ceiling. To infer the type of structural element, the highest features of entrenchment, such as notches in the canyon profile, are identified for each passage. The fractures visible above these features are the candidates for early guidance of

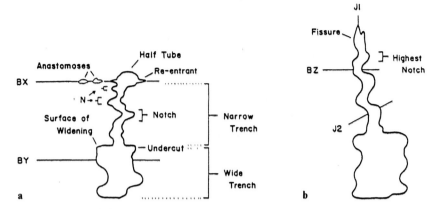

Fig. 8.6. a Cross-section of a canyon enlarged from a bed segment. The half tube and the anastomoses in the bedding plane *BX* are the earliest recognizable flow paths. They develop under phreatic conditions. Later entrenchment by vadose flow proceeded producing features such as the notches *N* and undercuts. **b** Cross-section of a canyon enlarged below an early fissure. Joint *J1* is exposed in the ceiling of the canyon. The bed parting *BZ* and lower joint *J2* lack dissolutional reentrants or anastomoses. They are located below the highest notch, which indicates the first free surface of vadose flow. These features indicate that joint *J1* guided the first flow path (Jameson 1985)

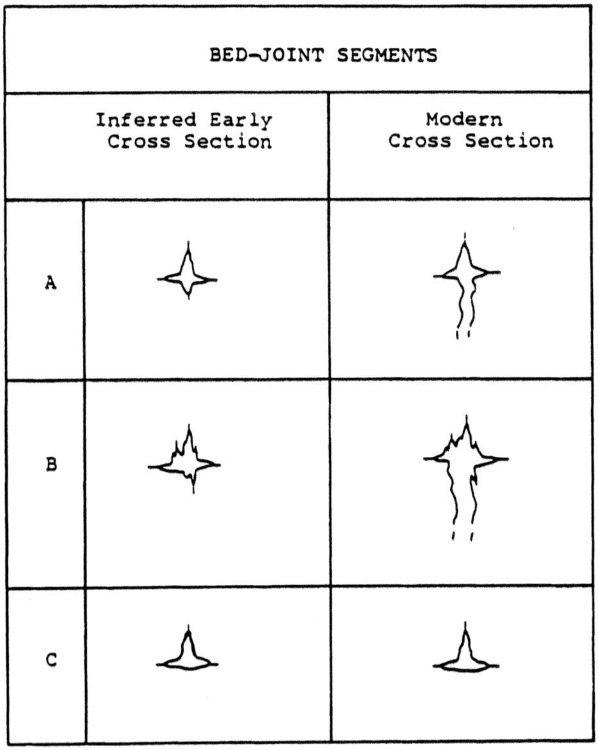

BED–JOINT SEGMENTS		
	Inferred Early Cross Section	Modern Cross Section
A		
B		
C		

Fig. 8.7. Typical cross-sections of passages in bed-joint segments. The inferred early cross-sections developed under phreatic flow conditions. Entrenchment occurs at free-surface flow of water-entrenched canyons into the floor of the phreatic conduits (**A,B**). Example **C** shows passages where the joint terminates at the bedding plane. These have been abandoned by water flow, thus preserving the phreatic features (Jameson 1985)

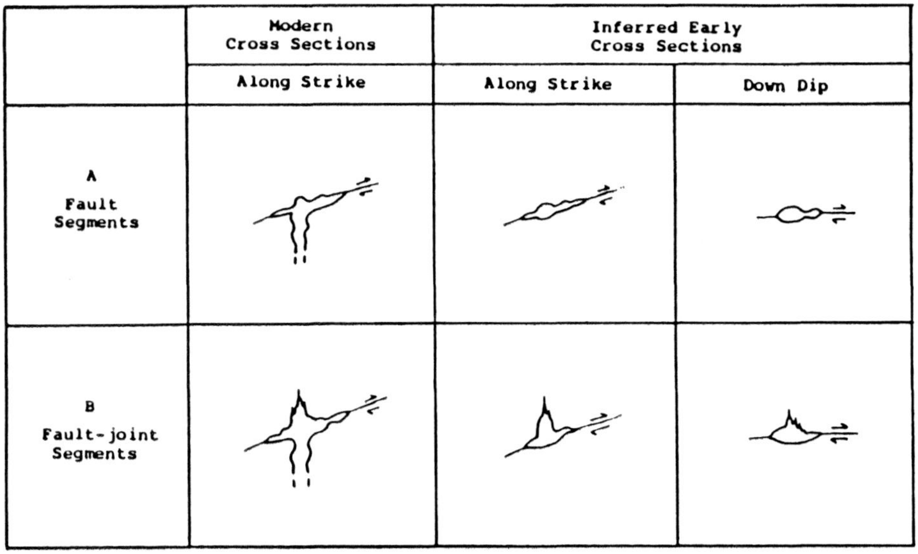

	Modern Cross Sections	Inferred Early Cross Sections	
	Along Strike	Along Strike	Down Dip
A Fault Segments			
B Fault-joint Segments			

Fig. 8.8. Typical cross-section of passages in fault and fault-joint segments. Inferred early cross-sections are shown for the strike and dip-oriented faults (Jameson 1985)

water flow. Figure 8.6 shows canyon cross-sections for a bed segment (a) and a joint segment (b). In both cases above the upper notch N, fractures (BX, J1) are clearly present, which must have guided flow before entrenchment. The identification of bed segments is further supported by the presence of anastomoses.

Similarly, interception segments, e.g. bed-joint or joint faults, can be analyzed. Figure 8.7 shows inferred early cross-sections of cave passages and the modern ones, created by later, downward entrenchment preserving remnants of the early profiles. Figure 8.7a shows a typical cross-section of bed-joint interception, whereas Fig. 8.7b is due to the interception of a single bed with several closely separated joints which are visible by the multiple spurs in the floor and the ceiling of the conduits. Figure 8.7c shows an interception, where the joint terminates at the bedding plane.

Similarly, Fig. 8.8 displaces cross-sections of fault segments and fault-joint segments, directed either along the strike or dip of the fault plane.

To infer the earliest recognizable flow path careful surveying is necessary, resulting in the preparation of a detailed base map, containing plan, profiles and cross-sections. From the morphological features structural segments are analyzed and the midlines of the early flow conduits are drawn. The result of such an investigation is shown schematically on a hypothetical canyon in Fig. 8.9. Alternately bed, joint or bed-joint elements are employed in establishing the early flow net. Figure 8.10 shows a three-dimensional representation of the early flow path in Snedegar Cave. The flow path from the right to left first follows a bed, then leaving it to be guided by a more favourable bed-joint intersection. After leaving this intercept the flow path again meanders along the bed. After branching, the pathways continue in the bed, leaving it eventually to follow bed-joint intersections or a fault.

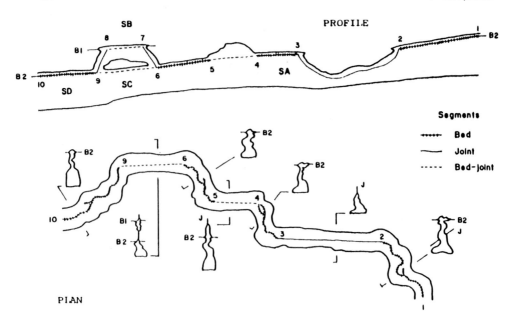

Fig. 8.9. Structural segments of a hypothetical canyon. The type of segments is indicated in the *profile* and also in the *plan*. The *plan* also shows the modern cross-section of passages from which the structural segments are inferred (Jameson 1985)

Then, it declines on joints until regions of major faults are encountered, where flow is directed by fault or fault-joint segments. This shows that, according to the hydrological setting, practically all types of fractures and fracture interceptions are used in the initiation of karst water flow. It is interesting to note the distribution of the different kinds of structural elements in relation to the total length surveyed and also to evaluate the mean lengths of the segments.

There are three joint sets in the cave area. Major joints are predominantly oriented between 60°N–75°E. Somewhat more abundant are joints with orientations between 30°N–45°E, which are minor joints. In the cave the set of major joints is mainly utilized to form structural segments. Roughly 30% of the segments are of this type, whereas the minor joint set contributes only about 2% of the comparatively short segments. Thirty percent of the segments are guided by bedding partings, and bed-joint elements amount to 25%. Fault and fault-joint segments finally contribute 5 or 4% respectively. The mean lengths of the segments are about 10 m for fault elements. Bed, joint, bed-joint and fault-joint elements, on the average, are somewhat shorter, on the average about 7 m.

This complex alternation between various structural elements is usual in many cave systems. As a further example, D.C. Ford and Ewers (1978) reported on the distribution of structural elements in Swildons Hole, England, where 50% of the known passage length is guided by bedding planes, 30% by joints and the remainder by faults or bed-joint intercepts.

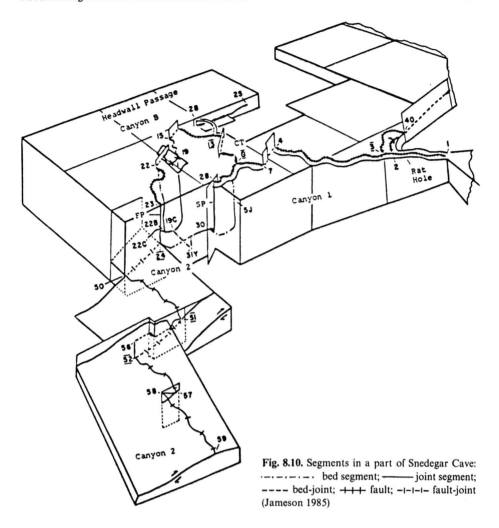

Fig. 8.10. Segments in a part of Snedegar Cave:
·—·—·—· bed segment; ————— joint segment;
———— bed-joint; +++ fault; –ı–ı–ı– fault-joint
(Jameson 1985)

There are extremes, however, and caves can be either controlled almost entirely by joints or, in the other extreme, by bed segments.

Powell (1977) has given a beautiful example of a cave in which all the passages are oriented in the direction of the joint strike. Figure 8.11a shows the map of this cave, which is entirely oriented along a joint system of two perpendicular joint sets. The east-west trending set consists of master joints, whereas the S–N trending set of cross joints is unique only to individual beds. The rose diagram in Fig. 8.11 includes 136 joint measurements. Figure 8.11b shows similar examples of caves in Europe, Schwäbische Ostalb, Germany (Bayer 1983), which show joint control also in dendritic cave patterns.

From this and other examples in Indiana caves, Powell (1977) concluded that permeability along bedding planes is not necessary for the development of cave systems. A sequence of strata, dissected by two sets of joints, master joints and cross

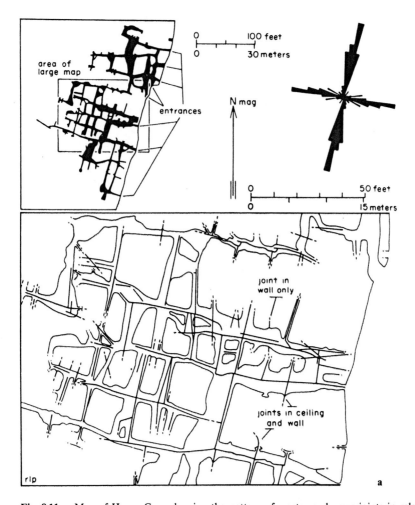

Fig. 8.11. a Map of Heron Cave showing the pattern of master and cross joints in relation to cave passages. The *rose histogram* shows the distribution of joint orientation (Powell 1977). **b** Maps of several caves from the Ostalb, Germany in relation to rose histograms of joints taken at the entrances of the corresponding caves. Note joint control also in dentritic cave patterns (Bayer 1983)

joints, is sufficient to form three-dimensional cave patterns, since flow of water in such a system of fractures is possible in both horizontal and vertical directions.

One method to clarify the influence of fractures is the comparison of the orientational patterns of fracture systems, measured at outcrops near the caves, to those orientational patterns of cave passages. This can be done by comparing rose histograms showing the directional distribution of fracture frequency to those showing the directional distribution of the lengths of cave passages. Furthermore, statistical methods can be used to determine average bearings within each quadrant of the rose diagrams. This method was used by Deike (1969). Figure 8.12 shows the

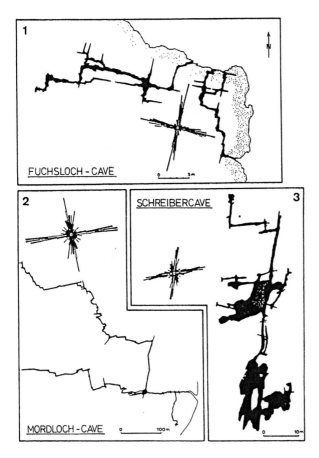

Fig. 8.11b

cumulative averages for cave passage orientation and joint orientation in relation to each other. Each point represents one cave and the corresponding joint system. The straight line would result if cave development depended solely on the abundance and orientation of fractures. The fact that many points lie very close to this line shows that this is true for many systems. The many points, however, deviating considerably show that other influences must also be of importance. One of these influences is due to different permeabilities of different joint sets.

This is also confirmed by investigations by Kastning (1984). Figure 8.13 shows orientation rosettes comparing the frequency of joints or photolineaments to the length of cave ensembles in the corresponding areas. The diagrams of the Langtry area show that only one set of joints is utilized for the flow of groundwater. This N–E bearing set is due to fractures which have been formed by extension and are assumed to be initially more open that the perpendicular set which was formed by compressional forces.

The examples of Bend Area and New Braunsfels Area show that all fracture systems are more or less evenly contributing to the development of the related karst

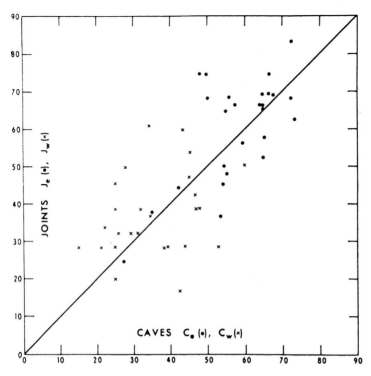

Fig. 8.12. Cumulative weighted averages of passage orientation (C_e, C_w) in relation to cumulative averages of joint orientation (Deike 1969)

systems. Similar findings are reported by Trudgill (1985) for caves in England. From the findings above one may conclude that karstification should be especially intensive in zones of high fracture concentration with fractures of sufficient extension. These zones can be identified as fracture traces (< 1 km) or lineaments (> 1 km) by areal photography (Parizek 1976, Wermund and Cepeda 1977). It has been found that aerial photography measurements and field measurement of joints show agreement, as can be seen by the comparison of the corresponding rose diagrams.

There is a clear relationship between fracture traces and development of permeability (Parizek 1976, Lattman and Parizek 1964). Highly fractured zones show enhanced karstification and control the location of caves and conduit systems. Therefore, karst features such as sinkholes, land subsidence and large karst springs occur favourably along zones of fracture concentration.

Despite the overwhelming importance of joints in the development of caves, there are caves which have developed almost entirely on bed partings. A prominent example is Mammoth Cave, Kentucky, USA. This cave is located in flat-lying rock formations with low gradients, $6-12$ m km^{-1}. Most of the passages of this 280-km-long cave system are concordant with the beddings of the rock, following the same individual bed of large distances up to several miles (Palmer 1977, 1981). The reason

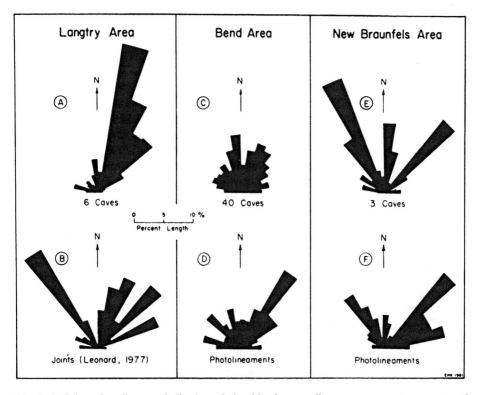

Fig. 8.13. Orientation diagrams indicating relationships between linear cave passage segments and mapped fractures and photolineaments for three areas in central Texas (Kastning 1984)

for this control is the poor jointing of these rocks. Only few joints cut across several beds. Only in thick-bedded layers, where large joints occur, do the passages abruptly change upwards or downwards by as much as 6 m. There are also shafts, mainly oriented on joints connecting the different levels of cave passages.

Because of the poor jointing, infiltrating water, descending through major joints or along outcrops of bedding partings, was forced to the bed partings as the only routes available. Even if joint frequencies are much higher, caves in flat-lying strata exhibit mainly bedding control (D.C. Ford 1971). This is explained as follows. In flat-lying rocks, for hydrological reasons, the predominant infiltration of water must be by way of joints. Joints, however, in contrast to beddings, are discrete features not extending individually to such a large dimension as beddings. Joints can be used as pathways of flow only through tortuous ways by many turns into different individual joints at their intersections. Bedding planes in flat-lying strata, however, represent continuous flow paths from possible places of water infiltration to outcrops where springs can emerge. Therefore, some bedding plane partings may represent a kind of short circuit to the high resistance joints and are therefore employed in establishing initial fracture conduits.

8.3 Karst Aquifers

In contrast to aquifers in homogeneous media, karst aquifers due to their inhomogeneous distribution of permeability are extremely complex. Groundwater flow
occurs according to two modes: turbulent conduit flow and laminar flow diffused
through the net of fine fractures. These two components of groundwater flow are
extremely different in the effectivity of groundwater transmission and groundwater
storage. Conduit flow transmits the majority of the water, whereas the net of
fractures and fissures stores most of it as true groundwater. Within a given rock mass
the distribution of both fissures and well-developed karst channels is extremely
inhomogeneous and anisotropic, which further enhances the complexity of those
systems.

Figure 8.14 gives a schematic vizualization of a karst aquifer. Precipitation (A)
infiltrates the limestone mass either directly on bare or soil-covered rock surfaces
(B), or it may first be diffused through highly permeable rock (C), where it is more or
less evenly distributed when entering into the limestone. This diffuse recharge
penetrates into the rock through small openings of the limestone and proceeds as
diffused flow along the joints, bedding planes and other favourable fissures. Flow is
governed by Darcy's law along the tortuous pathways prescribed by the complicated three-dimensional net of fracture planes.

If the rock is covered by impermeable shale (D), surface water is channelled into
small depressions and point inputs (PJ) are established. They may give rise to the
development of shafts and conduits, which transmit water effectively.

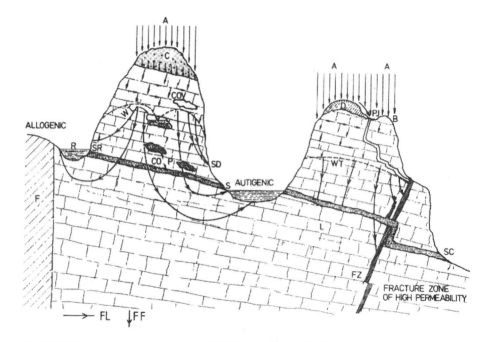

Fig. 8.14. Schematic visualization of a complex karst aquifer (see text)

Point inputs are also swallets situated at river sites (SR) where the river sinks into the ground. This often gives rise to large karst channels (cave streams) oriented down dip of the strata. There may also be conduits oriented along the strike parallel to the valley in our particular example. All of these conduits finally emerge as large karst springs (SC, S).

The zone of fracture concentrations (FZ) may be effective in altering the direction of karst channels.

Depending on the amount of recharging water, a water table or piezometric surface (WT) is established, the height of which fluctuates, thus defining a floodwater zone. Above the upper limit of this zone one defines the entirely vadose or unsaturated zone of the karst aquifer. The entirely phreatic or saturated zone lies below the lower border of the flood water zone. Note that vadose conduit channels in the region of this zone have significant influence on the location of the water table. It should be stressed here that the water table position changes on a short-term basis within the floodwater zone. This zone can extend several hundred metres (Bögli 1980). Thus, in the floodwater zone phreatic or vadose conditions exist alternately.

Karst channels in the vadose zone, where cave streams with a free surface exist, have a different morphology than those in the phreatic zone. Due to flow with a free surface the former develop as canyons, whereas the phreatic conduits, which are filled with water entirely, show elliptical or circular cross-sections.

In the vadose zone water infiltrates along the fractures (FF) following gravity until it reaches the phreatic zone, where it either may feed karst channels or proceed to the surface along the flow lines (FL), which give a crude vizualization of the average flow paths taken in the diffused aquifer part. Under favourable conditions these flow lines will concentrate to establish small springs (SD). As already mentioned in the introduction to Chapter 5, springs fed by diffuse flow differ largely in their geochemistry from those fed by conduits (Shuster and White 1971). Figure 8.15a gives a schematic diagram of the routes taken by flow in karst aquifers (Smith et al. 1976). The diagrammatic flowchart shows the many different routes, such as percolation input or input by swallets and the subsequent pathways taken by water precipitated to the escarpment. Figure 8.15b shows the different routes denoted in Fig. 8.15a on a schematic karst aquifer.

The complexity of karst aquifers is mirrored by the hydrographic response of the associated springs to flood pulses (White 1977a, Atkinson 1977, Smith et al. 1976).

We will first discuss the response of an idealized conduit stream, consisting of only one channel, to a flood pulse resulting from heavy precipitation during a short time interval, shown by the dashed area in Fig. 8.16. After a time lag there is a steep rise in the discharge of the spring. This is due to the fact that after a time lag, needed to channel the precipitation into the swallet, there is a rise of the water level at the entrance to the cave channel. If this channel is situated entirely in the phreatic zone, then there is an instantaneous response at the output due to instantaneous transmission of hydrostatic pressure. This high peak discharge is then followed by a rapid recession curve, which in the simplest case is a single exponential.

Associated to this hydrographic behaviour there is also a response in the chemistry of the discharging water. As soon as the water, which has filled the conduit

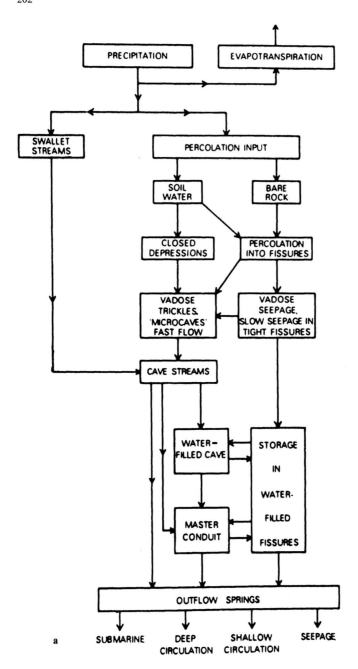

Fig. 8.15. a Possible routes taken by flow in a limestone aquifer (Smith et al. 1976). **b** Block diagram of limestone area showing drainage via caves, closed depressions and drainage into fissures (Smith et al. 1976)

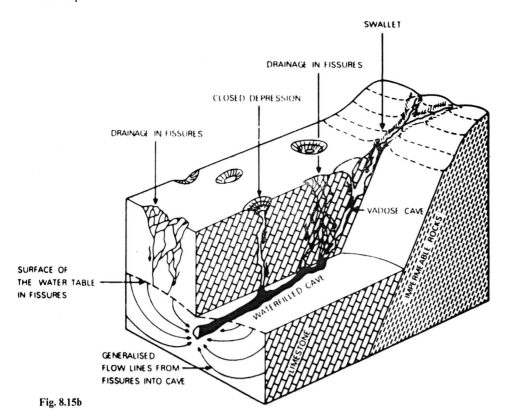

Fig. 8.15b

before the arrival of the flood pulse has been pushed out of this conduit, there is a decrease in hardness and pH due to the arrival of precipitated water. This can be used to determine the volume of the conduit. It is also possible to tag the flood pulse with tracers and measure the amount of discharged water from the onset of the hydrographic rise to the appearance of the tracer (Smith et al. 1976, Smith and Atkinson 1977).

For diffuse flow in a karst aquifer the response to a sudden recharge is much different. There is a large delay in time and a slow rise of discharge for springs fed by diffuse flow, as indicated in Fig. 8.16. The slow rise is due to the many alternative flow paths in the complex fracture system leading to different arrival times of the water at the spring. The decay follows an exponential and is extremely slow. The exponential constant k is related to the effective porosity and therefore to the storage capacity and the transmissivity of the fracture aquifer (Bear 1979).

This can be used to estimate the transmissivity of fracture aquifers from recession curves (Smith and Atkinson 1977). In a mixed aquifer, comprising conduit and fracture flow, the recession curve will be a superposition of two exponentials with k_c and k_d. The ratio k_c/k_d defines a decoupling between the two systems. With increasing karstification k_c increases as the transmissivity by channel conduits is increased. Thus, at least, part of the channel system in the flood zone will become

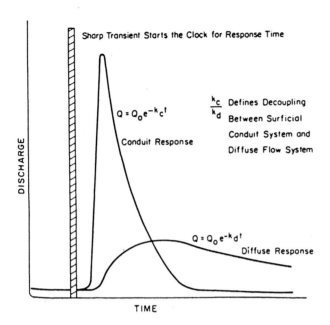

Sharp Transient Starts the Clock for Response Time

$\frac{k_c}{k_d}$ Defines Decoupling Between Surficial Conduit System and Diffuse Flow System

$Q = Q_0 e^{-k_c t}$

Conduit Response

$Q = Q_0 e^{-k_d t}$

Diffuse Response

DISCHARGE

TIME

Fig. 8.16. Schematic hydrographs for response of conduit springs and diffuse flow springs to a transient recharge event (White 1977a)

more tightly linked to the surface drainage system and the exchange between conduit system and the diffuse flow system will decrease.

From the diffuse recession curve by integration the total volume of diffuse flow storage can be determined.

The analysis of recession curves and flood response by Atkinson (1977) shows that the overwhelming part of groundwater is stored in the fissures and fractures constituting the diffuse flow system. Only 3 to 5% of the groundwater is stored in conduits.

By comparing the volume of groundwater with the volume of the rock containing the aquifer, an effective porosity of 1% is found, comprising all voids, fissures and channels in the karst aquifer. Similar values of effective karst porosity are given by Milanovic (1981). He reported on porosities ranging from 1 to 6% within one distinct region, showing that karst permeability is unevenly distributed. Similar values are also reported from the literature cited for various karst systems of the world by Bonnaci (1987), who gives a detailed discussion on this problem. Primary porosity in most of these karst systems seems to be small, amounting to only a few percent of the total effective secondary porosity created by dissolution processes.

The uneven distribution of permeability, which renders karst aquifers into such inhomogeneous complex structures, can be observed from water pressure tests in boreholes (Milanovic 1981) during drilling. The section of the borehole to be investigated is separated from the other parts of the borehole by rubber packers. Water under pressure is injected into the thus isolated section. The quantity of water in litres per minute lost into the rock at a given length (1 m) and a given injection pressure (0.1 atm) is then used as a measure for permeability. Typically, pressures

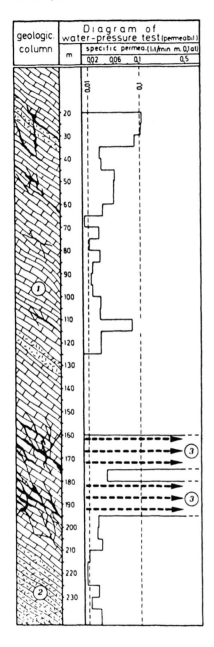

Fig. 8.17. Results of borehole investigations showing a graphical representation of the geological column in relation to measured permeability (Milanovic 1981)

of 10 atm and borehole sections of 5-m length were employed. Figure 8.17 shows a profile of such an investigation and a geological column showing the causal factors for the permeability observed.

Clearly, there are abrupt changes from impermeable to highly permeable zones (3), where the loss of water was so high that a pressure of 10 atm could no longer be sustained. The permeability in the upper zone is very unevenly distributed, ranging from impermeable to values between low permeability (0.01) to medium permeability (0.1).

This example also shows that in the early process of karstification selective mechanisms do exist, which from the many possible flow paths in the fracture system choose those which are the most favourable according to hydrological reasons (primary permeability, hydraulic gradients, etc.) and which can most easily be enlarged by dissolution of limestone.

8.4   Caves

Caves offer an overwhelming variety of morphological features on all scales. On the large scale the patterns of the caves are largely different, ranging from complex, three-dimensional, dentritic structures to regularly shaped, two-dimensional networks and a variety of mixtures between these two extremes.

Apart from the geological setting of the limestone with respect to synclinal or anticlinal structures and the dip and strike of the beds, much influence is exerted by the type of water input into the limestone mass. Palmer (1984b) has related different cave patterns to surface features of water input. Figure 8.18 shows these relations. In the case of authigenic recharge (i.e. water derived from the karstifying limestone

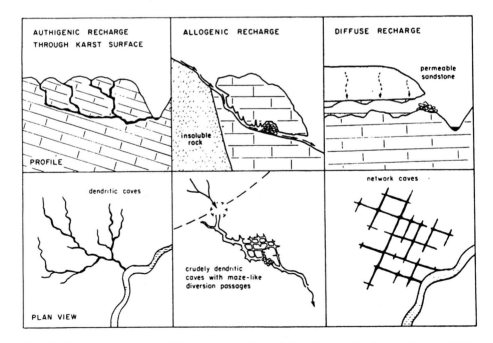

Fig. 8.18. Common patterns of solutional caves and their relationship to surface features (Palmer 1984b)

area) scattered to many points on the karst surface, point inputs of minor size from small catchment areas result. Each of these inputs might give rise to solutional enlargements of the fractures. Thus, dentritic caves result, similar to river networks, discharging at a few large springs. These caves are typical in soil-covered and bare limestone areas with no perennial surface streams.

In the case of allogenic recharge from water of non-carbonate catchment areas only a few, but significant point inputs exist where streams reach the border of the limestone region. Caves in such a situation often consist of only a single major conduit with a few minor tributaries (Palmer 1972). If such a master conduit is blocked by collapse or filling with sediments, water rises during flood periods, causing high hydrostatic pressures. Allogenic waters flowing into large karst channels maintain their high aggressivity over large distances of several tenths of kilometres (Dreybrodt 1988). Therefore, fractures and other openings, which are penetrated during flood periods due to the steep hydraulic gradients resulting from the corresponding high hydrostatic pressure, may be enlarged effectively and diversion passages are formed to bypass the blockage. Depending on the geological setting, they form braided, interconnected passages or a network of intersecting passages (Palmer 1975).

If water input is through highly permeable, non-carbonate caprocks, it is evenly distributed as diffuse flow to the limestone interface and water with similar hydrostatic conditions is available simultaneously at all the fissures and fractures regardless of their size. Thus, a more or less even enlargement is possible for many of them and network patterns result (Palmer 1975).

The initial stage of cave development starts after the incision of a valley into a limestone mass, which establishes a hydraulic gradient for flow from the upper limestone regions into the valley. Thus, water penetrates into the primary fissures, but most of it is still drained on the surface. The water table at this stage is high and fairly stable (Dreybrodt 1981a, Röglic 1965, D.C. Ford and Ewers 1978). In caves where the scale of folding of the rock mass is much larger than the dimensions of the cave system, the cave develops in a flat-lying or homoclinally tilted block. In such a structure flow along the dip of bedding planes is most likely and the earliest cave elements develop as dip tubes (D.C. Ford 1968, 1971, D.C. Ford and Ewers 1978). In flat strata these are subparallel bands of anastomoses oriented down dip with the largest tubes in the middle of the bands. Where the rocks dip steeply ($>5°$), anastomoses are replaced by an array of single channels down dip. As soon as the water table drops by the thus increased permeability, these tubes become important pathways of vadose flow and canyons oriented down dip start to be entrenched.

They guide water downstream to the phreatic zone, which at the water table is diverted along the strike. Therefore, in this region strike-oriented phreatic tubes develop which maintain their circular or elliptical shape even until they become dry. This situation is observed in Mammoth Cave (Palmer 1981, 1977). Most of the canyons in this cave are oriented down dip and finally merge in an abrupt transition into circular, phreatic tubes oriented along the strike. Figure 8.19 shows an excellent example. Boone Avenue is a canyon oriented along the axis of a gentle syncline. It guided water to Marion and Cleaveland Avenues, which are both strike-oriented phreatic tubes. After the water table dropped, the water was no longer diverted into

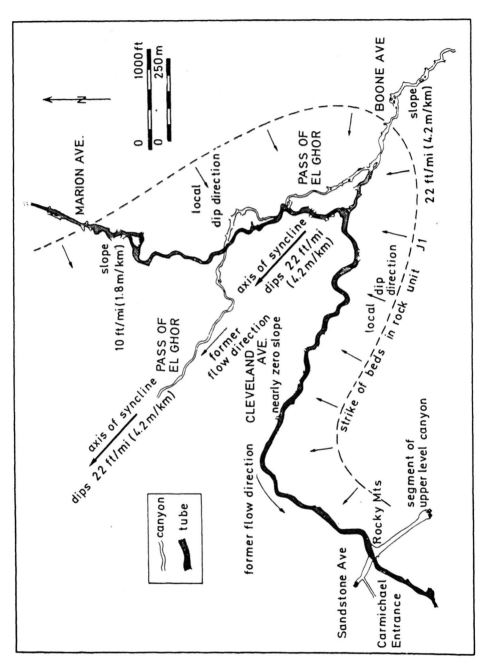

Fig. 8.19. Influence of geological structure upon trend and gradient of Cleveland Avenue and related passages in Mammoth Cave. Note that the passages along the dip are vadose canyons. Cleveland Avenue is strike-oriented and shows features of phreatic tubes (Palmer 1981)

the tubes along the strike but continued its way down dip, forming a lower level canyon, Pass of el Ghor.

To complete this section a further remark should be made. We have discussed here a few idealized examples, which under corresponding conditions can be observed in nature. In most cases, however, it is difficult to attribute those principles unequivocally to the structure of a cave. This is especially the case where the scale of folding matches that of the cave or is even smaller. Impermeable barriers in a limestone mass can divert flow and impose additional complexity to cave structures. Therefore, many of the basic principles described above are hidden in many caves.

8.5 Development of Caves

There have been many conflicting views on cave genesis among the early karst researchers. Three hypotheses on cave development have been advocated. They are illustrated in Fig. 8.20.

1. *Vadose Theory.* This theory implies that most of the cave volume is excavated by cave streams with a free surface. Water from a large point input flows down the vadose zone to the water table, which has been established by some fast previous processes in the pre-cave state (Martel 1921, Dwerryhouse 1907), which are not further specified.

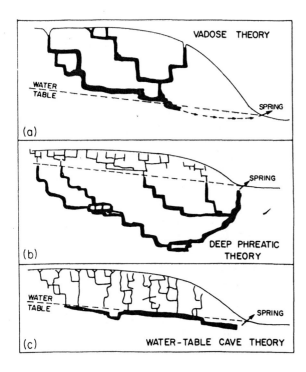

Fig. 8.20a–c. Diagrammatic long sections to illustrate cave system development as proposed by the vadose, deep phreatic and water-table theories (D.C. Ford and Ewers 1978)

2. *Deep Phreatic Theory.* If the development of permeability from the input to the resurgence of limestone-dissolving water is such that a stable water table is maintained over a sufficiently long period, most of the cave conduits will develop in the phreatic zone, far below the water table (Davis 1930, Bretz 1942).
3. *Water Table Theory.* Water percolates down the vadose zone to the water table. Then, most of its movement takes place along a shallow zone near the top of the phreatic zone. Therefore, the cave develops along this region (Swinnerton 1932, Rhoades and Sinacori 1941).

A review of these different approaches is given by Warwick (1962) and Powell (1975). Each of these conflicting theories is supported by field evidence. D.C. Ford and Ewers (1978) linked these different views into one genetic theory. They proposed a four-state model in which deep phreatic caves and water-table caves are end members. Which of these two types finally develops, depends on the frequency of fissures penetrable by aggressive waters. Vadose caves result from the drawdown of the water table in caves of primary phreatic origin, where the water follows early phreatic tubes.

This recent genetic theory, inferred from extensive field observations, summarizes the present knowledge derived from mostly geomorphological observations and gives the boundary conditions, which have to be observed, when trying to reconstruct cave genesis by physico-chemical modelling. Therefore, here we will present these findings in some detail.

Caves are defined by Ford and Ewers as integrated networks of solutional conduits with diameters larger than 5 mm, which transmit water from an input boundary to an output boundary. In conduits of these diameters flow generally is turbulent at all geological relevant hydraulic gradients (Dreybrodt 1981a). All other voids and fissures, which are too small for the occurrence of turbulent flow, are considered to be in a pre-cave state. This pre-cave state is the state of karst initiation. In the initial state the structure of the net of the many minute open fissures and fractures, which potentially might develop into pentrable fissures, defines the conditions of cave development. These early features cannot be seen in caves directly but at most can only be inferred by observation of geomorphological structures which have developed from them.

The earliest recognizable features are families of dip tubes in subparallel arrays which are entirely contained in single, individual bedding partings. They are directed down dip and show spacings ranging from several metres to tenths of metres. They can be observed in strike-oriented cave passages into which they merge. Figure 8.21 illustrates these dip tubes as they develop in strata with a dip larger than 5°. A series of dip tubes of diameters in the order of several centimetres are stacked one above the other in successive bedding planes. At dip angles below 5° the tubes are replaced by bands of anostomoses. Tubes in different beds are connected by joint chimneys which develop along the joints. This may happen if the joint is rendered sufficiently permeable and the hydraulic gradient along this fracture is favourable. Commonly, these joint chimneys transport water to the upper bedding plane. The lower dip tube plus the joint chimney are termed a simple, phreatic loop. These are the basic structural elements of integration of an early net of flow paths in the

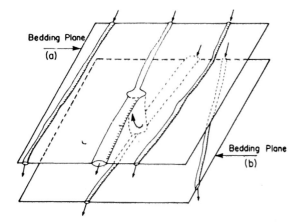

Fig. 8.21. Families of 'dip tubes' (oriented proximate to true dip) in successive, penetrable, bedding planes are the basic cave units from which many cave systems of explorable dimensions are composed. They are connected commonly by 'joint chimneys' through which groundwater flows from a lower plane (b) to a higher plane (a) (D.C. Ford and Ewers 1978)

phreatic zone. Furthermore, it is possible that neighbouring dip tubes in a distinct bedding plane can be connected by strike-oriented tubes (D.C. Ford 1968), providing a further structural complexity (not shown in the figure). As already stated all types of fractures can be employed to form flow paths which later can be inferred from the structural segments found in caves.

The frequency of fissures (i.e. their average number for a given length), which are penetrated by those early flow paths, is an important determinant for the further evolution of the cave. This fissure frequency can increase with time when more and more potentially favourable fractures and intersections develop into new tubes.

Figure 8.22a and b indicates the influence of fissure frequency on cave development in cases, where strata are flat-lying or homoclinally dipping. Four states of fissure frequency are defined.

In state 1, which represents the lowest value at which caves may develop, deep phreatic caves develop. The cave system develops along the first phreatic loops which, due to the few alternative pathways, originate far below the water table. The entire system of cave passages stays below the water table in this limiting case.

In state 4, which is the other extreme, fissure frequency is very high. In this case there are many alternative pathways, separated by relatively short distances. Thus, in the scale of these distances the rock acts as a highly permeable, porous medium (Long et al. 1982) and a continuous water table is established along which flow to the outlet proceeds. Enlargement of conduits by dissolution is now restricted to a shallow zone below the water table and an ideal water-table cave builds up. Support of this interpretation is given by results of Bedinger (1966), who simulated such a situation by using electric resistance networks.

Contemporaneously with the enlargement along the water table, vadose entrenchment in the zone above the water table proceeds by infiltration from the surface.

States 2 and 3 represent intermediate cases between the two extremes. Figure 8.22c shows the characteristics of these cave patterns in comparison to states 1 and 4. In state 2 a bathyphreatic (deep phreatic) cave develops, where at many points the

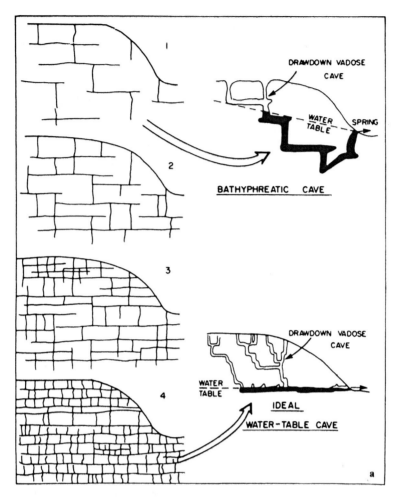

Fig. 8.22a–c. a The four states of fissure frequency that differentiate types of phreatic and water-table caves, here drawn for the case of flat-lying strata. Developed systems for *states 1* and *4* (the end members) are also shown. **b** The four states of fissure frequency drawn for the case of steeply dipping strata where the system outlets are in the general direction of true dip. Developed systems for *states 1* and *4* are shown. **c** Illustration of intermediate *states 2* and *3* between the extreme of *state 1* (bathyphreatic cave) and *state 4* (water-table cave) (D.C. Ford and Ewers 1978)

passages cut through the water table. Because of the still low fissure frequency, however, water flow follows the already excavated passages and is not able to penetrate through fractures along the water table. The state 3 caves are a mixture of phreatic loops and water-table parts.

In caves of types 2 and 3 gradational modifications may drive the geometry towards higher states. These modifications are:

1. Entrenchment of canyons along parts of the cave which reach above the water table.

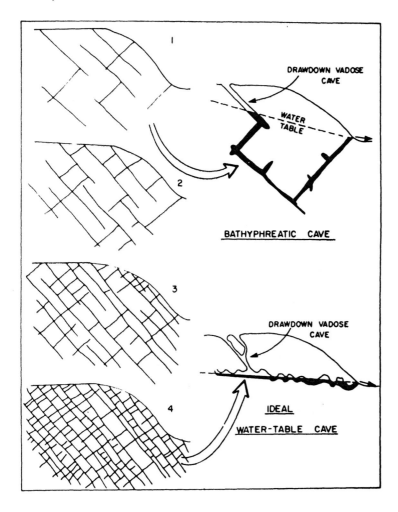

DRAWDOWN VADOSE CAVE

WATER TABLE

2

BATHYPHREATIC CAVE

3

DRAWDOWN VADOSE CAVE

4

IDEAL

WATER-TABLE CAVE

Fig. 8.22b

2. In the phreatic zone, if dissolution on the floor of the passage is prevented by sediments covering it, the upward-directed dissolution carves paragenetic passages towards the water table with vertical amplitudes as high as several tenths of metres.
3. Bypass tubes may shortcut phreatic loops under favourable conditions.

All these processes tend to establish a more gradual water table in concordance to more and more of the cave passages. In the examples of Fig. 8.22, it is a common feature that the resurgence of the caves remains at a fixed location at all times of the cave development.

This is not generally the case. Increase of fissure frequency as time elapses might produce new outputs to which the further development can adjust completely or in part. Allogenic influences, such as the incision of valleys, change hydraulic gradients and provide alternative routes along which new passages are created.

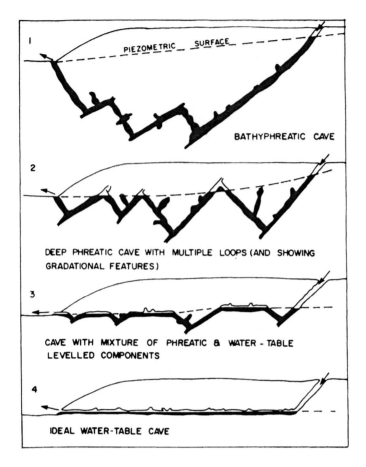

1

PIEZOMETRIC ___ SURFACE ___

BATHYPHREATIC CAVE

2

DEEP PHREATIC CAVE WITH MULTIPLE LOOPS (AND SHOWING
GRADATIONAL FEATURES)

3

CAVE WITH MIXTURE OF PHREATIC & WATER - TABLE
LEVELLED COMPONENTS

4

IDEAL WATER-TABLE CAVE

Fig. 8.22c

Figure 8.23 illustrates the development in such a case. At the initial state of karstification fissure frequency is low at state 1 and, according to the geological setting of the fractures, a first net of penetrable fissures is established. The water table in this net of fissures is practically at the surface of the rock (a). Thus, a state 1, deep phreatic loop develops (b). During that time the network of diffuse flow, presented by the potentially penetrable fractures, is enlarged by dissolution and the frequency of penetrated fissures increases, drawing down the water table and driving the cave into state 2 (c). When new outputs are formed because of lowering of the valley, the water table drops below the first loop and alternative flow paths below this loop are created. Because of the increased fissure density, the new phreatic loops show lower amplitudes but are more numerous. Therefore, they are subject to entrenchment and bypassing, which leads to a new state 3 system (d,e) below the first passages.

Upon renewed lowering of the output (f) this process is repeated and a cave system closer to state 4 originates. Increase of fissure frequency persists during all these phases as the depth of karstification increases. The fissure frequency in any of these states, however, diminishes with increasing distance from the water table until

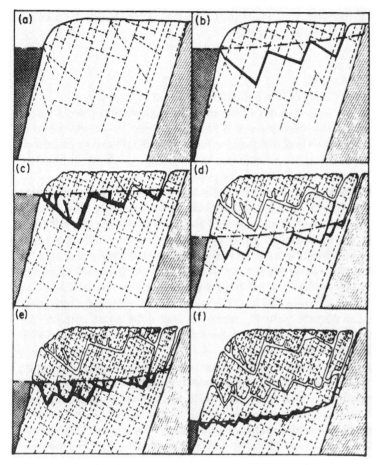

Fig. 8.23a–f. Illustration of the increase of fissure frequency in a limestone mass with the passage of time after onset of karstification and its effect upon the geometry of successive caves developing in a multi-phase system. The figure is hypothetical but based upon the situation in the central Mendip Hills, England (D.C. Ford and Ewers 1978)

practically tight rocks are encountered which prevent any flow of groundwater. On the other hand, the accessible array of potentially penetrable fractures below each of the water-filled conduits suffices for the creation of new phreatic loops.

Field evidence for this is given by Milanovic (1981). He observed the frequency of variation of karstification in an ensemble of 146 deep boreholes in the Dinaric karst, Yugoslavia, by classifying recorded permeabilities into five categories, from empty caves with dimensions of metres down to such low permeabilities which represent the initiation of karstification. From these data a coefficient of karstification is defined, which decreases with depth by an exponential with a mean length of 83 m in this special case.

Similar findings were reported by Burger and Pasquier (1984). They measured the hydraulic conductivity K, which is related to permeability, at various depths in the Swiss Jura near Neuchâtel. They also found a decrease of the hydraulic conductivity related to depth Z by an exponential

$$K = K \exp\left(-\frac{Z}{Z_K}\right). \tag{8.3}$$

Z_K is the mean depth of karstification and values of about 60 m resulted.

Karstification at the surface is most intense in the upper zone down to about 10 m. This zone is called the infiltration or b-zone by Jakucs (1977) who reported a depth of 15–20 m. This zone involves the formation of surficial karst features such as surface depression and solutional widening of fissure systems into kluftkarren, and it is gradually lowered by these erosional processes.

Although cave development follows the rules so far given, there are differences of cave genesis in flat-lying and steeply dipping rocks. A homoclinally tilted, steeply dipping rock at state 1 of fissure frequency will entrain water along the beddings deep into the rock, until favourable joints are encountered to guide the water upwards. This behaviour results when most of the employable bedding planes reach below the potential output. This creates deep phreatic loops, which are favoured in these situations. On the other hand, water-table caves are more common in flat-lying strata. Here, only joints or faults can lead the water to great depths. Since these fractures, however, are discrete features, the only continuous water paths to the output are supplied along the beddings and caves of the shallow phreatic type with low amplitudes of the loops will result.

The evolution of vadose caves is illustrated in Fig. 8.24. In the early state of karstification a water table close to the ground is established and an array of cave passages develops (a), which eventually drain the system to such an extent that the water table drops down to the level of the spring. This favours the sinking of allogenic streams using the routes of prior phreatic passages. These caves are defined as vadose drawdown caves (b). As a subsequent event the invasion of vadose caves will result, after primary inputs eventually have been blocked by sediments (c).

Vadose cave systems will develop favourably where relief is high and water input is due to streams originating from non-carbonate areas. This provides large amounts of highly aggressive water. Solution rates of these types of water of low calcium concentration are higher by an order of magnitude than those of water, which has drained through the infiltration zone, calcareous soils or small fissures (cf. Chap. 7) and thus has attained high calcium concentrations.

In areas of authigenic recharge through the karst surface much of the water is drained through fissures of relatively small size utilizing large parts of the vadose zone. Under these circumstances only a few and small streams, if at all, will flow on the surface and no significant point inputs arise from which vadose enlargement could be initiated. In such situations, apart from vertical shafts connecting different cave levels, the development of vadose cave systems of considerable size is much less probable.

The principles of cave development, as they are presented here, resolve many of the contradictions of the "classical" cave theories by showing that these are only special cases of different states or phases in the development of a cave system. It

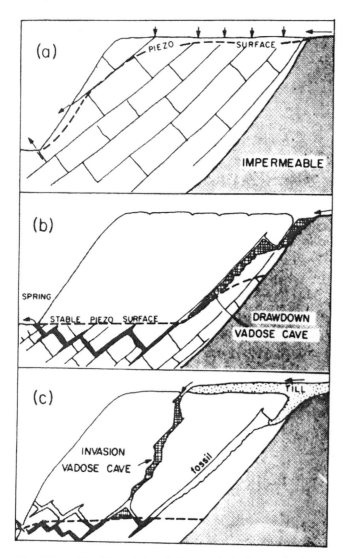

Fig. 8.24a–c. The differentiation of the two principal types of vadose caves. **a** The early condition of an unsignificant vadose zone. Groundwater drainage is channelled through an early array of phreatic tubes. **b** Due to volumetric expansion of the array, a vadose zone is created, in which streams greatly enlarge portions of the phreatic routes ('Drawdown Vadose Cave'). **c** Streams are introduced to an established vadose zone at new input positions and carve new routes that may utilize very little of the early phreatic array ('Invasion Vadose Caves') (D.C. Ford and Ewers 1978)

finally should be stated that only general principles have been discussed here, assuming relatively homogeneous limestone masses with respect to hydraulic permeability and geological structure, such as the system of fractures. If these are more complex, a variety of different cave geometries may result, rendering simple classifications much more difficult.

9 Models of Karst Development from the Initial State to Maturity

The evolution of secondary permeability in soluble rocks is determined by two fundamental factors. The first is the field of hydraulic gradients within the primary voids and partings of the rock which establishes flow of aggressive water from points of input to points of output. The second factor is the kinetics of the dissolution reactions. These determine to which extent the earliest initial flow paths are enlarged by dissolution. If saturation, with respect to calcite, is attained very quickly, dissolution can be active only at very short distances from the input, and only insignificant changes of the first flow channels result. On the other hand, at slow approach to equilibrium the earliest flow channels are modified by dissolution along their entire length. This changes the distribution of the pressure field and accordingly changes the flow fields. It is the purpose of this chapter to describe the evolution of karst according to these mutually interrelated factors.

9.1 Pressure Fields in Fractures and Their Influence on the Development of Early Channels

As we have seen in Chapter 8 flow paths develop favourably along partings, such as joints and bedding planes. Therefore, in this section we will discuss the pressure fields which are established in these early elements of permeability, once input and output points are defined.

The hydrodynamics of flow through a natural fracture is very often simulated by assuming a constant distance between the two confining planes and applying the laws of laminar flow (cf. Chap. 5) to this geometry. A real fracture, however, exhibits statistically varying distances between the confining planes, and also sites of contact.

Tsang and Witherspoon (1981) have shown that in this case for homogeneous flow in the x-direction, the volume rate is:

$$Q = \frac{W \rho g}{12 \eta} \langle d^3 \rangle_x^{1/3} \langle 1/d^2 \rangle_y^{-1} \cdot \frac{\Delta h}{L}, \tag{9.1}$$

where W is the width of the fracture, L its length along y and Δh the hydraulic head between $y = 0$ and $y = L$.

This expression is similar to Eq. (5.18) which gives the flow rate for an idealized slit fracture with distance d. In Eq. (9.1) d is replaced by two statistical averages of the distribution of distances; $\langle d^3 \rangle_x$ along the x-axis and $\langle 1/d^2 \rangle_y$ along the y-axis. Similar expressions result if one considers radial flow from one point input.

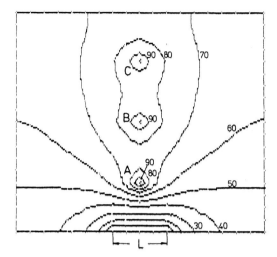

Fig. 9.1. Input-output geometry in a bedding plane. Input of water through water-leading joints is at points A, B and C. The joints have a leakage factor $t = 0$, i.e. they are sufficiently wide so as not to show flow resistance. Input head h at the joints is 100 m. The plane acts as a two-dimensional porous medium with transmissivity T. The boundaries are impervious except for that of the output region along L, where h = 0. Isolines of hydraulic head show the distribution of pressure. Flow is perpendicular to these lines (see text)

From these findings we conclude that flow in fractures is linearly related to pressure gradients and thus the fracture can be modelled by a two-dimensional porous medium obeying Darcy's law. In the following this approximation will be employed.

Figure 9.1 shows a bedding plane, where input of water flowing down joints is at points A, B and C. Output is established either along the line L or some points situated on this line. There are two ways of obtaining the pressure distribution in such a case. The first is by using electric analogues, such as an electrolytic tank or conductive paper (Bear 1972). This approach has been adopted by Ewers (1978, 1981) and Ewers and Quinlan (1981). The shortcoming of this approach is that one assumes hydraulic heads at the input points to be constant and all having the same values. In other words, this is equivalent to the assumption that the joints leading the water to the inputs are already sufficiently large, such that no pressure loss occurs along them.

An alternative way is to treat the problem numerically by solving the differential equation for the hydraulic head h with appropriate boundary conditions. This equation reads:

$$T \varDelta h + t(h_i - h) = 0 , \tag{9.2}$$

where T is the transmissivity of the confined, two-dimensional aquifer as represented by the parting; t is the leakage factor from the diffuse aquifer represented by the joints discharging into the bedding plane and is different from zero only at input points; h_i in the initial state is the position of the water table in the diffuse aquifer and is close to the ground. An easily applicable program for the solutions to this type of equation is given by Kinzelbach (1986).

Figure 9.1 shows solutions of Eq. (9.2) for the case $t = 0$ and the following boundary conditions. At points A, B, C water input is such that the hydraulic head remains constant at 100 m. Output is possible only along line L, where the pressure

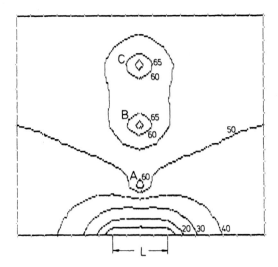

Fig. 9.2. Pressure field for the same situation as in Fig. 9.1, except for $t \neq 0$, $t = T \times 10^{-4}$. The distribution of pressure visualized by the isolines is similar to that in Fig. 9.1. The pressure at the input points is lower, however, due to the flow resistance of the joints leading the water to the inputs (see text)

head is kept at 0. All the other boundaries are impervious. The plotted contours are isolines of the hydraulic head h with increments of 10 m. According to Darcy's law, flow is along the gradient of h which is in the direction of the steepest descent perpendicular to the contour lines. In Fig. 9.1 flow velocity is largest from point A to the output line L, since the hydraulic gradient is steepest there. Flow from points B and C cannot proceed along the lines ABC to the output directly. It has to take detours along pathways with low hydraulic gradients. The flow lines can be constructed by using the fact that these lines are perpendicular to the contour lines of the hydraulic head. The detours, which flow from points B and C has to take, can also be visualized from the fact that flow from each input point extends into a well-defined, flow domain as illustrated in Chapter 5 (Figs. 5.14, 5.15).

Figure 9.2 shows the situation where at points A, B and C water input is through joints with a finite leakage factor t and where $h = 100$ m. The ratio is $T/t = 10^4$. Everything else remains as in Fig. 9.1.

In this case, due to the resistance exerted to flow by the joints, the hydraulic head at points A, B and C drops to about 65 m.

Again, the maximal flow velocity is from A to L, whereas flow from points B and C is much lower. The pattern of flow is changed only unessentially in comparison to Fig. 9.1. Therefore, in view of the idealizations in modelling flow through a bedding parting, it seems justifiable, to neglect the resistance of the input joints.

Figure 9.3 shows contour lines for a more complex situation. Points A to G are input points of constant hydraulic head at $h = 100$ m. Output is at the point indicated. As one can visualize from the figure flow velocities are highest from point C to output. There is also substantial flow from D to output. Flow from all the other points is slow, since it proceeds on detours along low hydraulic gradients.

The results of the calculations shown here and the many systematic electric analogue experiments by Ewers (1981) can be condensed in a simple general rule.

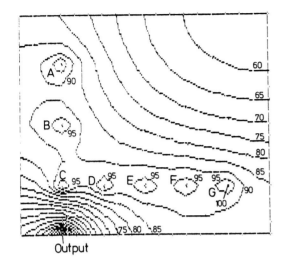

Output

Fig. 9.3. Isolines of pressure visualizing the distributions of hydraulic head for a complex input-output geometry. h = 0 at the output point; h = 100 m at points A-G (see text)

Provided that the heads at all input points are equal, flow velocity to the output is largest from that input with the shortest distance to the output. If there are several output points, also with equal head, flow proceeds mainly from the nearest input.

From this rule it is easy to envisage the development of cave patterns. As we will show later initial cave conduits develop most probably along the pathway of the steepest hydraulic gradient, where flow velocity is maximal. Therefore, after some time a region of high transmissivity will propagate from point C to the output 0 in the example of Fig. 9.3 Figure 9.4 shows in a sequence a–e, how the pattern of hydraulic head contour lines changes upon solutional enlargement of the parting along the direction of the steepest hydraulic gradient. A somewhat different geometry of inputs A–F has been chosen. Output is designated.

Figure 9.4a shows the initial state, where the hydraulic head at all input points is equal (100 m). At the output it is assumed to be zero.

Figure 9.4b shows how the hydraulic head field develops if transmissivity has become high along the line at A. At sufficiently high transmissivity this line becomes an isoline with h = 100 m. The total hydraulic head now drops to zero along the line A–0, thus increasing the hydraulic gradient. In Fig. 9.4c the initial channel has reached the output. At this moment the hydraulic head drops to zero along the line A–0, since to maintain the high hydraulic head of 100 m at practically infinitely high transmissivity would require an infinite amount of water feeding this channel. Therefore, the channel A–0 becomes vadose and now itself represents a suitable output. The nearest input point to this new output now is point B, and the steepest descent of the hydraulic head is along line AB. There is also a descent of head from point D to A because of the comparable distances. Figure 9.4d shows the situation after a channel has developed from B to A. The hydraulic gradients are now such that a new channel can develop from D to B and also from C to B. In Fig. 9.4e we illustrate the case where a channel has developed from D to B. As can be easily seen hydraulic heads are steepest from E to D and C to B, which are the directions of further passage development.

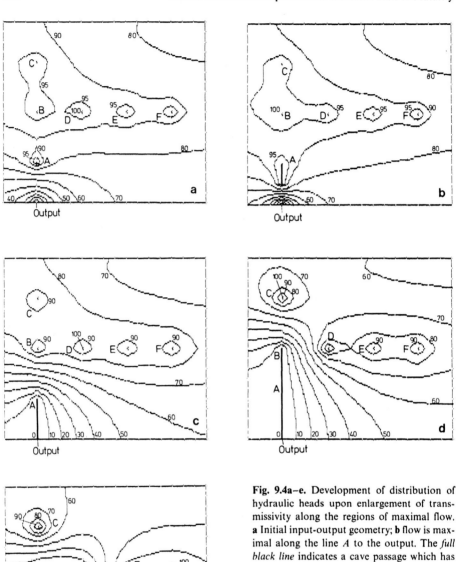

Fig. 9.4a–e. Development of distribution of hydraulic heads upon enlargement of transmissivity along the regions of maximal flow. **a** Initial input-output geometry; **b** flow is maximal along the line *A* to the output. The *full black line* indicates a cave passage which has developed to such an extent that T can be assumed T = 0. The pressure drop from 100 to 0 is now along a much shorter distance compared to the initial state **a**. Accordingly, flow increases from *A* to the output. **c** The cave passage has reached the output. Pressure therefore drops to zero along *A* to output. Pressure is redistributed with the largest gradients from *B* to *A*. **d** Analogously to the development of the first cave passage from *A* to output, a new passage has originated from *B* to *A*. The resulting new distribution of pressure shows the highest gradients along *BD* and *CB*. Each of these connections is a good candidate for the further development of passages. **e** Here, we have chosen connection *BD* for passage development. Candidates for further cave development are now *CB* and *DE*

By using the principle that initial channels always develop between input and output of the shortest distance, one can easily construe how cave patterns develop from a given configuration of inputs and outputs.

9.2   The Network Linking Models of Ewers

Ewers (1978, 1982) and Ewers and Quinlan (1981) have used the principles depicted above to suggest models for development of cavern conduits in different geological situations.

9.2.1   The Low Dip Model

In nearly horizontal carbonate sequences, input to bedding partings is mainly through joints or faults. Therefore, inputs may occur at many points on the surface. Furthermore, stream valleys are sites where water may enter directly into exposed bed partings.

Outputs are likely to occur along strike-oriented stream valleys lower than the input points. Where a water-guiding bed parting is exposed at the valley, resurgences are equally possible along distant lines of exposure. Ewers has summarized this geological setting into a simplified model, which is shown in Fig. 9.5.

Sinks (inputs) are modelled as linear arrays of points with equal heads. The discharge boundary is parallel to these arrays and each point of this boundary can be equally favourably employed as an output. This initial situation is represented in Fig. 9.5a. Input points 1 to 15 now all are equally distant from corresponding output points and flow from them experiences equal pressure fields. Therefore, initial channels develop as illustrated in Fig. 9.5b. Due to the fact that the transmissivity in a bedding parting is randomly distributed and not homogeneous, at sites of higher transmissivity flow velocities are larger than at sites of lower transmissivity. As a result the different flow channels propagate at different rates, some of them reaching first the discharge boundary (Fig. 9.5c). Thus, a new output configuration results and those channels which are closer to the new output points are directed towards these first channels which have reached the outputs. Other channels already close enough to the discharge boundary keep their direction of propagation and eventually reach the discharge boundary (Fig. 9.5d). Furthermore, the input points of the second row now start to develop channels towards those of the first row, which are already connected to the output boundary by channels (Fig. 9.5d).

Some of these channels as first reach points of the first array, which are already connected to the output. They then serve as new output points towards which channel development of their neighbours is directed (Fig. 9.5e). An important result of the combination of stochastic and deterministic principles is that inevitably flow domains and water divides develop, which constitute groundwater basins, completely isolated from each other. For instance, points 9 to 11 and 8' to 13' comprise one of these domains (Fig. 9.5e).

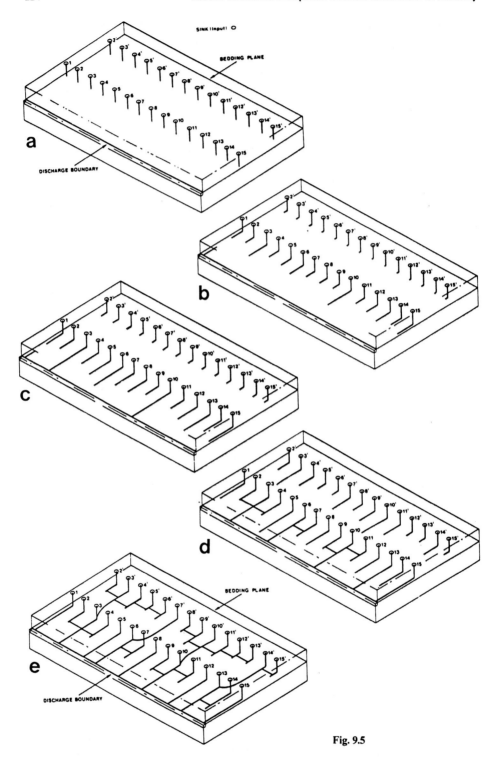

Fig. 9.5

To summarize, the low dip model shows some characteristic properties:

1. Because of the development of separated groundwater basins, one expects that the number of springs draining the area is inversely proportional to their discharge. In other words, there may be either many small springs or only a few but large ones draining the area.
2. Although each channel propagates from the input to the output, the final conduits integrate headwards from the resurgence opposite to channel propagation.
3. The conduits should develop into a trellised pattern, with one set of conduits oriented favourably into the dip direction and the other perpendicular along the strike. The dip-directed passages, in general, should be older than those in the strike direction.

A typical example for the low dip model is represented by the Mammoth Cave region (Ewers and Quinlan 1981, Ewers 1982). Various groundwater basins of varying size have developed in this area, supporting the low dip model (Quinlan et al. 1983). One example for point (3) is the Parker Cave system which shows a characteristically trellised pattern along the dip and strike with a morphology from which pre-dating of dip tubes to strike tubes can be inferred. Furthermore, joint control in this system is absent and all dip tubes have developed as described in Chapter 8.

9.2.2   The High Dip Model

In steeply dipping limestone strata, input of water is likely where the bedding plane intersects the surface. Inputs will be arranged along this line of intersection and may be due to strike-oriented streams. Output points occur either at an intersection of the bedding plane with the surface or at sites where joints, connecting this plane to the surface, have become sufficiently widened to guide the flow to the surface. Ewers' (1978) model is shown in Fig. 9.6a–e. In the initial state there are three input points along the intersection to the surface. The output point is lower by the elevation h and connected to the surface by a wide joint. Figure 9.6a shows also the flow lines directed towards the later spring.

Figure 9.6b shows the tubes (heavy lines) propagating along these flow lines. The channel from the input closest to the output reaches this point first (Fig. 9.6b,c). Figure 9.6d,e shows strike-oriented tubes integrating the shorter channels to the

---

Fig. 9.5a–e. The low dip model for cave development. **a** Initial situation with a linear array of input points (O) at equal head. The discharge boundary is assumed to be at h = 0. **b** Due to stochastic differences in transmissivity and solubility of the rock, passages to the output propagate with different speed. **c** The first of those passages have reached the discharge boundary. This produces significant redistribution of pressure between those points connected by the passages to the output and their nearest neighbours in the first and the second row of the input array. **d** Consequently, passages develop between these points, connecting points in the first row and initiating passages to grow from the second row towards the first row. **e** Finally, clusters of points are connected by passages. Pressure drop between these clusters is small and it is therefore unlikely that they will be connected by further passages. Thus, in this final state groundwater basins have developed (Ewers and Quinlan 1981)

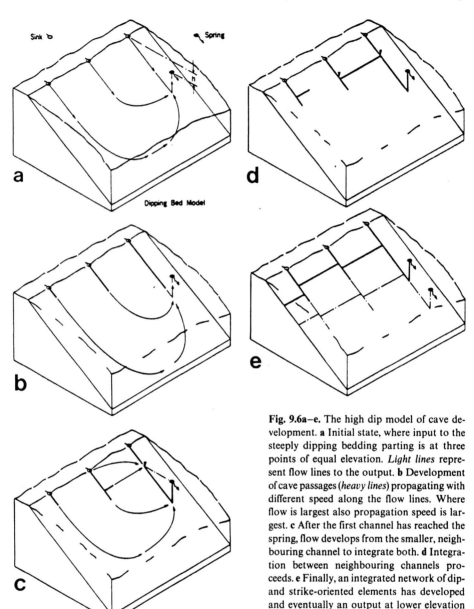

Fig. 9.6a–e. The high dip model of cave development. **a** Initial state, where input to the steeply dipping bedding parting is at three points of equal elevation. *Light lines* represent flow lines to the output. **b** Development of cave passages (*heavy lines*) propagating with different speed along the flow lines. Where flow is largest also propagation speed is largest. **c** After the first channel has reached the spring, flow develops from the smaller, neighbouring channel to integrate both. **d** Integration between neighbouring channels proceeds. **e** Finally, an integrated network of dip- and strike-oriented elements has developed and eventually an output at lower elevation will be employed (Ewers 1982)

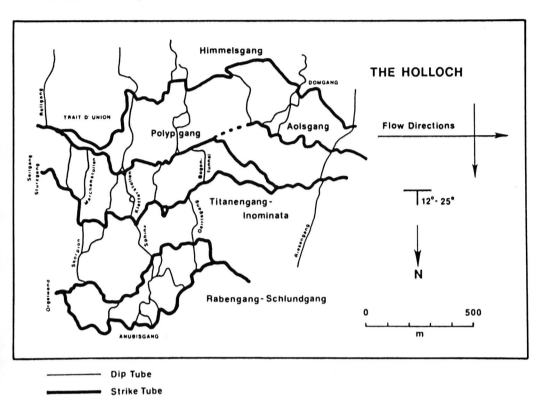

Dip Tube

Strike Tube

Fig. 9.7. Conduits of the middle portion of the Hölloch showing strike- and dip-oriented elements (Ewers 1978, 1982)

output. Finally, a new output, lower than the first one, may be established and the linking scheme to this new point will be repeated by new dip- and strike-oriented channels developing in response to this new output. The high dip model shares the characteristic properties (2) and (3) listed previously for the low dip model.

As an example for the high dip model Ewers (1978, 1982) referred to the Hölloch/Switzerland. Many passages in this 150-km-long cave have developed along bedding planes and are either dip- or strike-oriented as shown in Fig. 9.7. They show no joint control. All the tubes have developed under phreatic conditions as can be inferred from their morphology with elliptical or lenticular cross-sections. Canyons indicating vadose activity are almost absent.

9.2.3 The Restricted Input Model

In many geological settings input configurations may arise, in which a few inputs to a bedding plane are arranged in a linear array oriented perpendicular to the discharge boundary or collinear to an output point. This situation has been illustrated already in Fig. 9.1, where the initial hydraulic heads are shown.

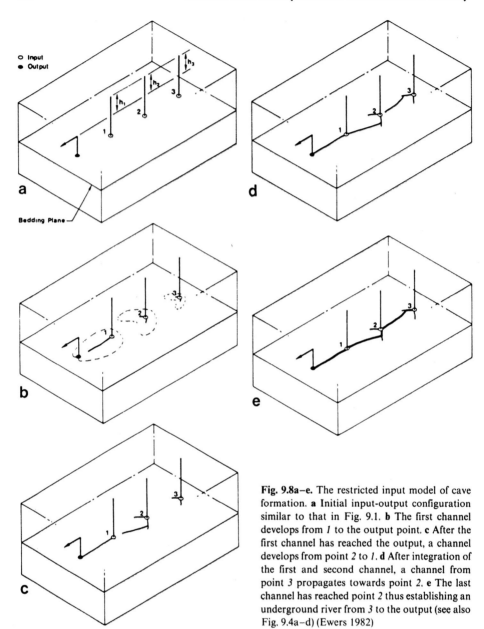

Fig. 9.8a–e. The restricted input model of cave formation. **a** Initial input-output configuration similar to that in Fig. 9.1. **b** The first channel develops from *1* to the output point. **c** After the first channel has reached the output, a channel develops from point *2* to *1*. **d** After integration of the first and second channel, a channel from point *3* propagates towards point *2*. **e** The last channel has reached point *2* thus establishing an underground river from *3* to the output (see also Fig. 9.4a–d) (Ewers 1982)

Such situations are likely where limestone is covered by an impermeable stratum. Inputs may be created where a stream by fluvial erosion breaches this stratum. Water is then guided by joints to the lower beddings.

The development of secondary cavern porosity is illustrated in Fig. 9.8a–e. Figure 9.8a depicts the initial input and output geometry. Output is via one point.

As can be seen from Fig. 9.2 the hydraulic gradient is large from point 1 to the output and correspondingly a channel propagates quickly to the output along this line. Small channels may develop along the flow lines directed away from inputs 2 and 3 (Fig. 9.8b). Once the major channel, originating from point 1 (Fig. 9.8c), reaches the output, a new channel originates from point 2 towards point 1. Figure 9.8d and e show how, by repetition of this process, a subsurface river is integrated. Note that propagation of the integrated river network is headwards, a principle which is operative in any model of cave development.

An illustrative example for the restricted input model is a cave which has developed in Cave Creek Valley in south-central Kentucky (Ewers 1982). Cave Creek Valley is a dry karst valley with several closed depressions, which has been cut into the impermeable caprock covering the area until the limestone was exposed. Thus, an array of inputs along this valley is established, draining allogenic surface water into the limestone mass.

The resurgence for these waters is at Cumberland River which at that point flows in a direction perpendicular to the valley axis. Below Cave Creek Valley a conduit pattern has developed which is directed along the valley axis and shows the basic characteristics of the restricted input model.

A final remark should be made. The three models discussed in this chapter are not different with respect to the processes of cave formation. What makes them different are the boundary conditions of input and output geometry. Using programs like those of Kinzelbach (1986), which can be easily carried out on personal computers, the field of hydraulic heads can be visualized for almost any type of input-output geometry. Furthermore, an inhomogeneous distribution of permeability can be implied. Thus, this technique may be a suitable tool to construe the evolution of a cave.

9.3 Dissolution Kinetics and the Concept of Penetration Length

In the discussion on the influence of hydraulic gradients on the development of the first initial karst channels we have assumed that these develop along pathways where flow velocity of calcite-aggressive water is maximal. In this section we will give a detailed treatment of this problem.

There are obviously two parameters determining whether a first small tube can develop.

1. The dissolution rate, which determines the annual retreat of the confining rock, has to be sufficiently large to excavate a channel to geologically reasonable dimensions in geologically reasonable times. There is much evidence that from the initiation of karst systems to their mature state, minimum times in the order of several 10 000 years are needed (Mylroie and Carew 1986, Bakalowicz 1982, White 1979).

2. Water entering a fracture loses its solutional aggressivity as time passes (cf. Eq. 7.44). Therefore, it can travel only a limited distance L_p until its solutional power has dropped to 10% of its initial value. Obviously, this distance is proportional to

the velocity of flow. If this penetration length L_p is in the order of the input-output distance or larger, solutional power is active over the entire path of flow and a solutional channel will be the result. In contrast, if L_p is small compared to the input-output distance, solutional excavation of a tube is only possible close to the input point and most of the parting along the flow path will remain unchanged. Therefore, in this case karstification is unlikely to evolve.

Dissolution rates and penetration lengths are counteractive since at large dissolution rates the solution becomes rapidly saturated and short penetration lengths result. In any case the interplay of dissolution rate and penetration length determines the time scale and the length scale of cave development.

9.3.1 The Concept of Penetration Length

To calculate the penetration length L_p one has to know the geometry of the conduit and also the dissolution rates as a function of the Ca^{2+} concentration c of the solution. Figure 9.9 shows a length fraction dx of an arbitrary conduit with the cross-sectional area A. Velocity of water flow in this channel is v. According to mass balance the amount of $CaCO_3$ dissolved per time unit from the walls over the length dx must be equal to the difference of the flux of Ca^{2+} into the conduit and the flux out of it. At the entrance the concentration is c and after length dx it has increased by dc due to dissolution from the wall. Therefore:

$$v \cdot A \cdot dc = F(c) P_r \cdot dx , \tag{9.3}$$

where P_r is the perimeter of the conduit and F(c) represents the dissolution rate. By integration one obtains:

$$x = \frac{v \cdot A}{P_r} \int_{c_0}^{c} \frac{dc}{F(c)} , \tag{9.4}$$

where c_0 is the concentration of Ca^{2+} at $x = 0$, the entrance of the conduit, and c is the concentration after the solution has travelled the distance x with the constant velocity v. From Eq. (9.4) the penetration length L_p can be derived by integrating to the value of c which corresponds to $F(c) = 0.1 F(c_0)$.

Several expressions for F(c) have been used in the literature to calculate L_p.

Weyl (1958) assumed that dissolution is determined entirely by molecular diffusion. By solving the transport equation for a laminar flow velocity distribution

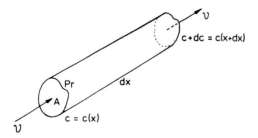

Fig. 9.9. Geometrical model of a flow conduit to calculate the penetration length L_p (see text)

in a plane parallel fracture and in a circular conduit, he obtained:

$$L_p^f = \frac{C_f \cdot v \cdot d^2}{D}; \qquad L_p^c = \frac{C_c \cdot v \cdot r^2}{D}, \tag{9.5}$$

where D is the molecular diffusion constant, d the aperture of the fracture and r the radius of the conduit. $C_f = 0.304$, $C_r = 0.572$ are factors resulting from the laminar velocity distribution. If one neglects the laminar velocity distribution, assuming plug flow, one obtains $C_f = 0.33$ and $C_c = 0.4$. This shows that neglecting velocity distribution is justifiable in problems of diffusional mass transport (cf. Eq. 7.37).

White (1977) has used experimental data of Plummer and Wigley (1976) to calculate L_p. Using a second-order rate equation (cf. Chap. 6, Fig. 6.30):

$$F = k_c(c_{eq} - c)^2, \tag{9.6}$$

one obtains by integration of Eq. (9.4):

$$L_p^f = 1.08 \frac{d \cdot v}{k_c c_{eq}}; \qquad L_p^c = 1.08 \frac{r \cdot v}{k_c c_{eq}}, \tag{9.7}$$

where the initial concentration at the input to the conduit is zero.

Note that Eq. (9.7) yields values which are smaller by a factor of four from those given by White. This is because we define L_p as the length after which the dissolution rates have dropped to 10%, whereas White defines the length after which 90% of saturation has been attained, or as one can see from Eq. (9.6), at 1% of the initial dissolution rate.

There is evidence that close to equilibrium a rate law of the fourth order may be valid (Plummer and Wigley 1976).

With

$$F = k_c'(c_{eq} - c)^4, \tag{9.8}$$

one obtains the penetration distances for fractures or circular conduits:

$$L_p^f = 0.77 \frac{d \cdot v}{k_c' c_{eq}^3}; \qquad L_p^c = 0.77 \frac{r \cdot v}{k_c' c_{eq}^3}. \tag{9.9}$$

From the experimental data of Plummer and Wigley (1976) one finds:

$$k_c = 30 \text{ cm}^4 \text{ mol}^{-1} \text{ s}^{-1} \quad \text{and} \quad k_c' = 1.2 \times 10^{13} \text{ cm}^{10} \text{ mol}^{-3} \text{ s}^{-1}.$$

The penetration distance for linear kinetics as discussed in Chapter 7 can be calculated from the linear rate law:

$$F = \alpha(c_{eq} - c) \tag{9.10}$$

by integration of Eq. (9.4) as:

$$L_p^f = \frac{v \cdot d}{\alpha} \cdot 1.15; \qquad L_p^c = \frac{v \cdot r}{\alpha} \cdot 1.15. \tag{9.11}$$

Fig. 9.11. Dissolution rate of a calcite- ▶ aggressive stagnant solution in an fracture conduit with $d = 0.01$ cm for second-order kinetics (Eq. 9.6) and fourth-order kinetics (Eq. 9.8) as a function of time. In both cases $c_0 = 0$, $c_{eq} = 2 \times 10^{-3}$ mmol cm^{-3}. Note the logarithmic scale. $F_0 = F(0)$

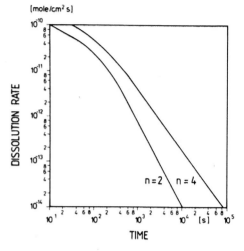

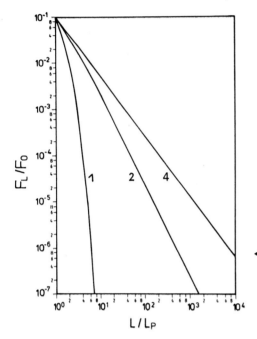

◀ **Fig. 9.10.** Ratio of the dissolution rate to the initial dissolution rate for first- second- and fourth-order kinetics, as designated on the curves, as a function of travelled distance L in units of the respective $L_p^{(i)}$, $i = 1, 2, 4$. Note the logarithmic scale

One comment should to be made here. The penetration distances L_p for different types of kinetics are not directly comparable. Figure 9.10 shows the relative change F_L/F_0 of the dissolution rates as a function of the distance travelled. This distance is measured in multiples of the corresponding L_p. From this figure it becomes evident that first-order kinetics, with an exponential approach to equilibrium, lose dissolution power very fast. For this type of kinetics dissolution rates drop by a factor of ten for each increment L_p of flow path. This is not the case for higher order reactions. Table 9.1 visualizes this. With increasing order of reaction the solution can travel longer distances until the dissolution rates have dropped to a given fraction. This behaviour is due to the time dependence of higher order chemical kinetics and has been discussed in Chapter 4. Figure 9.11 illustrates the time dependence of higher order reaction rates and is equivalent to Fig. 9.10 as lengths L and time t are related to each other by velocity v.

From Eqs. (9.5), (9.7), (9.9) and (9.11) two important conclusions can be drawn:

1. Penetration lengths depend on the rate law of dissolution. A change in dissolution kinetics as the system approaches equilibrium can be of utmost importance for initiation of karst development. At Ca^{2+} concentrations below approximately

Table 9.1. Travel distances of calcite-aggressive water in units of penetration lengths $L_p^{(i)}$ until the solution rates have dropped to $F = F_0 \cdot 10^{-n}$, for first; second- and fourth-order kinetics (the initial concentration $[Ca^{2+}]_i = 0$) F_0 is the initial rate at $c = 0$

F/F_0	$L/L_p^{(1)}$	$L/L_p^{(2)}$	$L/L_p^{(4)}$
10^{-1}	1	1	1
10^{-2}	2	4.5	6.6
10^{-3}	3	15	38
10^{-4}	4	49.5	217
10^{-5}	5	158	1216
10^{-6}	6	449.5	6840
10^{-7}	7	1581	38465

60 to 90% of saturation there is a linear rate law, as discussed in Chapter 7. Closer to saturation dissolution rates drop sharply and eventually may be described by rate laws of higher order. In this region, due to the small dissolution rates, high penetration lengths are then encountered. Slow dissolutional excavation of first initial channels proceeds until their dimensions have become sufficiently large to keep the solution undersaturated approximately below 60%. From this point linear dissolution kinetics take over and determine the further evolution of the karst system.

2. Penetration lengths are related linearly to flow velocity. In laminar flow, as exists in the initial state of karstification, therefore calcite-aggressive water with the highest flow velocity can penetrate large distances into the rock until it loses its solutional power. This justifies the assumptions made above that cave development proceeds favourably along pathways of the steepest hydraulic gradient.

To estimate how far an aggressive solution can penetrate into a fracture, it is important to know the dissolution rate after distance X. Using the data of Plummer and Wigley (1976) for fourth- and second-order kinetics and those of Buhmann and Dreybrodt (1985b) (cf. Chap. 7), we therefore calculate the distance X water can travel in a joint of initial aperture d subject to a given hydraulic gradient J, until its solution rates have dropped to 10^{-12} mmol cm^{-2} s^{-1} corresponding to a widening of the fracture to 2×10^{-6} cm year^{-1}. As typical for karst water we assume an equilibrium concentration of 2×10^{-3} mmol cm^{-3} and $\alpha = 2.5 \cdot 10^{-5}$ cm s^{-1}. For $d = 0.01$ cm and $J = 0.1$ and laminar flow (cf. Eq. 5.17) and Table 9.1

$$X^{(1)} = 1.76 \text{ m}, \qquad X^{(2)} = 22.7 \text{ m}, \qquad X^{(4)} = 131 \text{ m}.$$

If one assumes that the calcite concentration of the solution is zero when entering the fracture, the dissolution will proceed in first-order kinetics only a few metres. At about 60% of saturation second-order kinetics will be rate-determining

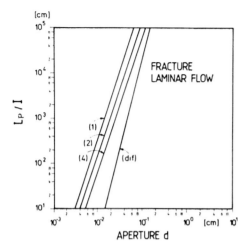

Fig. 9.12. Ratio of penetration length to hydraulic gradient for laminar flow in plane parallel fractures of aperture d for first-, second-, and fourth-order kinetics and the case of diffusion control. In all cases one assumes $c_0 = 0$. For first-order kinetics $\alpha = 2.5 \times 10^{-5}$ cm s^{-1}

until after about 10 m travel distance there is a change to slow dissolution, fourth-order rate, which is still sufficient to excavate a channel of 0.05 cm width in approximately 25 000 years.

Note that the fourth-order dissolution rates decrease slowly. After a travel distance of 65 m the fourth-order rate is $1.5 \times 10^{-5} F_0$, at 131 m $5.2 \times 10^{-6} F_0$, and at 360 m it is $1.5 \times 10^{-6} F_0$.

An important result of this analysis is that in the initial state of karstification, where conduit dimensions are below 0.05 cm, first-order penetration lengths are less than 20 m even at hydraulic gradients of 0.1. Therefore, water flowing in aquifers fed by diffuse flow of this type should always be close to saturation as observed by Schuster and White (1971) for springs (cf. Chap. 5).

To summarize the data on penetration length Fig. 9.12 gives the ratio L_p/J as a function of fracture aperture d for first; second- and fourth-order kinetics, as well as for the diffusion limited case.

9.3.2 Penetration Lengths in Turbulent Flow

In the initial state of fracture widening flow is laminar. As soon as a critical fracture aperture is achieved flow becomes turbulent. For a circular conduit this limit is reached for the Reynolds number $N_{Re} \approx 2000$. This corresponds to a critical radius of $r_c = 1$ cm for an hydraulic gradient $J = 10^{-3}$, $r_c = 0.5$ cm for $J = 10^{-2}$, and $r_c = 0.2$ cm for $J = 10^{-1}$ (Dreybrodt 1981a). Similar values have been given by Wigley (1975a), Atkinson (1968) and White and Longyear (1962).

In the region of these conduit dimensions penetration lengths for first-order kinetics are in the order of several kilometres even for low hydraulic gradients ($J = 10^{-3}$), as can be seen from Fig. 9.12. One can therefore assume that in most cases the solution will be sufficiently undersaturated for first-order kinetics to prevail.

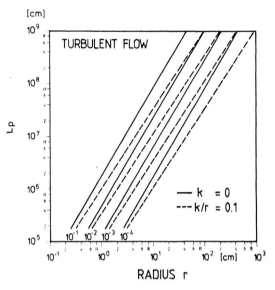

Fig. 9.13. Penetration lengths for first-order kinetics in turbulent flow. *Full lines* refer to hydrodynamically smooth tubes, *dotted lines* to rough tubes with k/r = 0.1 (cf. Chap. 5). The *numbers* on the curves designate the hydraulic gradient

As turbulence occurs the dissolution rates increase as discussed in section 7.5. The upper limit, which can be reached, is for the case of a vanishing diffusion boundary layer and is discussed in Chapter 7. Furthermore, the flow velocity in turbulent flow depends on the square root of the hydraulic gradient (cf. Eq. 5.34).

These two effects determine the penetration distance in turbulent flow, which is given by:

$$L_p^{turb} = \frac{v \cdot r}{\alpha^t} \cdot 1.15 ,$$ (9.12)

$$v = \sqrt{4grJ} \left[-2\log\left(\frac{k}{7.4r} + \frac{2.51v}{\sqrt{16gt^3 J}}\right) \right] ,$$

where α^t is the kinetic first-order rate constant and is tabulated in Table 7.1 and 7.2 for the open and closed system respectively. Since this value represents an upper limit, the penetration distance calculated by use of this value gives a lower limit.

For first-order kinetics Fig. 9.13 presents penetration distances for various hydraulic gradients for smooth tubes and tubes of roughness k/r = 0.1. All the curves are drawn for values of $r > r_c$. In all cases penetration lengths are greater than 1 km. Once conduits of 10-cm radius are excavated, even at lowest hydraulic gradients, penetration lengths are in the order of 10 km. As a consequence water flowing in conduits of this size remains well below saturation and first-order kinetics determine dissolution. This is in agreement with the observations of Shuster and White (1971), who found that springs fed by conduits remain undersaturated with respect to calcite at saturation ratios Ω between 0.14 to 0.51 (cf. Chap. 5).

9.4   Karst Development from the Initial State to Maturity

Onset of karstification is triggered by entrenchment of a valley which provides the hydraulic gradient for the flow of water through primary fissures. The apertures of these fissures are distributed statistically. Apertures range from 10^{-3} cm up to 10^{-2} cm (Tsang 1984). Motyka and Wilk (1984) measured the widths of fissures in karstified Triassic rocks in the surroundings of Olkusz, Poland in underground mine workings located 150 m belowground and also on the surface. The most frequent widths of fissures was found to be 200 μ.

To model the evolution of the earliest karst channels in these fractures we assume each fracture to be of a constant average aperture over a long distance L. The pressure drop from the input of water to its outlet at distance L from the input is Δp.

Flow in this plane parallel fracture of initial aperture d is laminar. The volumetric flow rate per unit width of the fracture is given by:

$$Q = vd . \tag{9.13}$$

The penetration lengths $L_p^{(1)}$, $L_p^{(2)}$ and $L_p^{(4)}$ given by Eqs. (9.11), (9.7) and (9.9) are directly proportional to Q. If one assumes the width of the fracture to be constant, transverse to flow then the penetration lengths $L_p^{(i)}$, i = 1, 2, 4 remain constant along the entire flow path, since Q is constant according to the equation of continuity. This is important when d varies along the flow path, because it simplifies mathematical analysis considerably. In terms of Q, one has:

$$L_p^{(1)} = 1.15\frac{Q}{\alpha}; \qquad L_p^{(4)} = 0.77\frac{Q}{k_c'c_{eq}^3} . \tag{9.14}$$

The spatial dependence of the dissolution rate can now be calculated by integration of Eq. (9.4). For first-order kinetics one obtains:

$$F(x) = F_0^{(1)}\exp\left(-\frac{2.31x}{L_p^{(1)}}\right), \tag{9.15}$$

where $F_0^{(1)}$ is the initial rate of dissolution at c = 0 and is given as (cf. Chap. 7):

$$F_0^{(1)} = \alpha c_{eq} . \tag{9.16}$$

After the distance $L_p^{(1)}$ the dissolution rate drops to 10% of $F_0^{(1)}$ and the concentration c is 90% of the saturation concentration c_{eq}. We assume that at this concentration dissolution kinetics change to the fourth-order type as given by Eq. (9.8).

Thus, at x = $L_p^{(1)}$ water with concentration $0.9c_{eq}$ dissolves limestone by fourth-order kinetics and the dissolution rate is given by integration of Eq. (9.4) with the lower limit $c_0 = 0.9c_{eq}$:

$$F = F_0^{(4)}\left(\frac{4.62}{L_p^{(4)}}\cdot x + 1000\right)^{-4/3} , \qquad x \geq L_p^{(1)} . \tag{9.17}$$

Using these rates in a numerical iterative procedure, one can calculate the aperture as a function of distance x from the input and time t after first onset of dissolution as:

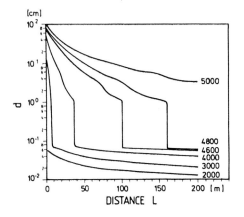

Fig. 9.14. Enlargement of a fracture with initial width of 10^{-2} cm and a length of 200 m by calcite-aggressive water. The *numbers* on the curves give the time in years after the onset of enlargement (see text) $J = 0.1, \alpha = 2.5 \cdot 10^{-5}$ cm s^{-1}

$$d(x, t + \Delta t) = d(x, t) + \gamma \cdot F^{(i)}[x, L_p^{(i)}(t)] \cdot \Delta t \,, \qquad (9.18)$$

where γ is a factor to convert the expression $F(x, L_p)$ into cm year^{-1} and Δt is a suitable time increment and is measured in years.

Note that in Eq. (9.18) L_p is dependent on time, since with the widening of the fracture the flow rate Q increases. Therefore, after each time step Q(t) has to be calculated by:

$$Q(t) = \frac{\Delta p}{R(t)}, \qquad (9.19a)$$

with the flow resistance R(t):

$$R(t) = \sum_n \frac{\Delta x}{d^3(n \cdot \Delta x, t)} \cdot \frac{12\eta}{\rho g} \qquad (9.19b)$$

where Δx is an increment along x.

The penetration distances are either $L_p^{(1)}$ or $L_p^{(4)}$ depending on which dissolution kinetics $F^{(i)}$, $i = 1, 4$ is effective.

Figure 9.14 represents fracture apertures as functions of distance for various times after onset of dissolution. As a typical concentration $c_{eq} = 2 \times 10^{-3}$ mol l^{-1} has been assumed. The initial fracture aperture at $t = 0$ is $d = 0.01$ cm and the length $L = 200$ m. The hydraulic gradient is 10^{-1}, $\alpha = 2.5 \cdot 10^{-5}$ cm s^{-1}.

At times below 3000 years $L_p^{(1)}$ is so small that practically fourth-order kinetics govern dissolution along the entire length.

The curve, after 2000 years have elapsed, is a characteristic example. The fracture is now widened by only a small amount of several 10^{-2} cm. Note that in Fig. 9.14 the apertures are plotted on a logarithmic scale. As the apertures increase in time the first-order penetration length becomes sufficiently large such that dissolution can now proceed on the much larger first-order rates. This is observed after 3000 years. Due to the large rates, a channel of 10-cm dimension at the entrance and of 0.1 cm at a distance of 10 m is created. Then dissolution changes into fourth-order kinetics and dissolution proceeds at a slow path along the remaining length of the fracture. With increasing time $L_p^{(1)}(t)$ increases as does the distance up to which fast dissolution occurs. Finally, $L_p^{(1)}(t)$ becomes larger than the fracture length. This occurs between 4800 and 5000 years. After 5000 years $L_p^{(1)}$ is so large that dissolution rates are close to $F_0^{(1)}$ in Eqs. (9.15) and (9.16) and are almost constant along x. Thus, the channel widens uniformly with a retreat of bedrock of about 0.05 cm year^{-1}.

Some comments should be given at this point. We have assumed that the dissolution rates in first-order kinetics, as given in Eq. (9.16) exhibit a kinetic constant α independent on d. This is approximately true for values of d $\leq$ 1 cm (cf. Eq. 7.36). Therefore, in our calculations the first-order dissolution rates at d > 1 cm have been chosen too high by up to one order of magnitude. This, however, does not influence the results for two reasons:

1. The flow resistance R is determined by the narrow part of the channels and the influence of the too large apertures in the wide regions is negligible.
2. The penetration lengths in our approximation are lower than those with correct dissolution rates. Thus, the times at which the channel breaks through at the outlet are estimated too high.

Since the aim of our considerations is to estimate orders of magnitude, we have chosen the approximation above with the important result that at geologically reasonable initial conditions, such as aperture, hydraulic gradients and lengths of fractures, first flow channels do develop to reasonable dimensions in reasonable times.

To illustrate the limit of lower dissolution rates we have calculated fracture apertures as they develop in time, if first-order dissolution rates are lowered by a factor of ten compared to the value used in Fig. 9.14. This is shown in Fig. 9.15. The curves are similar to those in Fig. 9.14. Breakthrough is achieved earlier but is in the same time scale. The real profiles are in between those of Fig. 9.14 and Fig. 9.15. To illustrate the flow rates which are carried through the fractures with increasing time

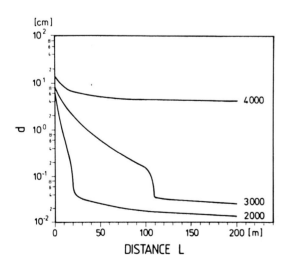

Fig. 9.15. Enlargement of the fracture at various times as in Fig. 9.14 except for a smaller kinetic constant α in the first-order kinetics of dissolution (see text). $\alpha = 2.5 \cdot 10^{-6}$ cm s^{-1}

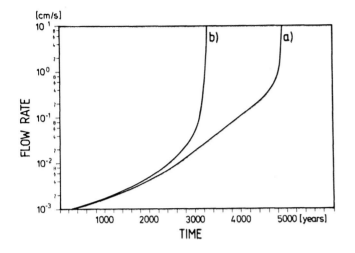

Fig. 9.16. Flow rate Q through the aperture, enlarging as a function of time. *Curve a* refers to the situation in Fig. 9.15, *curve b* to that in Fig. 9.14 respectively. Breakthrough times can be defined at the steep rise of the curves

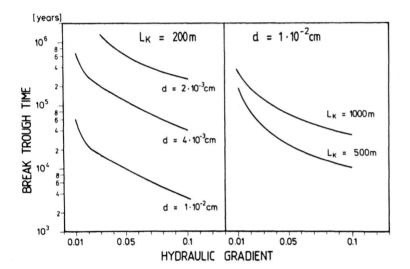

Fig. 9.17. Breakthrough times as can be read from breakthrough curves as in Fig. 9.16 as a function of the hydraulic gradient. *Left:* The length of the aperture is 200 m, curves are for different initial apertures d as designated. *Right:* Initial aperture is 1×10^{-2} cm, curves are for fracture lengths of 500 and 1000 m respectively (see text). $\alpha = 2.5 \cdot 10^{-6}$ cm s^{-1}

Fig. 9.16 shows breakthrough curves in a logarithmic scale as a function of time for both limits of first-order rates. Breakthrough is achieved when the first-order penetration lengths are in the order of the fracture length. The times when this happens can be taken as characteristic times for the evolution of first channels. Figure 9.17 gives these times for various hydraulic gradients and initial fracture d for fractures of 200 m length. For apertures of 10^{-2} cm, for all hydraulic gradients given, first channels of sizes up to a few millimetres develop in the time scale of several 1000 to 10 000 years over distances of as much as 200 m. Initial fractures with apertures of less than 4×10^{-3} cm practically have no chance to develop first karst channels. Therefore, in the initial state only few fractures of sufficient aperture are selected. Those of them with the largest hydraulic gradient will be the sites where first, secondary permeability is created. After breakthrough of these channels is achieved, they develop with fairly uniform dissolution rates, corresponding to retreat of bedrock in the order of several 10^{-3} cm year^{-1}. Therefore, in a geologically very short time of 1000 years they achieve dimensions of about 1 cm and turbulent flow occurs. Upon onset of turbulent flow penetration lengths of first-order kinetics are in the order of 1 km and therefore dissolution proceeds by first-order kinetics with the kinetic constant α^t of turbulent flow (cf. Chap. 7). Retreat of confining rock is now in the order of several 10^{-2} cm year^{-1}. Thus, intermediate channels develop into mature cave passages of 1 m diameter in time scales of 10 000 years.

To summarize the consequences of our calculations we can state that cave evolution can be visualized at three stages. The initial phase results from dissolution close to saturation in the inhibited fourth-order kinetics region. Even though dissolution rates are small they are maintained over large distances of several 100 m. The slow increase in fracture aperture yields higher penetration lengths until first-order kinetics with much higher dissolution rates become operative at the channel entrance. This again enhances penetration length until it finally reaches the order of the input-output distance. The minimal channel dimension is then in the order of 10^{-1} cm. The time needed to complete this phase amounts to several 10 000 years. It should be stressed that higher order dissolution kinetics with low rates close to equilibrium are absolutely necessary for the initiation of karstification. First-order kinetics with the kinetic constant α, as obtained from experiments far from saturation, exhibit such low penetration lengths that initiation of karstification could never take place. We have calculated, by using first-order kinetics exclusively, the breakthrough time for fractures of 20-m length, 0.01-cm aperture and hydraulic gradients of 0.1. We have found that this time exceeds 10^8 years.

After breakthrough has occurred penetration lengths are in the order of several 100 m (cf. Fig. 9.12) and there is a short intermediate state until turbulence occurs. Then penetration distances are in the order of several kilometres and dissolution increases due to eddy diffusion. To develop to maturity with underground channels of several metres the system needs a time of several 10 000 years in agreement with field observations. Saturation lengths amount now up to 100 km even at the lowest hydraulic gradients ($J = 10^{-4}$) as can be seen from Fig. 9.13. This last value gives the correct scale in lateral dimensions of large karst systems.

9.5 Geomorphological Theories from the Viewpoint of the Mathematical Model

Conceptual mathematical models can provide insight into processes of karstification, when their results are interpreted in terms of models inferred from geomorphological observations. Two conditions have to be met. Geomorphological models have to be in agreement with the physical, chemical and mathematical principles of karstification. On the other hand, the results of mathematical models have to be concordant with the geomorphological observations, which provide the only firm basis from which models can be tested. It is therefore of interest to discuss the comprehensive, four-state model of D.C. Ford and Ewers (1978) (cf. Chap. 8) with respect to the results of the last section.

The first recognizable early flow paths are dip tubes as shown in Fig. 8.21. These are oriented down dip and should be the result of the initial phase of karstification. The mathematical model predicts initial channels in the direction of the steepest hydraulic gradient in those bedding partings, which are sufficiently wide. Thus, it explains the dip orientation of the tubes. Furthermore, the fact that only a few of many bedding partings are selected as sites of dip tube development can now be understood. The geomorphological term of a fracture penetrable by water can now be rendered precise as a fracture with an aperture in the order of 0.01 cm subject to an hydraulic gradient in the order of 10^{-2} to 10^{-1}. Hydraulic gradients of this magnitude should be common in the first stage of karstification, when the water table is close to ground and thus driving head is equal to the relief from the position of the water table to the lowest outputs possible. D.C. Ford (1980), from field experience, tentatively gave values of minimum hydraulic gradients required to initiate karst water circulation. For bathyphreatic caves in state 1 and multi-loop phreatic caves these values are approximately 3×10^{-2}.

If primany fissure frequency in a rock mass is high, there may be many fissures which are selected to develop early channels and early flow is spread throughout the rock mass. Therefore, in this situation enterable caves may not develop and thus also typical karst landforms are missing. The threshold in massiveness of beddings desirable to develop mature cave systems has been estimated by D.C. Ford (1980) to be at least 40 cm between each parting plane penetrable by groundwater.

From the results of breakthrough times in Fig. 9.17 one can see that a minimum time is needed for the first initial stage of karstification. This time is in the order of 10 000 years in agreement with observations (Mylroie and Carew 1986, Bakalowicz 1982, White 1979, Droppa 1966). It implies, however, that fractures of sufficiently wide initial apertures are present ($d_o \approx 0.01$ cm). If this is not the case longer times will be necessary with an upper limit of 10^6 years. This upper limit is given by assuming that fissures with apertures of less than 20 μ are not penetrable at all (Bocker 1969).

One important feature of the four-state model is the increase of frequency in penetrable fractures as karstification proceeds. This can also be explained from the mathematical model. With establishment of the first solutional "caves" with diameters of several millimetres in the intermediate state, i.e. after break-through of the first tubes, hydraulic gradients change considerably. This has two effects:

1.   According to Ewers' (1982) network-linking model new favourable hydraulic gradients arise along fissures which have not yet been selected. New dip tubes will develop in these fractures and will be integrated into the cave system.

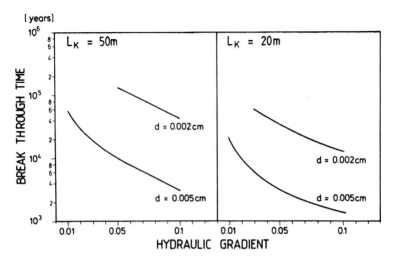

Fig. 9.18. Breakthrough times as a function of the hydraulic gradient for short fractures of 50- and 20-m length respectively and initial apertures of 2×10^{-3} and 5×10^{-3} cm (see text). $\alpha = 2.5 \cdot 10^{-6}$ cm s^{-1}

2. Due to the variety of boundary conditions, which change permanently upon breakthrough of initial channels, there may be possible input and output points along short fractures connecting different parts of passages already excavated.

Figure 9.18 shows breakthrough times for short fractures of 20 and 50 m with small initial apertures depending on the hydraulic gradients. As can be seen from these data widening of even narrow fissures of 0.005 to 0.002 cm proceeds in time scales of 10 000–100 000 years. Thus, additional porosity is created by widening of many small fissures. Since all these features integrate into one common network, the rock is rendered more and more permeable and a well-defined water table as in a porous material is finally established. In this state-4 situation dissolution will proceed primarily in its shallow region and an ideal water-table cave will be the result.

There has been much debate on the role of mixing corrosion on the initiation of karstification. Mixing corrosion was first introduced by Bögli (1963, 1964) to explain initial enlargement of small fractures. He found it necessary to draw onto such a mechanism, since at that time it was assumed, according to evidence given from Weyl's (1958) work, that karst water becomes entirely saturated after travelling short distances in narrow fissures.

Thrailkill (1968) has shown that the undersaturation, which on average results from mixing of two representative saturated solutions, is sufficient for the excavation of caves.

Dreybrodt (1981b, 1988) has suggested that mixing of water occurring at intersections of fractures might create first, secondary porosity along these lines of intersections, which will initiate karst development.

White (1984) has doubted the exclusive role of mixing corrosion, since there are many examples of covered karst where such a mechanism does not seem to be likely.

Instead, he pointed out that the discrepancy between the dissolution rates predicted by the PWP equation and those observed experimentally indicates a region of greatly inhibited dissolution rates.

In view of our considerations above, this regime of inhibition is sufficient for karst initiation. Mixing corrosion therefore must no longer be regarded as a necessity for karst initiation. Caves can well be initiated without this mechanism. Still, it may play an important supporting role in the development of first, secondary permeability. From the data of Plummer and Wigley (1976) one finds that dissolution with fourth-order kinetics dominates at concentrations above 90% saturation. This is also the region of undersaturation resulting from mixing corrosion. Thus, mixing corrosion might enhance solutional processes along water divides, where mixing of the different groundwaters occurs by dispersion. It can also be effective where solutional power in the inhibited regime has been exhausted at distances far from the input, which exceed the lengths along which inhibited dissolution can proceed.

9.6   Experimental Models

To simulate the development of first, secondary permeability by the growth of initial tubes in laboratory experiments, Ewers and Quinlan (1981) and Ewers (1982) used blocks of paris plaster as soluble material. By creating a parting of known aperture between the soluble plaster block and a transparent insoluble lower boundary, a laboratory model of a bedding plane parting was modelled. The surface of the plaster block was prepared by casting it from a sheet of abrasive paper (100 grit) to obtain replicable roughness.

This corresponds to an average initial aperture of 2×10^{-2} cm. Point inputs and point outlets for water were provided for various boundary conditions.

One example is shown in Fig. 9.19 right side. Input is at one point in the middle of the boundary to the right. There is one output site at the left boundary. At the start of the experiment an hydraulic gradient of 0.1 is applied to the liquid at the input. Small solutional tubes begin developing in the direction of streamlines, thus changing the distribution of the pressure field.

As the solutional channels approach the output point, the hydraulic gradients in the not yet enlarged parts of the fracture become steeper. Finally, the first tube breaks through the boundary, drawing most of the liquid from the other parts of the fracture. This behaviour is in accordance to the principles derived in Section 9.1.

The experiment can also be used to visualize more complex input-output geometries, such as those of the high dip model, low dip model and restricted input model. Ewers has performed such experiments and has found them to be in agreement with the models derived in Section 9.2.

As an example Fig. 9.19, left side, gives the experimental results for an input geometry corresponding to the high dip model. There are four input points with equal hydraulic head and one output point.

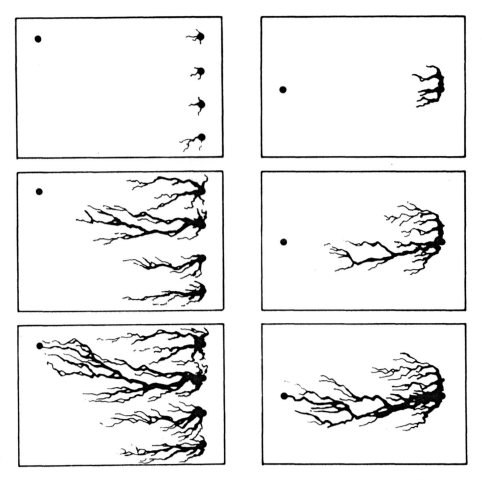

Fig. 9.19. Two different input-output geometries in the experiment of Ewers (1981). *Right*: Here, there is one single input point and one output point. As time passes channels gradually develop which finally reach the output point. *Left*: A more complex configuration with a linear input array and one output point (after Ewers 1982)

After 15 h primary tubes develop along the steepest hyraulic gradient quite evenly. They eventually branch into smaller secondary tubes. After 60 h one of the primary channels has gained advantage over the others and finally reaches the boundary as the first tube. The output during this process is shown by Fig. 9.20. There is a slow increase of output initially. As the first tube approaches the output boundary discharge increases rapidly and rises steeply at breakthrough. Figure 9.20 also shows a theoretical breakthrough curve as discussed below.

Although Ewers' experiments prove the principles of the influence of pressure fields on the development of early channels, it has shortcomings:

1. The dissolution process of paris plaster is entirely by diffusion and is therefore different from the chemical kinetics governing calcite dissolution.

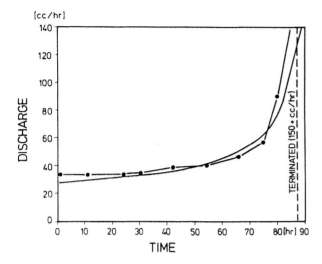

DISCHARGE

TIME

TERMINATED (150 ≐ cc/hr)

Fig. 9.20. Measured discharge through the output in Ewer's experiment (see Fig. 9.19, left) compared to the theoretical curve calculated by assuming that dissolution is controlled by diffusion (see text). The theoretical curve was scaled with one fitting parameter to the experimental data

2. The diffusion-limited saturation length $L_p^{(d)}$ in Ewers' experiments is in the order of 10 cm (coefficient of diffusion D of Ca^{2+} is 0.78×10^{-5} cm² s⁻¹). This is in the same order as the input-output distances of maximal 28 cm and does not apply to karst systems where input-output distances are by several orders of magnitude larger than penetration distances.

Nevertheless, Ewers' results can be used to test some principles of the calculations presented above (Sect. 9.4). The flux of Ca^{2+} for diffusion-limited dissolution is given by applying Eq. (3.23) to Ewers' experimental geometry:

$$F = c_{eq} \cdot \frac{2D}{a} \cdot \exp\left(-\frac{\pi^2 D \cdot t}{4a^2}\right) = \frac{2D}{a} c_{eq} \cdot \exp\left(-\frac{x \cdot 2.31}{L_p^{(d)}}\right), \qquad (9.20)$$

where a is the aperture of the experimental fracture (a = 0.02 cm) and $x = v \cdot t$.

Note that the flux depends on the aperture a of the fracture, in contrast to dissolution of calcite for the range of apertures used in the experiment. Therefore, the plaster model is not similar chemically to dissolution of calcite. The penetration length $L_p^{(d)}$ is given by:

$$L_p^{(d)} = 0.93 \cdot \frac{va^2}{D} . \qquad (9.20a)$$

Using Eq. (9.20) in combination with Eqs. (9.18) and (9.19), we can compute the discharge as a function of time. The results are given in Fig. 9.20 as the full line and are in remarkably good agreement to the experimental data, represented as dots.

9.7 Karst Processes on the Surface

Although formation of karst caves, serving as potent conduits for runoff, is most specular, it represents only an almost insignificant site of karst erosion. Only a few percent ranging from 0.05% to 0.5% of the total amount of limestone removed by solution from the landscape result from water-filled conduits (Gunn 1986, Atkinson and Smith 1976, Smith and Atkinson 1976). 31%, however, is removed in the diffused aquifer consisting of small fissures in the bedrock and about 67% is dissolved in the upper surface zone of about 10-m depth, i.e. in the soil and the subcutaneous zone. Accordingly, this upper weathered layer of rock beneath the soil is a highly permeable zone, which overlies a region of low permeability, since fissures tend to close with increasing depth (Williams 1983). Thus, percolating water is not transmitted directly to the phreatic zone. It can be stored in a perched aquifer from which lateral flow occurs at sites of high vertical permeability in the underlying rock, as presented by widened master joints. These eventually develop into solution dolines or shafts leading subsurface runoff directly into caves. In this section we will discuss the formation of dolines from the viewpoint of dissolution kinetics and also the rates of karst denudation.

9.7.1   The Formation of Dolines and Shafts

Karstic closed depressions are mainly solutional in origin (Williams 1983). Figure 9.21 shows a schematic sequence of the solutional development of dolines. In the initial state some major joints reaching down to the phreatic zone of an already karstified rock mass may be employed to transmit vertically some of the surface and subsurface water. Hydraulic gradients in this situation will be high, i.e. up to about 0.5. With increasing time the joints will be opened by dissolution. A small surface depression will result and as a consequence this funnel will attract more water. This again increases the area of depression. From this model it is plausible that dolines will be situated above cave conduits or at sites where high vertical permeability and high hydraulic gradients are available. Many joints in these zones can be enlarged, at the same time contributing to the considerable sizes of the dolines, which eventually collapse.

   Thus, dolines play a considerable role in the formation of karst landscapes (Jennings 1985, Sweeting 1972). In this context we want to discuss the time scale of doline development from the basis of the solutional enlargement of fissures. We assume the distance to the phreatic zone to be in the order of 100 m. Hydraulic

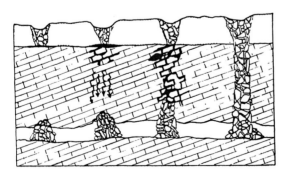

Fig. 9.21. Formation of a doline located above a vadose section of a cave (Jennings 1985)

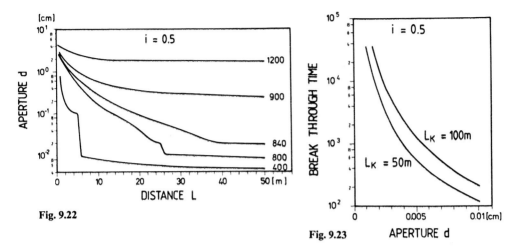

Fig. 9.22

Fig. 9.23   APERTURE d

Fig. 9.22. Enlargement of a fracture with conditions favourable to forming dolines at different times after onset of solution. Initial fracture width is $4 \cdot 10^{-2}$ cm. Due to the geological situation, as schematized in Fig. 9.21, the hydraulic gradient is high (0.5) and the length is in the order of several 10 m

Fig. 9.23. Breakthrough times for short fractures (50, 100 m) subjected to high hydraulic gradients as a function of the initial aperture of the fracture. Time scale extends from 10^2 up to 10^4 years, thus giving the range of time to form a doline $\alpha = 2.5 \cdot 10^{-6}$ cm s^{-1}

gradients are high, i.e. approximately 0.5. Joint widths employed vary from 0.002 to 0.01 cm.

The development of a karst channel vertically downward over a total length of 50 m, as calculated by our model, is shown in Fig. 9.22. Here, we have assumed an initial joint opening of 0.004 cm and an hydraulic gradient of 0.5. The figure shows a logarithmic plot of the joint widening as a function of depth for various times. After 840 years breakthrough is achieved with a minimal channel dimension of 0.2 cm. After this time solutional widening is by first-order kinetics and proceeds quickly. To visualize the time scale for the origin of solution dolines Fig. 9.23 shows breakthrough times for fracture lengths of 50 and 100 m, subjected to an hydraulic gradient of 0.5, as a function of the initial fracture aperture. Due to the large hydraulic gradient, even narrow fissures can be utilized in the establishment of vertical permeability in times of no more than 10 000 years. Fissures of 0.005 cm are enlarged in extremely short times of a few 100 years.

Further results, which are applicable to doline formation at a lower hydraulic gradient, are given in Fig. 9.18. For high hydraulic gradients, $i \geq 0.5$, and short fracture lengths, $L_k \leq 20$ m, breakthrough times of several 10 years result from our calculations. Recently Gunn and Gagen (1987) have observed formation of sinkhole like features, which have developed in abandoned limestone quarries within these length and time scales. This gives confidence into the concepts of our model.

Where vadose flow of water descends through solutionally enlarged joints, shafts will develop as an important pathway of vertical transmission of ground-water. Shafts show vertical walls and cut through the bedding planes. Their walls show sharp encarvements, similar to karrens, due to the thin layer of water streaming down the shaft. A detailed description of the morphology and role in groundwater hydrology of vertical shafts has been given by Bruckner et al. (1972).

One important feature of many shafts is their location at the edges of noncarbonate caprocks. As a consequence vadose waters entering this type of shaft are highly aggressive. Their chemistry and flow properties have been investigated by Bruckner et al. (1972). The Ca^{2+} concentration of vertical shaft waters was found to range from 2.5×10^{-4} mol l^{-1} up to 1.5×10^{-3} mol l^{-1} with a saturation index, $\log \Omega$, ranging from -0.5 to -3.0. The carbon dioxide pressures with which such waters would be in equilibrium are in the order of up to about 5×10^{-3} atm. Water flow down the vertical wall is laminar with the relation between velocity and thickness δ of the water sheet given by (Bird et al. 1960):

$$v = \frac{g\rho\delta^2}{3\eta} .$$
(9.21)

Measurement have shown values of δ ranging from $\delta = 0.12$ cm down to $\delta = 0.05$ cm.

These data enable us to estimate the lateral enlargement of shafts. Since there is no appreciable change in the hardness of waters when descending to the bottom, the $H_2O-CO_2-CaCO_3$ system can be assumed under open system conditions to be in equilibrium to CO_2 with a partial pressure of 5×10^{-3} atm. Figure 7.5 gives the dissolution rates for these situations. For the thickness of the laminar flowing water sheets between $\delta = 0.1$ down to 0.05 cm dissolution rates are independent of δ, as can be seen from Fig. 7.5. At $10°C$ the average dissolution rates are 1×10^{-8} mol cm^{-2} s^{-1}. This corresponds to a widening of the shaft diameter of 2×10^{-2} cm $year^{-1}$. Thus, a shaft with 2-m diameter may be formed within 10 000 years.

9.7.2 Karst Denudation

Karst denudation rates are defined as the annual removal of limestone rock from a carbonate area and are measured in m^3 km^{-2} $year^{-1}$ of the limestone area. This value corresponds to an average lowering of the area by 1 mm in thousand years (1 mm/ka). Both units are commonly used in the literature.

A most commonly used expression, which relates annual precipitation P, evapotranspiration E and the mean hardness H of an average of at least 50 samples of runoff water, to the denudation rate DR is a modification of Corbel's (1959) equation:

$$DR = \frac{(P - E)H}{1000 \cdot \rho} \cdot \frac{1}{f} ,$$
(9.22)

where P and E are in mm $year^{-1}$, H in mg $CaCO_3$ per litre and ρ is the density of limestone; f is the fraction of the limestone area in the total catchment. This expression implies that no allogenic surface water enters the catchment and thus

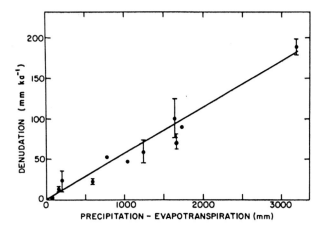

runoff is only derived from meteoric water. Figure 9.24 gives some reported denudation rates as a function of $(P - E)$ as summarized by White (1984). The data points can be fitted by a linear relation $DR = 0.049 \, (P - E) + 6.3$ with a correlation coefficient $r = 0.92$.

This linear relation can be justified also by theoretical considerations. Denudation rates can be derived by measuring the total amount of $CaCO_3$ removed from a catchment area A. If runoff $Q(t)$ is known as a function of time, one obtains:

$$DR = \int_0^t Q(t)H(t)dt \cdot \frac{1}{1000A \cdot f \cdot \rho \cdot t} , \qquad (9.23)$$

In Canadian karst Drake and D.C. Ford (1981) have found an inverse relation between Q and H given by:

$$H(t) = aQ(t)^{-1} + b . \qquad (9.24)$$

From this one obtains:

$$DR = \frac{1}{1000A \cdot f \cdot \rho}(a + b\bar{Q}) , \qquad (9.25)$$

with $\bar{Q}$ as the average annual runoff. $\bar{Q}$ is given by mass balance as:

$$\bar{Q} = P - E - \Delta S , \qquad (9.26)$$

where S is the amount of groundwater stored in the aquifer and ΔS its change. If observation time t is sufficiently long, ΔS will average to zero and $\bar{Q}$ averaged over a sufficiently long time will be given by $P - E$. This, however, implies that recharge of the catchment is entirely authigenic.

A maximum denudation rate can now be calculated, if one assumes $f = 1$ and that runoff is saturated with respect to calcite (White 1984). The equilibrium concentration of Ca^{2+}_{eq} is given by Eq. (2.35c) in Chapter 2. Using Eq. (9.22) one obtains with $H = c_{eq}$ converting mol l^{-1} in Eq. (2.35c) into mg l^{-1}:

$$DR = \left(\frac{K_1 K_c K_H}{4K_2 \gamma_{Ca} \gamma_{HCO_3}^2}\right)^{1/3} P_{CO_2}^{1/3}(P - E) \cdot \frac{100}{\rho}.$$ (9.27)

Equation (9.27) contains no adjustable parameters and combines the three climatic variables temperature, precipitation and CO_2 pressure.

Figure 9.25 gives maximal denudation rates calculated from Eq. (9.27) for temperatures of 5°, 10° and 25°C and three regions of P_{CO_2}; $P_{CO_2} = 10^{-3.5}$ atm, representative for bare rock; $P_{CO_2} = 10^{-2.5}$ for normal soils usual in temperate climates; and $P_{CO_2} = 10^{-1.5}$ for tropical organic-rich soils in dependence on $(P - E)$.

It is easily seen from these results that the direct influence of temperature is small; most of the variation of the denudation rate stems from CO_2 pressure.

The average P_{CO_2} in soils depends on biological activity and therefore on climate and on temperature. A simple biologically reasonable model to estimate CO_2-pressure in the soil has been given by Drake and Wigley (1975) and Drake (1980) as:

$$\log P_{CO_2}^* = -2 + 0.04T ;$$ (9.28a)

$$P_{CO_2} = [(0.21 - P_{CO_2})/0.21] P_{CO_2}^* .$$ (9.28b)

$P_{CO_2}^*$ is a potential CO_2 pressure due to the biological activity in the soil dependent upon temperature T in °C. P_{CO_2} is the actual CO_2 pressure in the soil and the second equation expresses the inhibition of soil respiration resulting from decreasing concentration of oxygen as CO_2 pressure increases.

Drake (1983) and Drake and Ford (1981) have shown that by using this expression to calculate the equilibrium concentrations of calcite of groundwater in the closed and open systems gives good agreement to the average Ca^{2+}-concentrations observed in different climates.

The combination of Eqs. (9.28) and (9.27) can therefore be used to predict maximal denudation rates.

Figure 9.25 shows the theoretical curves calculated for various P_{CO_2} and temperatures and for comparison, gives also the observed data from Fig. 9.24 and also data from Atkinson and Smith (1976) which are separated into three sets related to arctic, temperate and tropical climate. There is general agreement showing that these considerations give the correct order of magnitude.

Recently Gunn (1986) has commented the model of Drake and Wigley (1975) and Drake (1980) as expressed by Eqs. (9.28). He argues that the variability of soil CO_2 pressure is so large, even in repeated measurements at one single site, that the definition of averages is not meaningful, since also little is known about the shape of the probability distributions.

On the other hand, Gunn recognizes that in the application of this model to the hardness of groundwater, P_{CO_2} in Eqs. (9.28) are inferred to be the same as P_{CO_2} of saturated water samples with the given hardness. The variability of the hardness of groundwaters in the arctic and alpine, the temperate and the tropical climates, however, shows an approximately Gaussian distribution where average values are well defined (Smith and Atkinson 1976). To reconcile this contradiction, Gunn suggests that groundwaters are in fact in equilibrium with ground-air CO_2. The

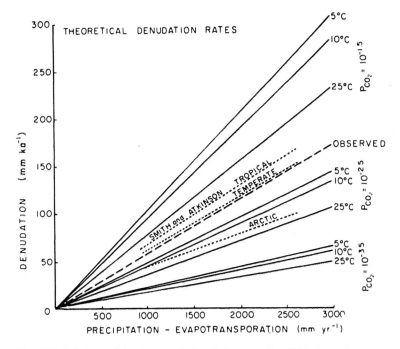

Fig. 9.25. Solutions of the theoretical denudation equation (9.27) for various temperatures and CO_2 pressures. Observed denudation rates from Fig. 9.24 and from Smith and Atkinson (1976) are also shown (White 1984)

concept of ground-air CO_2 was first introduced by Atkinson (1977) as the CO_2 content of the air in the unsaturated epikarstic zone, which results from microbial decay of organic matter in joints and fissures. At a depth of a few meters temperatures are almost constant and a constant supply of debris could well result in a more or less constant rate of CO_2-production. In this situation, Eqs. (9.28) might apply.

Thus Gunn concludes that Drake's model predicts P_{CO_2} of the ground-air, which shows a high degree of temperature dependence but only low seasonal fluctuations in agreement to Atkinson's (1977) predictions.

9.7.3 Denudation Rates on Bare Rock at the Surface and in Caves

The lower curves in Fig. 9.25 give an estimate for denudation rates of meteoric water on bare rocks. There are two methods to obtain such data. One method, which gives long-term averages, is the use of pedestals which have formed on a bare limestone surface beneath glacial erratic blocks. These rocks have protected the underlying limestone against removal by dissolution, thus forming pedestals the height of which is a measure of denudation since the time of ice retreat. Jennings (1985) summarized some of these data from the literature. Values of surface lowering range from 15 to

40 mm/ka. This is in the range predicted by Eq. (9.27) as shown by the lowest set of lines in Fig. 9.25.

Short-term measurements of denudation on bare rock surfaces, either in caves or on karst areas, can be performed by use of the micro-erosion meter (Trudgill et al. 1981). A micrometer probe is mounted to a rigid plate, which is seated at three points to steel studs imbedded into the rock. Measuring periodically the distance of the rock to the studs gives directly the rates of rock retreat.

Such measurements have been performed by Forti (1984) in the karst of Triest (Italy). In the Triestine karst average denudation rates are 20 mm/ka at a precipitation of 1442 mm year^{-1} and in the region of Monte Canin the values are about 30 mm/ka at an annual precipitation of 2800 mm. These values are lower than the corresponding theoretical values of 30 and 50 mm/ka in Fig. 9.25 respectively. This may be due to the fact that water does not attain equilibrium as it seeps quickly from the surface into the rock.

Jennings (1985) has reported additional data from the recent literature. Values range from 5 to 17 mm/ka on bare rocks. These data can also be compared with the values obtained from the dissolution rates discussed in Chapter 7. Bauer (1964) has observed that meteoric waters running down limestone surfaces remain far from equilibrium. Therefore, first-order kinetics of dissolution are present. Thus it is possible to estimate values from the model represented in Sect. 7.3. The dissolution rates for $P_{CO_2} = 3 \times 10^{-4}$ atm are given in Fig. 7.7 for open system conditions and laminar flow. Average dissolution rates for sheets of water with a depth in the order of several tenths of a millimetre are in the order of 5×10^{-9} mmol cm^{-2} s^{-1}. This corresponds to 5×10^{-3} cm year^{-1} if water runs down the surface continuously. Assuming rainfall to occur only during 20% of the time, an annual retreat of bedrock of about 10^{-3} cm year^{-1} or 10 mm/ka will result, which gives a correct order of magnitude compared to the data observed.

The intensity of corrosion in caves has been measured by various authors using different methods. One which has become more commonly applied is the use of limestone standard tablets. These are circular discs of limestone from a given area. They are dried and weighed. Then they are placed at a site of active calcite dissolution and the dissolution rates are determined by weight loss. Droppa (1986) has reported such measurements in cave streams of the Demänova Cave. In the stream entering the karst area corrosion intensity was between 107–145 mm/ka in winter, when water temperature was low at about 1°C. In summer much higher values were observed, up to 320 mm/ka. The Demanova spring showed corrosion intensities lower by about a factor of ten.

From this one may estimate an average denudation rate in the cave conduits in the order of 0.015 cm year^{-1}. The average amount of aggressive CO_2 of the water when entering the karst system was from 7.7 mg l^{-1} to 3.7 mg l^{-1}. In spring, due to consumption by calcite dissolution, this value dropped to 3.9–1.9 mg l^{-1}. This corresponds to CO_2 pressures in equilibrium with the solution of about 10^{-3} atm up to 4×10^{-3} atm.

Assuming flow to be fully turbulent, maximal dissolution rates can be estimated from Fig. 7.26 to be at about 10^{-7} mmol cm^{-2} s^{-1} for highly undersaturated water with about 5×10^{-4} mol l^{-1} Ca^{2+} concentration. This relates to a denudation rate

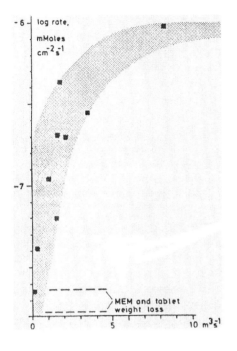

Fig. 9.26. Dissolution rates (*full squares*) in a phreatic cave conduit as a function of discharge (multiplied by a factor of 10, see text) The *dotted lines* show long term averages obtained by micro-erosion-meter (*MEM*) and weight loss of marble tablets for comparison (Lauritzen 1986)

of 0.1 cm year^{-1}. It is reasonable, however, to assume that only at very high discharges diffusion boundary layers are to be neglected. The minimal rate expected, if this is not the case, is about 0.02 cm year^{-1}. Thus, the calculations of dissolution rates as represented in Chapter 7 give reasonable results.

The influence of hydrodynamic flow conditions on the denudation rate in phreatic cave conduits was shown by Lauritzen (1986). By measuring calcium concentrations at the sink and the resurgence of a 580-m-long, isolated phreatic conduit, and also the discharge of the resurgence, the total amount of dissolved limestone was computed. The surface of the conduit was estimated from a cave survey. Figure 9.26 shows the dissolution rates as a function of discharge. Note that the data given by full squares have to be multiplied by a factor of 0.1 due to an underestimation of the rough surface of the conduit by a factor of ten (Lauritzen 1987, personal communication). There is an increase of dissolution rates up to a maximal value of 10^{-7} mmol cm^{-2} s^{-1} (0.1 cm year^{-1}) at high discharge. This is expected from the results of Chapter 7, since the diffusion boundary layer decreases with flow velocity.

Minimal dissolution rates at turbulent flow, according to Fig. 7.24, should be lower by a factor of seven compared to the maximal values. The data of Fig. 9.26 show a much larger decrease, but due to the large errors in the rate determination at low rates, a correct comparison is not possible. In comparison, micro-erosion-meter and standard-tablet data, which also have been measured over periods of 5 years are given. They give an average rate, where the rates at low discharge are favoured by a high statistical weight of about 2×10^{-2} cm year^{-1}. This puts the time

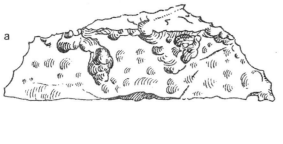

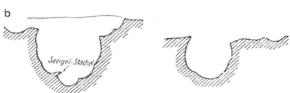

Fig. 9.27. a An originally triangular limestone block has been eroded by corrosion and shows solutional channels (see text). **b** Profile of solutional channels of maximum diameter 3 cm. The protruding spine of a sea urchin shows that the channels result from dissolution (Spöcker 1950)

scale for the development of the cave conduit to an average diameter of 6.5 m from the onset of turbulent flow at 1-cm diameter, into the region of 20 000 years.

Further denudation rates have been investigated by Gascoyne (1981) by determining the age of speleothems which are in a fixed position relative to local base levels of vadose cave streams. If, for instance, the basal age of such a speleothem is 100 000 years at an elevation of 2 m above base level of a stream in a cave passage, the averaged entrenchment rate is inferred to be 20 mm/ka. This is a lower limit since it is assumed that the stream has always taken the same route and that the speleothem has started to grow upon onset of entrenchment. For caves in a temperate climate, values of entrenchment rates between 20 to 80 mm/ka are reported. In a tropical climate they are higher, from 130 to 200 mm/ka.

Finally, an interesting observation of Spöcker (1950) should be given, since he provided also the chemical data of calcite-aggressive waters from which dissolution rates can be calculated and compared to observation. When repairing the 50-year-old casting from limestone blocks (Malm β) of a well, situated in sandstone near the town Lau in Germany, he found that the original, triangular limestone blocks had been heavily corroded. Figure 9.27 shows a drawing of such a block, which had been below the water table for 50 years. The block had lost 23% of its original weight and showed corrosion channels with a depth of 2.5 cm. One of these channels contained a protruding spine of a sea urchin. This shows that erosion was entirely by dissolution. The chemical data of the water are given as follows: pH $\approx$ 6, free $[H_2CO_3^*] \approx 7 \times 10^{-4}$ mol l^{-1} and $[Ca^{2+}] = 2.7 \times 10^{-4}$ mol l^{-1}. Discharge of the well is about 4 l s^{-1} and temperature 10°C. Assuming that dissolution has proceeded with fully turbulent flow, these data are sufficient to obtain dissolution rates from the equilibrium calculation of Chapter 2 and then by using the PWP equation (cf. Chap. 6, Eq. 6.12). The result thus derived is a removal rate of 1.5 mm year^{-1}. The observed rate amounts to 0.5 mm year^{-1}. In view of the fact that the PWP equation is accurate only within a factor of two, and that the diffusion boundary layer has been neglected, this has to be regarded as good agreement.

To summarize the results, we would like to state that sufficiently far from equilibrium calcite dissolution rates are well described by first-order kinetics with the kinetic constant α as derived in Chapter 7. Therefore, this model can be applied to natural processes and gives results which are sufficiently accurate for geological applications.

10 Precipitation of Calcite in Natural Environments

CO_2 flux into or out of water influences the saturation state with respect to calcite. The result is either dissolution or precipitation of calcite. Outgassing of CO_2 can lead to precipitation of calcite in surface streams, which are fed by karst springs, as observed by Jacobson and Usdowski (1975) and Michaelis et al. (1984). Other studies show that the states of the saturation index Ω up to ten can be maintained without precipitation (Suarez 1983, Herman and Lorak 1987). In this chapter we will discuss these processes with respect to precipitation kinetics of calcite and its dependence on the flow properties of the precipitating solution.

Most spectacular is the deposition of calcite in caves forming a variety of speleothems, such as stalagmites, stalactites, flowstone, draperies and rimstone dams. This deposition usually results from thin sheets of stagnant or flowing water entering the vadose part of the cave from joints and fissures with a high load of dissolved $CaCO_3$. Once these waters have entered into the vadose cave supersaturation is established quickly due to outgassing of CO_2 and deposition of calcite is effected from highly supersaturated solutions. One particularly simple type of speleothem, which can be used to determine deposition rates, is the regular stalagmite and we will focus mainly on this topic because of its importance in inferring paleo-climatic data from the analysis of its age and growth rates.

10.1 Speleothems: Growth and Morphology

Rainwater entering into a karst system at sites, where the ground is covered by vegetated soil, can quickly absorb large amounts of CO_2. Due to the fast reaction of first-order kinetics, it is capable of dissolving limestone and approaches towards equilibrium. This may occur already in the upper parts of the soil under open system conditions if percolation is sufficiently slow. Otherwise, the dissolution process can proceed in the small vertical joints under closed conditions. Finally, the calcareous solution emerges somewhere in a vadose part of the cave either as drops of water hanging on the cave roof for some time until they drip to the ground or as dripping or trickling water. When the dripping or trickling water impinges on the surface, it quickly (by outgassing of CO_2) attains equilibrium with the CO_2 pressure existing in the cave atmosphere (Dixon and Hands 1957). Therefore, the calcareous solution becomes supersaturated, if its initial calcium concentration, when entering the cave, is higher than the calcium concentration, which would be in equilibrium with the CO_2 pressure in the cave atmosphere.

This of course is only possible, if the P_{CO_2} in the soils considerably exceeds that in the cave. CO_2 contents in soil atmospheres vary by some orders of magnitude, ranging from 0.1% up to 10% (Trudgill 1985). In tropical soils high CO_2 concentrations of several percent are encountered. Temperate climate zones generally show CO_2 concentrations of several tenths of a percent. There are also large seasonal variations in CO_2 content since the source of soil CO_2 is of biological origin. This fact is used to draw information on the age of speleothems according to the radio carbon method (Franke 1951, Geyh 1970, Hendy 1969, Wigley 1975b) and also to relate variations of the isotopic ratios $^{14}C/^{12}C$ and $^{13}C/^{12}C$ to changes in palaeo-climate (Goede and Harmon 1983). CO_2 in the cave atmosphere ranges from 3×10^{-4} atm up to 3×10^{-3} atm (White 1976), depending on the ventilation in the cave.

Figure 10.1 summarizes the physical pathway of calcite-depositing solutions from the atmosphere to the cave, whereas Fig. 10.2 gives the chemical evolution of the water on the pathway of Fig. 10.1. In Fig. 10.2 the curve for aragonite solubility is also shown, which sometimes is also deposited in speleothems, especially at high supersaturation and in a warm climate.

10.2   Stalagmites

10.2.1   Morphology of Stalagmites

The variety of shapes which are sculptured as speleothems by dripping, trickling and flowing water are overwhelming and at a first glance one would reject the idea that even the most simple shapes could be described by mathematical models. At a second glance, however, one realizes that there are a few particular formations of quite regular form, which can be found in almost any cave. One of these is the regular stalagmite, which forms a column with a height of up to a few metres and of almost constant diameter. Diameters of these stalagmites range from about 3 cm up to several 10 cm. These regularly shaped stalagmites show one general principle of growth, which was first described by Franke (1965) as follows: "When a drop of solution reaches the flat ground it spreads out radially. Precipitation is greatest at the centre since the solution has lost some of it calcareous excess. In this way a thin layer of calcium carbonate is precipitated in the shape of a disc, thick at the centre and thinning out at the edges. This is enhanced by the next and following drops, so that a series of dome-shaped layered hoods build up one above the other." Finally, a stalagmite shows a stable form on its top, which does not change with time, provided that all parameters determining its growth remain constant. This growth principle can be easily envisaged by cutting a regular stalagmite into two halves along the growth axis. In many cases growth lines, representing former shapes of the stalagmite, can be seen which are schematically shown in Fig. 10.4a for a stalagmite growing on flat ground. The lines have been calculated by a computer simulation, which will be discussed below. There is one important fact on this type of growth: After the equilibrium shape has been attained *all* parts of the stalagmite grow with

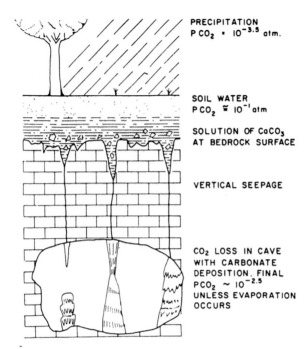

PRECIPITATION
P CO$_2$ $\cdot$ 10$^{-3.5}$ atm.

SOIL WATER
P CO$_2$ $\cong$ 10^{-1} atm

SOLUTION OF CaCO$_3$
AT BEDROCK SURFACE

VERTICAL SEEPAGE

CO$_2$ LOSS IN CAVE
WITH CARBONATE
DEPOSITION. FINAL
PCO$_2$ $\sim$ 10$^{-2.5}$
UNLESS EVAPORATION
OCCURS

Fig. 10.1. Model of calcite deposition in caves. For discussion, see text (White 1976)

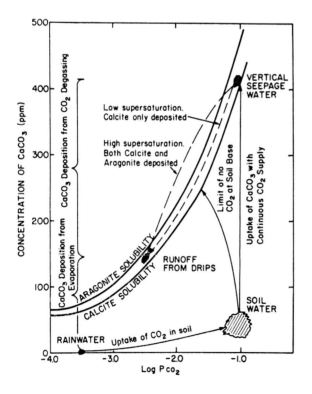

Fig. 10.2. Changes in the chemical composition of water (rain) entering into the soil where CO_2 uptake is effected. During vertical seepage in the rock $CaCO_3$ is dissolved under open or closed system conditions. Supersaturation in the cave develops by outgassing of CO_2 and further by evaporation of water (White 1976)

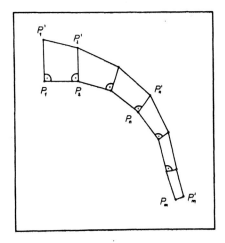

Fig. 10.3. Geometrical construction of the growth of stalagmites (Dreybrodt and Lamprecht 1981)

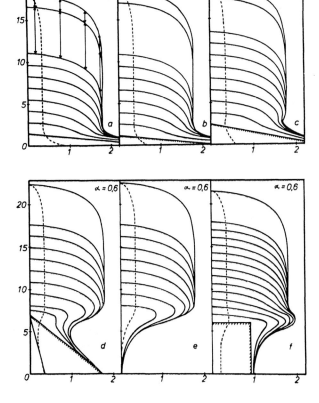

Fig. 10.4a–f. Growth of a regular stalagmite calculated by computer runs. **a** Growth on an initially flat surface; **b, c, d** growth on inclined surfaces with increasing inclination. **e** Growth at a cave wall; **f** growth at the edge of a rock. Note that in all cases an identical final shape is achieved and that Franke's principle of growth is well reproduced; $\alpha = 1/\lambda$ (Dreybrodt and Lamprecht 1981)

an *equal* vertical growth rate. This makes these speleothems ideally suited to determine growth rates, since it is sufficient to know the growth rate at the top of the stalagmite to provide interpretable data.

Dreybrodt and Lamprecht (1981) have visualized stalagmite growth by computer simulation and have found Franke's principle of growth verified by computer experiment. The principles of the calculations are shown in Fig. 10.3 as a geometrical construction.

It is assumed that the fluid forming the stalagmite spreads out radially and that calcite deposition rates decrease monotonically with increasing radius. Furthermore, growth is always perpendicular to the surface already existing.

The surface at a given time t represented by the polygon P_1 to P_m thus can be used to construct the surface at a later time $t + \Delta t$, represented by polygon P_1' to P_m' in the following way:

1. From each point P_n a vertical is drawn.
2. One assumes P_1 to be the point where dripping water arrives at the speleothem. The length of the flow path of the calcite-depositing solution is then the distance $P_1 P_n$ along the polygon $P_1 P_m$.
3. Assuming that the deposition rate is a monotonically decreasing function $R(P_1 P_n)$ each new point P_n' of the surface at $t + \Delta t$, is found by plotting the distance $R(P_1 P_n) \cdot \Delta t$ along the vertical lines from P_n to find P_n'.

In the computer calculations we have used:

$$R = R_0 \exp\left(-\frac{P_1 P_n}{\lambda}\right), \tag{10.1}$$

where R_0 was chosen as 0.1 cm/ΔT and ΔT is the time necessary for growth of the stalagmite by 0.1 cm. λ is a constant.

Figure 10.4a–f shows results of these calculations. Note that the axes are scaled in centimetres and that the abscissa is exaggerated by a factor of five. The dotted lines show the real shape at the end of the computer run. The distances between the lower lines correspond to $15 \cdot \Delta T$ in Fig. 10.4a–e, and $10 \cdot \Delta T$ in Fig. 10.4f.

Figure 10.4a shows the growth on an initially flat surface. An equilibrium shape is quickly achieved after the stalagmite has grown to a height of about twice its diameter. Figure 10.4b–d shows growth on inclined surfaces. With increasing angle of inclination there is an incision at the base of the stalagmite. As the speleothem grows, however, the same equilibrium shape is reached as in Fig. 10.4a. This shows that the final shape does not depend on the initial surface from which the stalagmite grows. This is also shown in the extreme case of Fig. 10.4e, where growth is at a perpendicular cave wall. Finally, Fig. 10.4f shows the case of growth at an edge of rock. All the shapes which are shown in Fig. 10.4 can be observed in caves, thus providing confidence to the simulation model.

The radius of the stalagmites in these calculations turns out to be equal to λ. If the amount of solution dripping to the stalagmite increases, one expects the velocity of radial spread to increase, which leads to an increase in λ and therefore in radius. This is shown by Fig. 10.5 where two stalagmites with $\lambda = 3.33$ cm and $\lambda = 2.5$ cm on an inclined surface are shown.

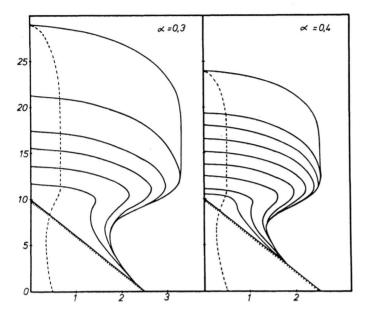

Fig. 10.5. Stalagmite growth on an inclined surface with a different supply of solution, $\alpha = 1/\lambda$ (Dreybrodt and Lamprecht 1981)

Therefore, a change in water supply will increase or decrease the radii of stalagmites as shown in Fig. 10.6a,b. In Fig. 10.6a λ is changed during the growth of a stalagmite from 2.5 cm to 1.66 cm, simulating a decrease in water supply. This leads to a decrease in stalagmite radius. Similarly, in Fig. 10.6b the radius increases with increasing water supply, giving a clublike shape. Conelike shapes of stalagmites, as they are frequently observed in caves, therefore can be interpreted to result from continuously decreasing water supply to the speleothem.

Sometimes one observes stalagmites which are extremely twisted columns, although the diameter remains fairly constant. These are found especially on clay grounds where the speleothem is not firmly anchored and may be tilted. Thus, after tilting of an already grown specimen, the drop impinges to a new point and the shape is changed as shown in Fig. 10.6. Here, the axis has been tilted by 10°, which produces a twisted, transitional shape.

To summarize, stalagmites under stable conditions of growth finally grow to an equilibrium shape independent of the initial surface onto which growth started. This equilibrium shape grows upward by a lateral translation. There are only two conditions for this:

1. Growth rates decrease with increasing distance from the drip point as a monotonic function, reaching the value zero at some finite distance l_0 from the centre.
2. Growth is directed perpendicular to the surface at any point, which implies that growth rates are defined in this direction.

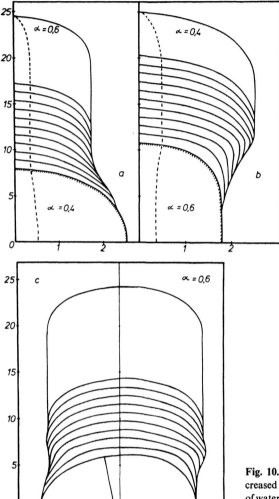

Fig. 10.6a–c. Growth of a stalagmite **a** upon increased supply of water; **b** with a decreasing supply of water; **c** after inclination of the initial growth axis (Dreybrodt and Lamprecht 1981)

Using these two principles, a strict mathematical proof can be given on the existence of an equilibrium shape and its upward shift by lateral translation.

Figure 10.7 shows a stalagmite surface $y(x, t)$ at time t. The distance of point P from the apex along the curve $y(x, t)$ is $l(x, t)$ which is a function of time t and distance x from the axis. In a first step one shows that $l(x, t)$ becomes independent of time, i.e. $l(x, t)$ is a bounded function in t and approaches a limiting value $l(x)$ for large times. We do so by assuming the contrary. Then $l(x, t)$ would be a function which is unbounded in t. In any case $l(x, t)$ must be a function increasing monotonically in t for each value of x. Therefore:

$$l(x, t) \le l(x, t + \Delta t) \tag{10.2}$$

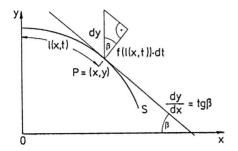

Fig. 10.7. Representation of the relation of growth rate dy of a stalagmite to the growth rate $f\,dt$ perpendicular to its surface

for any value of t and $\Delta t > 0$. This statement follows from the condition that growth is perpendicular to the surface and growth rate decreases with increasing distance x from the axis. If now $l(x, t)$ were an unbounded function in t, one can define an arbitrary value $l_c > l_0$, such that after the time t_c the relation:

$$l(x, t_c) \geq l_c \quad \text{for} \quad t > t_c \tag{10.3}$$

holds for each value of x.

Now the growth rate $f(l)$ is a monotonically decreasing function of l with a lower limit of zero (there can be no negative growth rate), which is attained for $l \geq l_0$. Therefore, from Eq. (10.3) a time t_0 exists such that:

$$l(x, t_0) \geq l_0 \quad \text{for} \quad t \geq t_0 \tag{10.3a}$$

for all values of x.

Since $f(l) = 0$ for $l \geq l_0$, this means that growth of a stalagmite would stop after time t_0.

This clearly is a contradiction since the stalagmite does grow for infinite times as long as a supersaturated solution is supplied. Therefore, the assumption that $l(x, t)$ is unbounded in t must be wrong, which means:

$$l(x, t) \leq l(x) . \tag{10.4}$$

Thus (at least, asymptotically in time) $l(x, t)$ approaches a time-independent value $l(x)$. One can now assume that this is accomplished with sufficient accuracy after time t_0. The computer simulations show that this is reasonable. Therefore, we have to a sufficient degree of accuracy:

$$l(x, t) = l(x) \quad \text{for} \quad t \geq t_0 . \tag{10.4a}$$

From this one can now deduce a differential equation for the shape $y(x, t)$.

The vertical growth rate can be calculated from Fig. 10.7 as:

$$dy(x, t) = \frac{f[l(x, t)]}{\cos \beta} \cdot dt ; \qquad \frac{\partial y(x, t)}{\partial x} = \tan \beta . \tag{10.5}$$

From this by using a trigonometric formula to convert $\cos \beta$ to $\tan \beta$, one obtains:

$$\frac{\partial y(x, t)}{\partial t} = f[l(x, t)] \cdot \sqrt{1 + \left[\frac{\partial y(x, t)}{\partial x}\right]^2} . \tag{10.6}$$

The length $l(x, t)$ is given by standard formula as:

$$l(x, t) = \int_0^x \sqrt{1 + \left[\frac{\partial y(x, t)}{\partial x} \right]^2} \, dx \, . \tag{10.7}$$

Equations (10.6) and (10.7) can be used to calculate the stalagmite shape for any time provided the function $f(l)$ is known. It is this function which contains the physics and chemistry of the process.

In the next step one can now prove that the derivative in Eq. (10.7) depends only on x and not on t after a sufficiently long time t_0. This is trivial since we have already shown that for this time $l(x)$ becomes independent of t, Eq. (10.4). Therefore, also the derivative with respect to x in Eq. (10.7) becomes independent of t. Thus, we have from Eq. (10.6):

$$\frac{\partial y(x, t)}{\partial t} = f[l(x)] \sqrt{1 + \left[\frac{\partial y(x)}{\partial x} \right]^2} = v \neq v(x) \, . \tag{10.8}$$

The solution $y(x, t)$ of this equation represents the equilibrium shape of the stalagmite and has the property that:

$$y(x, t) = y(x) + v \cdot t \, . \tag{10.9}$$

If v were a function of x, $l(x, t)$ would be unbounded in t as can be seen by inserting into Eq. (10.7), in contrast to Eq. (10.4). Thus v is constant. This, however, means that the stalagmite grows vertically, where v is the growth velocity, without changing its shape, which is given by $y(x)$.

10.2.2 Growth Rates of Stalagmites and Related Diameters

When a drop falls onto the top of a stalagmite, it is spread radially and forms a thin layer of thickness δ at the apex. From this thin layer calcite is deposited under open system conditions. According to Eq. (7.38) deposition rates are given by:

$$F = \alpha([Ca^{2+}]_{eq} - [Ca^{2+}]) = -\alpha \cdot c \, . \tag{7.38}$$

They are converted to growth rates in cm year^{-1} by:

$$\frac{dy}{dt} = v = 1.174 \times 10^6 \cdot F \, ,$$

where F is in mmol cm^{-2} s^{-1}.

To calculate growth rates of the stalagmite it is sufficient to know δ the thickness of the water layer at the top of the speleothem and to calculate the rates at this point. This was first done in a general way by Curl (1973). To define the problem the following parameters are used:

Drops of volume V drip to the stalagmite with time t' elapsing between two subsequent drops. The saturation excess of these drops is c. Now if a drop falls into the liquid layer, part of this layer or all of it will be removed and is replaced by the liquid of the new drop. Thus, mixing occurs and the concentration in the newly formed layer is:

$$c' = (1 - \Phi)c + \Phi c_0, \tag{10.10}$$

where Φ is a mixing coefficient and c is the concentration of the layer after time t'. If $\Phi = 1$, the new drop replaces the old layer completely. For $\Phi = 0$ the new drop does not mix at all with the layer and flows away without changing its concentration. This means that there is no supply of dissolved calcium carbonate and consequently no growth. Thus, $\Phi = 0$ has to be excluded from our considerations.

The change of excess saturation in time is now given by:

$$\delta(dc/dt) = -\alpha c . \tag{10.11}$$

The solution of this equation is:

$$c(t) = c' \exp(-t/T_d), \qquad T_d = \delta/\alpha \tag{10.12a}$$

with

$$c' = (1 - \Phi)c' \exp(-t'/T_d) + \Phi c_0 . \tag{10.12b}$$

The time history of the supersaturation excess given by Eq. (10.12) is illustrated in Fig. 10.8. In the uppermost drawing the decay time $T_d \ll t'$. Therefore, the super-saturation excess approaches zero before a new drop is supplied. This is called the low flow regime. In the high flow regime the time t' between two drops is short $t' \leq T_d$ and the excess remains high close to c_0 as shown in the lowest part of the figure. The intermediate region is shown in the middle section of Fig. 10.8.

The deposition rate at any moment is calculated by use of Eqs. (10.11) and (10.12):

$$\delta \frac{dc}{dt} = \frac{\alpha \Phi c_0 \exp(-t/T_d)}{[1 - (1 - \Phi)\exp(-t'/T_d)]} . \tag{10.13}$$

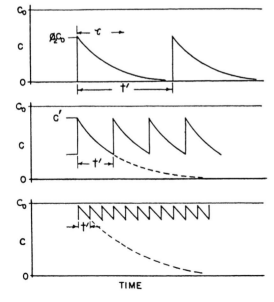

Fig. 10.8. Time histories of the super-saturation of the solution at the apex of a stalagmite for different drip times t'. Top: $t' \gg T_d$; low flow regime; middle: $t' \approx T_d$; intermediate region; bottom: $t' \ll T_d$; high flow region (Curl 1973)

From this by averaging over time period t' the average growth rate is found:

$$\frac{d\langle c\rangle}{dt} = \frac{\delta\Phi c_0\,[1 - \exp(-\,t'/T_d)]}{t'[1 - (1 - \Phi)\exp(-\,t'/T_d)]}\,,\tag{10.14}$$

where $d\langle c\rangle/dt$ is the average deposition rate in mmol cm^{-2} s^{-1}.

In the equilibrium state of stalagmite growth by definition the solution flowing away from the speleothem has lost its supersaturation completely. Therefore, the average amount of calcite is deposited on its area A is:

$$\frac{Vc_0}{t'} = \frac{d\langle c\rangle}{dt}\cdot A = \frac{d\langle c\rangle}{dt}\,\pi\frac{d_s^2}{4}\tag{10.15}$$

where d_s is the diameter of the stalagmite. From this one finds a relation between the diameter d_s and the flow rate of solution to the stalagmite, given by the parameters V and t':

$$d_s^2 = \frac{V\cdot[1 - (1 - \Phi)\exp(-\,t'/T_d)]\cdot 4}{\delta\Phi[1 - \exp(-\,t'/T_d)]\cdot\pi}.\tag{10.16}$$

In the low flow region, where $t' \gg T_d$, the diameter d_s becomes a minimum:

$$d_s^{min} = \sqrt{\frac{4V}{\delta\Phi\pi}}.\tag{10.17}$$

In the high flow region $t' \ll T_d$, and one obtains by expansion of the exponential:

$$d_s = \sqrt{\frac{4V}{\pi\alpha t'}}.\tag{10.18}$$

In the low flow region the diameter is given by δ and the volume V of the drop. The kinetics of deposition are of no concern since there is sufficient time for the solution to lose its supersaturation. The value of δ can be determined experimentally by placing a piece of water-absorbing paper of known surface area on various parts of the stalagmite and measuring the weight difference between the dry and wet paper (Dreybrodt 1980). An average value of $\delta = 0.01$ cm was obtained. The volume of a drop dripping from a stalactite is estimated by Curl (1972) to be at least 0.075 cm^3. Here, we use a somewhat higher value of $V = 0.1$ cm^3. From this the minimum diameter of the stalagmite is 3.6 cm. This is in good agreement with observations in nature.

In the high flow regime the diameter is determined by the kinetic constant α and by the flow rate V/t' of the water supply. Using the results shown in Figs. 7.9 and 7.10, Dreybrodt and Franke (1987) have calculated growth rates and diameters of regular stalagmites by using Eqs. (10.14) and (10.16) with $\Phi = 1$. The results are shown in Fig. 10.9. The growth rate W is in cm year^{-1} and depends heavily on temperature in the high flow regime, i.e. $t' = T < 2000$ s. For very high drip rates, $T < 100$ s the concentration in the layer stays practically constant (cf. Fig. 10.8) as also the growth rate. The diameters increase with flow rate and for $T < 10$ s they are given by Eq. (10.18).

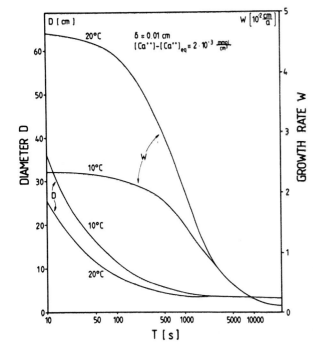

Fig. 10.9. Growth rate W and diameter D of a stalagmite at $10°$ and $20°C$ as a function of time $T = t'$ between two drops of solution (Dreybrodt and Franke 1987)

It should be noted here that even in the case of high flow we assume the water layer to be in laminar flow. This is reasonable for Reynolds numbers up to 1000 for flow of films on the vertical walls of the stalagmite (Kay and Nedderman 1985). On a stalagmite with 10-cm diameter a Reynolds number of 1000 would require a water supply of 100 cm³ s⁻¹ corresponding to a drip rate of 1000 drips s⁻¹. Thus, regular stalagmites always grow with laminar flow conditions.

Maximum deposition rates on a stalagmite are achieved for $\Phi = 1$ and t' $\ll$ T$_d$. Thus, the maximal growth rate in cm year⁻¹ is obtained from Eq. (10.14) by:

$$v = -1.174 \times 10^6 \, \alpha c_0 = -\alpha \cdot 1.174 \times 10^6 ([Ca^{2+}] - [Ca^{2+}]_{eq}). \qquad (10.19)$$

The values of the kinetic constants α and $[Ca^{2+}]_{eq}$ can be taken from the results in Chapter 7 (Figs. 7.9 and 7.10). Figure 7.9 shows the deposition rates for various thickness δ of the water layer and for $10°$ and $20°C$. P_{CO_2} of the solution is 3×10^{-4} atm, i.e. atmospheric CO_2 pressure. Note that the influence of δ on the kinetic constant α is insignificant, justifying the above assumption of α as a constant. There is a large temperature dependence however. Figure 7.10 shows precipitation rates for $10°C$ and $\delta = 0.01$ cm for various external CO_2 pressures as they are observed in caves. Note that α is also independent of external P_{CO_2}. The deposition rates, however, become lower due to decreasing supersaturation. Average calcium concentrations in drip water is about 2×10^{-3} mol l⁻¹ with large variations up to 4×10^{-3} mol l⁻¹ (Moore 1962, Sweeting 1972, Gams 1968).

Maximal deposition rates on stalagmites can now be read from Figs. 7.9 and 7.10. At $10°C$ the maximal rate is about 0.05 cm year⁻¹, which increases to about

0.1 cm year^{-1} at 20°C in warmer climates. Maximal growth rates of up to 0.09 cm year^{-1} have been found by ^{14}C dating of stalagmites for samples in Belgium (Gewelt 1986). In France growth rates of up to 0.06 cm year^{-1} were observed by Labeyrie et al. (1967) and Duplessy et al. (1972). In Germany Franke and Geyh (1970) obtained rates of up to 0.05 cm year^{-1}. In general, however, most of the rates observed are lower by a factor of about ten.

Harmon et al. (1975) observed rates by using 230Th/234U methods. The samples with ages between 25 0000–250 000 years showed average growth rates 0.004 cm year^{-1} (Mexico) down to 0.0004 cm year^{-1} (Kentucky).

Further measurements have been made by direct observation (Bögli 1980, Moore and Sullivan 1978). In a few cases very high rates of up to 0.5 cm year^{-1} have been ovserved, which possibly are due to some special conditions and do not seem to be representative. A possible explanation may be that the growth of these specimens proceeds under turbulent flow, which would explain the highly increased rates (cf. Figs. 7.11, 7.29). Considering the many parameters, such as drip rates, calcium concentration, temperature and mixing coefficient Φ, such large variations are reasonable.

According to Fig. 10.9 there is a correlation between the growth rate and the diameter of the stalagmite. Stalagmites with diameters exceeding 30 cm are characterized by high drip rates and accordingly show maximal growth rates. This has been found by Hohmann (1979). The growth rates found by ^{14}C dating of a huge stalagmite of 4.90-m height and a diameter of 1.30 m at its base ranged from 0.065–0.071 cm year^{-1}. This agrees well to that predicted by Fig. 10.9. In the same cave a neighbouring smaller stalagmite of 20-cm diameter showed growth rates of 0.02 cm year^{-1}.

10.3   Calcite Precipitation in Surface Streams

There are only a few studies on calcite precipitation in surface streams. The first comprehensive study was published by Jacobson and Usdowski (1975) who observed considerable calcite precipitation on the Westerhof Bach, a small stream in West Germany, 30 km north of Göttingen. This small stream of 265-m length has no tributaries. Therefore, no changes in its chemistry result from mixing of different waters. This is an important condition when studying precipitation processes in natural streams. The spring is located in the Triassic upper Muschelkalk, which is predominantly limestone. It is underlain by Middle Muschelkalk, which is dolomite with gypsum lenses. The water emerging from the spring appears to circulate through both strata. Figure 10.10 shows the profile of the stream. Water samples have been taken at stations 1 to 9 and observations of the stream water chemistry have been conducted since 1974 up to now. Characteristic profiles of temperature, pH, Ca concentration, concentration of bicarbonate, the P_{CO_2} pressure in equilibrium with the solution and the saturation index $\log S_c = \log \Omega$ are also given in Fig. 10.10.

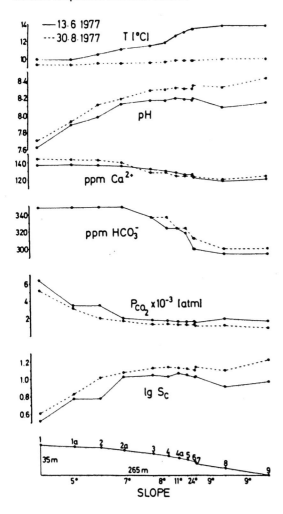

Fig. 10.10. Variations of temperature, pH, [Ca^{2+}], [HCO$_3^-$], p$_{CO_2}$ and log S$_c$ along the profile of the Westerhof stream. *Below*: The elevation profile of the stream and the location of stations *1* to *9*, where the water samples were taken for analysis (Usdowski et al. 1979)

There is relatively fast degassing of CO_2 from the stream water, the P_{CO_2} of which drops from 6×10^{-3} atm to about 2×10^{-3} atm from station 1 to 2a. Correspondingly, pH and log S_c increase. Ca^{2+} and HCO_3^- concentrations, however, remain constant, indicating that no precipitation of calcite occurs. Once log $S_c > 1$ is achieved the Ca^{2+} and HCO_3^- concentrations drop and heavy calcite deposition proceeds. It is especially high between stations 6 and 7. A small waterfall of about 2 m height is situated at this location and flow is supercritically turbulent. This enhances deposition of calcite.

Similar observations have been made by Herman and Lorak (1987). They also found CO_2 outgassing and accordingly an increase in pH and log S_c in a stream in the United States. The first visible precipitation of calcite occurs at the vertical cliff of a 21-m-deep waterfall, which is covered by travertine. At this point log S_c has reached a value of 1.20. This high value is maintained along the further course of the stream, where a continuous decrease in the Ca^{2+} and HCO_3^- concentration is found. Thus, as a general rule it appears that precipitation in surface streams is only possible if $\log S_c > 1$. Suarez (1983) and Troester and White (1986) reported on surface rivers with $0.6 \leq \log S_c \leq 0.8$ (Colorado River) and $\log S_c < 0.9$ (three tropical streams in Puerto Rico). Nowhere in these rivers was there any detectable precipitation of calcite.

Michaelis et al. (1984) found that calcite deposition in the Güterstein River, located in the Schwäbische Alb, Germany, occurred for values above $\log S_c = 0.8$. Thus, the threshold for calcite deposition may vary, but should be close to about 1. There may be, however, other reasons preventing calcite precipitation, even at high saturation levels. The Urach Waterfalls, located in the vicinity of the Güterstein River, are known to have deposited calcite over long periods, as can be seen by their tuffa formations. At present, no deposition is observed, although $1.2 \leq \log S_c \leq 13$. As a reason for this behaviour Michaelis et al. reported on concentrations of phosphate ions of 0.3 ppm, which act as inhibitors (cf. Chap. 7). In all the cases discussed above, $CaCO_3$ precipitation is of anorganic origin without assistance of biological processes, which in many cases are of overwhelming importance.

This is supported by experiments on a calcite-precipitating stream which were performed by Dreybrodt et al. at the Westerhof Bach. They measured calcite precipitation rates by immersing limestone samples of a rectangular flat plate shape into the stream. The precipitation rate was calculated from the known surface and the increase in weight after a few days. The limestone samples had been taken from the Westerhof region. Additionally, white Carrara marble samples were used. At the same time, water samples were taken at the location where the limestone tablets were situated. They were analyzed for Ca^{2+}, Mg^{2+}, pH, alkalinity, Na^+, K^+, Cl^- and SO_4^{2-} concentrations. From these parameters P_{CO_2}, HCO_3^- and CO_3^{2-} were calculated. These data are sufficient for the calculation of calcite deposition rates, provided the thickness ε of the diffusion boundary layer is known (cf. Chap. 7), which influences precipitation rates significantly (see also Fig. 7.28).

To estimate the thickness ε the following procedure was adopted. Samples of monocrystalline gypsum with a similar shape as the limestone samples were immersed close to the limestone samples. Dissolution rates of these samples were measured according to weight loss. Dissolution of $CaSO_4$ is known to be limited by diffusion. Therefore, dissolution rates can be described by:

$$F_d = \frac{C_{eq} \cdot D}{\varepsilon},$$
(10.20)

where F_d is the dissolution rate in $g\ cm^{-2}\ s^{-1}$ and C_{eq} is the equilibrium concentration of gypsum (1.93 g l^{-1} at 10°C). Using this method values of ε in the range from 0.01 cm up to 0.03 cm were found and used for the calculation of precipitation rates.

Figure 10.11 shows rates observed at several times, for stations 1 to 9, including long-term runs, where the samples stayed immersed for 143 days.

The full lines represent precipitation rates calculated from the chemical data at the start of the experiment (open crosses). The calculations have been run for several values of ε as denoted on the curves. This was done since the determination of ε

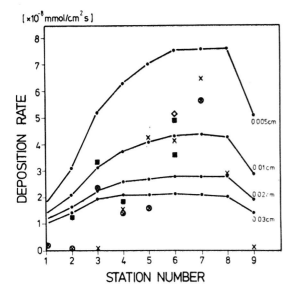

Fig. 10.11. Deposition rates measured at various stations in the Westerhof stream. Different *symbols* represent the period of time the samples were immersed in the stream: ■ 134 days, ◇ 11 days, ⊗ 14 days, × 3 days. The *full lines* are linear interpolations between the points which represent the deposition rates calculated from the chemical composition of the water at the corresponding station. *Numbers* on these lines designate the thickness ε of the diffusion boundary layer used in the calculation

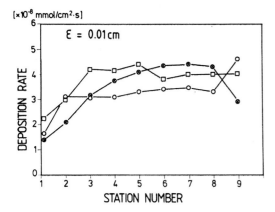

Fig. 10.12. Deposition rates calculated with $\varepsilon = 0.01$ cm for various chemical compositions observed in the stream at the corresponding station for different times of observation during the period from 1977 to 1986. Note that the calculated values are in the range of those observed (Fig. 10.11)

implies only a short time span and there might be variations. At least the lower and upper curves give conservative lower and upper limits. It turns out that the precipitation rates are well within these boundaries. Figure 10.12 shows rates calculated for various chemical compositions measured at different times. The boundary layer thickness here has been assumed as $\varepsilon = 0.01$ cm. The figure shows that variations in deposition rates due to different chemical compositions amount to a scattering of about $\pm 20\%$. Thus, the largest variations are due to changes of hydrodynamic flow conditions. The theory predicts precipitation rates of $1-2 \times 10^{-8}$ mmol cm^{-2}s^{-1} for station 1 and 2, where $\log S_c \leq 0.8$. These are not observed as the system is in a region of inhibited slow kinetics, where $\log S_c < 1$.

Summarizing these experiments shows that calcite precipitation occurs at solutions sufficiently far from equilibrium and follows first-order kinetics as described in Chapter 7.

In a recent investigation, Schulz (1986) has reported on calcite depositions in Roman aqueducts to Cologne, Germany. By statistical analysis of the number of growth layers visible on a given sample he derived growth rates of about 0.1 cm year^{-1}. This agrees well to the rates observed in the Westerhof stream and also to those predicted by Fig. 7.28.

11 Conclusion and Future Perspective

In this book we have presented only one view of karst systems by discussing the physics and chemistry of dissolution of limestone and its consequences for the development of karst. One important outcome of this procedure is the recognition of important external and internal parameters, which determine karst processes. Internal parameters are, for example, the dimensions of karst channels or fissures, their connectivity, and hydraulic gradients which determine the type of flow, i.e. laminar or turbulent. As has been shown in Chapter 7, the type of flow is of utmost influence to the dissolution rates of limestone and thus to the further development of the karst system.

External parameters are dependent on the climatic regime of the region, which in turn determines the amount of water infiltrated and its content of carbon dioxide, which is regulated by the biological activities in the soil and by the vegetation. These parameters also have important influence to the dissolution rates of limestone.

Internal and external factors are related to each other. High biological productivity of carbon dioxide can enhance karstification which consequently has an impact to the vegetation and thus to CO_2 production. Karst-ecosystems are therefore complex and comprise many feed-back loops.

Physical and chemical processes are only a few of the links of the mutual interrelations in this multi-loop system. The factors by which these processes are determined can result from multivariate effects.

Biological processes on karstification can be indirect. In this case, the biological activity produces a chemical composition of the solute entering into the fissures, which then acts according to physical and chemical laws, as have been described in this book.

There is, however, also a direct attack of microorganisms on limestone, which is by far more complex and not easy to describe (Golubic and Schneider, 1979).

The roots of arboraceous and other plants etch tortuous and branching channels into the rock, thus changing the permeability of the upper karst zone. Jakucs (1977) reported on such channels reaching down to 25 m into the rock and constituting "something like a large-scale skeletal sponge" in the upper karst zone. Especially under tropic conditions, these processes are extremely fast and a few years suffice to change a hairline crack into a channel with a diameter of a few tens of cm.

These examples show that, although understanding the physical and chemical principles of abiotic calcite dissolution can contribute valuable information and conclusions upon many karst processes, it may not necessarily explain many other features of more complex origin. Nevertheless, to disentangle the many complex mechanisms acting in karst systems, it is of importance that each scientific discipline first describes corresponding different mechanisms in its own intradisciplinary approach. Only with as much knowledge as possible of this type can an interdisciplinary approach be taken to contribute to a much deeper understanding of this synergetic multi-loop feedback system. This is the context within which the results discussed in this book should be seen.

As has been discussed in Chapter 9.4, two differing kinetic regimes in calcite dissolution are important in the process of karstification. Once karst channels have developed to dimensions of a few cm, CO_2 containing water can travel distances in the order of a few km without reaching saturation. Therefore first-order kinetics determine dissolution of limestone.

In the initial state of karst, however, karst water is close to saturation and slow high-order kinetics control dissolution. As we have shown, this dissolution regime is responsible for the onset of karstification. Little is known, however, of the detailed mechanisms of this type of dissolution. Future research has to focus on this problem, which is difficult to treat because of the long-time experiments involved. Furthermore, the kinetics in this region is expected to depend very strongly on the chemical composition of the dissolving rocks. Magnesian calcites may show different behaviour compared to pure limestone. Thus there is a wide field of research open to clarify these questions.

Once more details are known about these slow kinetics, triggering the first water pathways, it may be possible to work on realistic models, by which time scales for the creation of first water pathways can be estimated sufficiently accurately to be of importance in the construction of dams in karst areas and in the utilization of karst water.

In 1980 the workshop on Karst Water Research Needs has discussed future research topics to the understanding of karst (Yevjevich, 1981). One of the topics in karst development is "dynamics of carbonate solution and precipitation within underground karst water systems". The expected output was described as follows:

"The most important results will be then the development of methods of predicting, where and how intense the chemical solution will be, and where and under what conditions the chemical precipitation of bicarbonates will take place undergound, with the corresponding changes in transport and water storage characteristics. In fact, any new knowledge on dynamics and regularities in complex karst undergound chemistry will produce better tools for explaining features and the time evolution of karst water systems".

I hope that this book makes a first step into this direction.

References

Atkinson TC (1968) The earliest stages of underground drainage in limestone: a speculative discussion. Proc Br Speleol Assoc 6:53–70

Atkinson TC (1977) Diffuse flow and conduit flow in limestone terrain in the Mendip Hills. J Hydrol 35:93–110

Atkinson TC (1977a) Carbon dioxide in the atmosphere of the unsaturated zone: an important control of groundwater hardness in limestones. J Hydrol 35:111–123

Atkinson TC, Smith DJ (1976) The erosion of limestones. In: Ford TD, Cullingford CHD (eds) The science of speleology. Academic Press, London New York pp 151–177

Back W, La Moreaux PE (ed) (1983) V.T. Stringfield Symp: Proc Karst Hydrol. J Hydrol 43:287–312

Bakalowicz M (1982) La genèse de l'aquifer karstique vue par un geochimiste. Reun Monogr Sobre Karst-Larra 82:159–174

Bauer (1964) Kalkabtragungsmessungen in den österreichischen Kalkhochalpen. Erdkunde 18:95–102

Baumann J (1986) Dissolution kinetics of calcite in porous media. Rep Geol Paleon Inst Univ Kiel (FRG) 15

Bauman J, Buhmann D, Dreybrodt W, Schulz HD (1985) Calcite dissolution kinetics in porous media. Chem Geol 53:219–228

Bayer HJ (1983) Beispiele zur tektonischen Höhlenanlage aus der Schwäbischen Ostalb (Württemberg). Höhle 34:93–99

Bear J (1972) Dynamics of fluids in porous media. Am Elsevier, New York.

Bear J (1979) Hydraulics of groundwater. McGraw Hill, New York, p 55

Bedinger MS (1966) Electric-analog study of cave formation. Natl Speleol Soc Bull 28:127–132

Beek WJ, Mutzall KMK (1975) Transport phenomena. Wiley, New York

Berner RA, Morse JW (1974) Dissolution kinetics of calcium carbonate in sea water IV: Theory of calcite dissolution. Am J Science 274:108–134.

Berner RA, Wilde P (1972) Dissolution kinetics of calcium carbonate in sea water I: Saturation state parameters for kinetic calculations. Am J Science 272:826–839.

Bircumshaw LL, Riddiford AC (1952) Transport control in heterogeneous reactions. Q Rev Chem Soc 6:157–185

Bird RB, Stewart WE, Lightfoot EM (1960) Transport phenomena. Wiley, New York

Blackwell RI, Rayne IR, Terry WM (1952) Factors influencing the efficiency of miscible dislacements. J AIME 217:1–8

Bocker (1969) Karstic water research in Hungary. IAHS Bull 14:61–70

Bögli A (1963) Ein Beitrag zur Entstehung von Karsthöhlen. Höhle 14:63–68

Bögli A (1964) Mischungskorrosion: Ein Beitrag zum Verkarstungsproblem. Erdkunde 18:83–92

Bögli A (1969) Neue Anschauungen über die Rolle von Schichtfugen und Klüften in der karsthydrographischen Entwicklung. Geogr Rundsch 59:395–408

Bögli A (1980) Karst hydrology and physical speleology. Springer, Berlin Heidelberg New York

Bonacci O (1987) Karst hydrology. Springer series in physical environments 2. Springer, Berlin Heidelberg New York

Bretz JH (1942) Vadose and phreatic features of limestone caves. J Geol 50:675–811

Bruckner RW, Hess JW, White WB (1972) Role of vertical shafts in the movement of groundwater in carbonate aquifers. Groundwater 10 (6):9.

Buhmann D (1984) Ein neues kinetisches Modell zur Berechnung der Lösung and Abscheidung von Kalk bei Verkarstungsprozessen und seine experimentelle Bestätigung. PHD Thes, Univ Bremen (FRG)

Buhmann D, Dreybrodt W (1985a) The kinetics of calcite dissolution and precipitation in geologically relevant situations of karst areas, 1. Open system, Chem Geol 48:189–211

Buhmann D, Dreybrodt W (1985b) The kinetics of calcite dissolution and precipitation in geologically relevant situations of karst areas, 2. Closed system, Chem Geol 53:109–124

Buhmann D, Dreybrodt W (1987) Calcite dissolution kinetics in the system $H_2O-CO_2-CaCO_3$ with participation of foreign ions, Chem Geol 64:89–102

Burger A, Pasquier F (1984) Prospection a captage d'eau par forages dans la vallée de la Brevine (Jura Suisse) In: Burger A, Durbertret L (eds) Hydrology of karstic terrains vol 1 Unesco pp 145–149

Burton WK, Cabrera N, Frank FC (1951) The growth of crystals and the equilibrium structure of their surfaces, R Soc London Philos Trans A 243:299–358

Capellos C, Biekski BHJ (1972) Kinetic systems, Wiley, New York

Carslaw HS, Jaeger IC (1959) Conduction of heat in solids, Clarendon, Oxford

Compton RG, Daly PJ (1984) The dissolution kinetics of iceland spar single crystals, J Colloid Interface Sci 101:159–166

Corbel J (1959) Erosion en terrain calcaire, Ann Geogr 68:97–120

Cornet L, Lewis WN, Kapesser R (1969) The effect of surface roughness on mass transfer to a rotating disc, Trans Inst Chem Eng 47:T222–226

Crank I (1975) The mathemathics of diffusion, Clarendon, Oxford

Curl RL (1972) Minimum diameter stalactites, Natl Speleol Soc Bull 34:129–136

Curl RL (1973) Minimum diameter stalagmites, Natl Speleol Soc Bull 35:1–9

Cvijic J (1924/26) Geomorfologija I e II, Beograd, Yugoslavia

Dahmke A, Matthess G, Pekdeger A, Schenk D, Schulz HD (1986) Near surface geochemical processes in quaternary sediments. J Geol Soc, London 143:667–672

Darcy H (1856) Les fontaines publiques de la ville de Dijon. Dalmont, Paris

Davies CW, Jones AL (1955) The precipitation of silver chloride from aqueous solutions Part 2: Kinetics of growth of seed crystals. Trans Faraday Soc 51:812–817

Davis SN (1930) Origin of limestone caverns. Geol Soc of Am Bull 41:475–628

Davis SN (1968) Initiation of groundwater flow in jointed limestone. Natl Speleol Soc Bull 28:111–117

Degens ET, Kempe S, Spitzy A (1984) Carbon dioxide: A biogeochemical portrait. In: The handbook of environmental chemistry Vo. 1, part C, Springer Berlin pp 127–215

Deike RG (1969) Relations of jointing to orientation of solution cavities in limestones of central Pennsylvania. Am J Sci 267:1230–1248

Dixon BE, Hands GC (1957) Desorption and absorption of gases by drops during impact. J Appl Chem 7:342

Drake JJ (1980) The effect of soil activity on the chemistry of carbonate groundwater. Water Resour Res 16:381–386

Drake JJ (1983) The effect of geomorphology and seasonality on the chemistry of carbonate groundwater. J Hydrol 61:223–236

Drake JJ, Ford DC (1981) Karst solution: a global model for groundwater solute concentrations. Trans Jpn Geomorph Un 2:223–230

Drake JJ, Wigley TML (1975) The effect of climate on the chemistry of carbonate groundwater. Water Resour Res 11:958–962

Dreybrodt W (1980) Deposition of calcite from thin films of calcareous solutions and the growth of speleothems. Chem Geol 29:89–105

Dreybrodt W (1981a) Mixing corrosion in $CaCO_3-CO_2-H_2O$ systems and its role in karstification of limestone areas. Chem Geol 32:221–236

Dreybrodt W (1981b) Kinetics of dissolution of calcite and its application to karstification. Chem Geol 31:245–269

Dreybrodt W (1988) The kinetics of calcite dissolution and its consequences to karst evolution from the initial to the mature state. Natl Speleol Soc Bull (in press)

Dreybrodt W, Buhmann D (1988) Dissolution and precipitation rates of calcite influenced by boundary layers in turbulent flow. Submitted to Chem Geol

Dreybrodt W, Franke HW (1987) Wachstumsgeschwindigkeiten und Durchmesser von Kerzenstalagmiten. Höhle 38:1–6

Dreybrodt W, Lamprecht G (1981) Computer-Simulation des Wachstums von Stalagmiten. Höhle 31:11–21

Droppa A (1966) The correlation of some horizontal caves with river terraces. Stud Speleol 1:186–192

Droppa A (1986) L'influence des saisons de l'année sur la dénudation karstique. Proc 9th Int Congr Speleol 1:237–241

Duplessy IC, Lalon C, Delibrias G, Nguyen HV (1972) Datations et études isotopiques de stalagmites. Application aux paleo temperature. Ann Speleol 27:445–464

Dwerryhouse AR (1907) Limestone caverns and potholes and their mode of origin. J Yorks Rambler Club 2:223–228

Erga O, Terjesen SG (1956) Kinetics of the heterogeneous reaction of calcium bicarbonate formation, with special reference to copper ion inhibition. Acta Chem Scand 10:872–875

Ewers RO (1978) A model for the development of broad-scale networks of groundwater flow in steeply dipping carbonate aquifers. Trans Cave Res Group Great Britain 5:121–125

Ewers RO (1982) Cavern development in the dimensions of length and breadth. Ph D Diss, McMaster Univ, Hamilton, Ontario

Ewers RO, Quinlan JF (1981) Cavern porosity development in limestone: a low dip model from Mammoth Cave, Ky. Proc 8th Int Speleol Congr, Bowling Green, USA, pp 721–731

Ford DC (1968) Features of cavern development in Central Mendip. Trans Cave Res Group Britain 10:11–25

Ford DC (1971) Geologic structure and a new explanation of limestone cavern genesis. Trans Cave Res Group Great Britain 13:81–94

Ford DC (1980) Threshold and limit effects in karst hydrology. In: Coates DR, Vitek ID (ed) Threshold in geomorphology. Allen and Unwin, London Boston Sydney pp 345–362

Ford DC, Drake JJ (1980) Spatial and temporal variations in karst solution rates: the structure of variability. In: Thorn CD (ed) Space and time in geomorphology. Allen and Unwin, London Boston Sydney pp 147–170

Ford DC, Ewers RO (1978) The development of limestone caves in the dimensions of length and depth. Can J Earth Sci 15:1783–1798

Ford TD (1971) Structures in limestone affecting the initiation of caves. Trans Cave Res Group Great Britain 13:65–71

Ford TD (1976) The geology of caves. In: Ford TD, Cullingford CHD (eds) The science of speleology. Academic Press, London New York pp 11–60

Forti F (1984) Messungen des Karstabtrags in der Region Friaul-Julisch-Venetien (Italien). Höhle 35:135–139

Franke HW (1951) Alterbestimmungen an Kalzit-Konkretionen mit radioaktivem Kohlenstoff. Naturwissenschaften 38:527–528

Franke HW (1965) The theory behind stalagmite shapes. Stud Speleol 1:89–95

Franke HW, Geyh MA (1970) Zur Wachstumsgeschwindigkeit von Stalagmiten. Atompraxis 16:46–48

Gams J (1968) Über die Faktoren, die die Intensität der Sintersedimentation bestimmen. Proc 4th Int Cong Speleol in Yugoslavia 1965, 3:107–116. Ljubljana 1968

Garrels MP, Christ CL (1965) Solutions, minerals, and equilibria. Harper and Row, New York

Gascoyne M (1981) Rates of cave passage entrenchment and valley lowering from speleothem measurement. In: Beck BF (ed) Proc 8th Int Cong Speleol 1:99–100

Gewelt M (1986) Datations 14C de concrétions de grottes Belges: vitesses de croisance durant l'Holocène et implications paléochimatiques. In: Paterson K, Sweeting MM (ed). New directions in Karst. Geo Books, Norwich, England pp 293–321

Geyh MA (1970) Zeitliche Abgrenzung von Klimaänderungen mit 14C Daten von Kalksinter und organischen Substazen. Betr Geol Jahrb 98:15–22

Goede A, Harmon RS (1983) Radiometric dating of Tasmanian speleothems—evidence of cave evolution and climatic change. J Geol Soc Aust 30:89–100

Golubić S, Schneider I (1979) Carbonate dissolution. In: Trudinger PA, Swaine DJ (eds) Biogeochemical cycling of mineral-forming elements. Elsevier, Amsterdam pp 107–129

Gregory DP, Riddiford AC (1956) Transport to the surface of a rotating disc. J Chem Soc 3:3756–3764

Gunn J (1986) Solute processes and Karst landforms. In: Trudgill ST (ed) Solute processes. Wiley, Chichester New York pp 363–437

Gunn J., Gagen P (1987) Limestone quarrying and sinkhole development in the English Peak District. In: Beck BF, Wilson WL (eds) Karst hydrogeology: Engineering and environmental applications A.A. Balkema, Rotterdam, Boston.

Hanshaw BH, Back W (1979) Major geochemical processes in the evolution of carbonate-aquifer systems. J Hydrol 61:1–355

Harmon RS, Thomson P, Schwarcz HP, Ford DC (1975) Uranium-series dating of speleothems. Natl Speleol Soc Bull 37:21–33

Harned HS, Hamer WJ (1933) The ionization constant of water. J Am Chem Soc 55:693–695

Harris JF, Taylor GL, Walper JL (1960) Relation of deformational fractures in sedimentary rocks to regional and local structures. Bull Am Assoc Petrol Geol 44:1853–

Hendy CH (1969) The use of C14 in the study of cave processes. In: Olssen IU (ed) Radiocarbon variations and absolute chronology, Proc 12th Nobel Symp Uppsala, Wiley, New York

Herman JS (1982) The dissolution kinetics of calcite, dolomite and dolomite rocks in carbon dioxide water system. Ph D Thes Pennsylvania State University

Herman JS, Lorak MM (1987) CO_2 outgassing and calcite precipitation in Falling Spring Creek, USA. Chem Geol 62:251–262

Hohmann W (1979) Zum Wachstum kolozäner Großstalagmiten in der Knitterthöhle bei Lethmate/Sauerland — und zur Methodik der Sinterproben-Entnahmen durch Kernbohrungen. Dortmunder Beitr Landeskd 13:45–63

House WA (1981) Kinetics of crystallization of calcite from calcium bicarbonate solutions. J Chem Soc Faraday Trans 1, 77:341–359

Hydraulic Research Station (1969) Charts for the hydraulic design of channels and pipes. 3rd edn, H.M.S.O., London

Inskeep WS, Bloom PR (1985) An evaluation of rate equations for calcite precipitation at p less than 0.01 atm and pH greater than 8. Geochim Cosmochim Acta 49:2165–2180

Jacobson RL, Usdowski E (1975) Geochemical controls on a calcite precipitating spring. Contrib Mineral Petrol 51:65–74

Jakucs L (1977) Morphogenetics of karst regions. Hilger, Bristol

Jameson RA (1985) Structural segments and the analysis of flow paths in the North Canyon of Snedegar Cave, Friars Hole Cave System, W.V. Thesis, MS, West Virginia Univ, Morgantown

Jennings JN (1978) Limestone solution on bare karst and covered karst compared. Trans Brit Cave Res Assoc 5:215–220

Jennings JN (1985) Karst geomorphology. Blackwell, Oxford

Jordan PC (1979) Chemical kinetics and transport. Plenum, New York

Kastning EH (1984) Hydrogeomorphic evolution of karsted plateaus in response to regional tectonism. In: Fleur RG (ed) Groundwater as a geomorphic agent. Allen and Unwin, London Boston Sydney

Katzer F (1909) Karst and Karsthydrographie. D.A. Kajon, Sarajevo, Yugoslavia

Kay JM, Nedderman RM (1985) Fluid mechanics and transfer processes. Cambridge Univ Press, Cambridge

Kazmierzak TF, Thomson MB, Nancollas GH (1982) Crystal growth of calcium carbonate: a controlled composition study. J Phys Chem 86:103–107

Kempe S (1979a) Carbon in the freshwater cycle. In: Bolin B, Degens ET, Kempe S, Ketner P (eds) The global carbon cycle, Scope 13 Wiley, New York pp 317–342

Kempe S (1979b) Carbon in the rock cycle. In: Bolin B, Degens ET, Kempe S, Ketner P (eds) The global carbon cycle, Scope 13 Wiley, Chichester pp 343–377

Kern DM (1960) The hydration of carbon dioxide. J Chem Educ 37:14–23

Kielland J (1937) Individual activity coefficients of ions in aqueous solution J Am Chem Soc 59:1675

Kinzelbach W (1986) Groundwater modelling. Elsevier, Amsterdam New York

Kiraly L (1975) Rapport sur l'état actuelle des connaissances dans le domaine des caractères physiques des rockes karstiques. In: Burger A, Dubertret L (eds) Hydrogeology of karstic terrains. IAH, Paris pp 53–67

Labeyrie I, Duplessy IC, Delibrias G, Letolle R (1967) Etude des températures des climats anciens par la mesure de l'oxygène 18, du carbon 13 et du carbon 14 dans les concretions des cavernes. In: Radioactive dating and methods of low-level counting, IAEA Vienna pp 153–160

Landolt-Börnstein (1969) Zahlenwerte and Funktionen aus Physik, Chemie, Astronomie, Geophysik und Technik, 6th edu, part 5, vol a. Springer Berlin Heidelberg New York

Lasaga AC (1981) Rate laws of chemical reactions. In: Lasaga AC, Kirkpartick (eds) Reviews of mineralogy and kinetics of geochemical processes. Mineral Soc Am, Washington DC

Lattman LH, Parizek RR (1964) Relationship between fracture traces and the occurrence of groundwater in carbonate rocks. J Hydrol 2:73–91

Lauritzen SE (1986) Hydraulics and dissolution kinetics of a phreatic conduit. Proc 9th Int Congr Speleol 1:20–22

Lerman A (1979) Geochemical processes. Water and sediment environments. Wiley, New York

Levich VG (1962) Physiochemical hydrodynamics. Prentice Hall, Englewood Cliffs

Li YH, Gregory S (1974) Diffusion of ions in sea water and in deep sea sediments. Geochim Cosmoschim Acts 38:703–714

Loewenthal RE, Marais GVR (1978) Carbonate chemistry of aquatic systems: theory and applications, vol 1. Ann Arbor Science, Ann Arbor, Mich

Long JCS, Remer CR, Wilson CR, Witherspoon PA (1982) Porous media equivalent for networks of discontinuous fractures. Water Resourc Res 18:645–658

Luikov AV (1968) Anaytical heat diffusion theory. Academic Press London New York

Martel EA (1921) Noveau traité des eaux souterraines. Edition Doin, Paris

Michaelis I, Usdowski E, Menschel G (1984) Kinetische Faktoren der $CaCO$ –Abscheidung und Fraktionierung von C und C. Z Wasser-Abwasserforsch 17:31–36

Milanovic PT (1981) Karst hydrogeology. Water Resourc Publ, Littleton, Col USA

Moore GW (1962) the growth of stalactites. NSS Bull 24:95–106

Moore GW, Sullivan GN (1978) Speleology. Zephyrus, Jeaneck

Morse JW (1974) Dissolution kinetics of calcium carbonate in sea water III: a new method for the study of carbonate reaction kinetics. Am J Sci 274:97–107

Morse JW (1983) The kinetics of calcium carbonate dissolution and precipitation. In: Reeder RI (ed) Reviews in mineralogy, vol 11: Carbonates: mineralogy and chemistry, Mineral Soc Am, Washington

Morse JW, Berner RC (1979) Chemistry of calcium carbonate in sea water. In: Jenne EA (ed) Chemical modeling in aqueous systems, Am Chem Soc, Washington

Motyka I, Wilk Z (1984) Hydraulic structure of karst-fissured Triassic rocks in the vicinity of Olkusz (Poland) Kras i speleologia 14(5):11–24

Mullin JW (1972) Crystallization. Chem Rubber, Cleveland, Ohio

Mylroie JE, Carew JL (1986) Minimum duration for speleogenesis. 9th Congr Int de Espeleol Barcelona 1:249–251

Nancollas GH, Reddy MM (1971) The crystallization of calcium carbonate II: calcite growth mechanism. J Coll Interface Sci 37:824–830

Nancollas GH, Kazmierczak TF, Shuttringer E (1981) A controlled composition study of calcium carbonate crystal growth: the influence of scale inhibitors. Corrosion–NACE 37:76–81

Nernst (1904) Theorie der Reaktionsgeschwindigkeit in heterogenen Systemen. Z phys Chem 47:52–55

Nielsen AE (1964) Kinetics of precipitation. Pergamon Oxford

Palciauskas VV, Domenico PA (1976) Solution chemistry, mass transfer, and the approach to chemical equilibrium in porous carbonate rocks and sediments. Geol Soc Am Bull 87:207–214

Palmer AN (1972) Dynamics of a sinking stream system: Onesquetaw cave, N.Y. Natl Speleol Soc Bull 34:89–110

Palmer AN (1975) The origin of maze caves. Natl Speleol Soc Bull 37:56–76

Palmer AN (1977) Influence of geological structure on groundwater flow and cave development in Mammoth Cave National Park, USA. In: Tolson JS, Doyle FL (eds) Karst hydrology, UAH Press, Huntsville, Alabama pp 405–414

Palmer AN (1981) A geological guide to Mammoth Cave National Park. Zephyrus, Teaneck

Palmer AN (1984a) Recent trends in karst geomorphology. J Geol Educ 32:247–253

Palmer AN (1984b) Geomorphic interpretation of karst features. In: La Fleur RG (ed) Groundwater as a geomorphic agent. Allen and Unwin, Boston pp 174–209

Parizek RR (1976) On the nature and significance of fracture traces and lineaments in carbonate and other terranes. In: Karst hydrology and water resources, vol 1. Karst hydrology water resources publication. Fort Collin, Colorado

Pfeffer KH (1978) Karstmorphologie. Wiss Buchges Darmstadt

Pielsticker KH (1970) Jahresschichten in Anschliffproben von Höhlensinter. Aufschluss 21:211–213

Pleskov YV, Filinovskii VY (1975) The rotating disc electrode. Studies in Soviet Science. Consultants Bureau, New York London

Plummer LN, Busenberg E (1982) The solubilities of calcite, aragonite, and vaterite in CO_2–H_2O-solutions between 0 and 90°C and an evaluation of the aqueous model for the system $CaCO_3$–CO_2–H_2O. Geochim Cosmochim Acta 46:1011–1040

Plummer LN, Mackenzie FT (1974) Predicting mineral solubility from rate data: application to the dissolution of magnesian calcites. Am J Sci 274:61–83

Plummer LN, Wigley TML (1976) The dissolution of calcite in CO_2-saturated solutions at 25°C and 1 atmosphere total pressure. Geochim Cosmochim Acta 40:191–202

Plummer LN, Wigley TML, Parkhurst DL (1978) The kinetics of calcite dissolution in CO_2–water systems at 5°C to 60°C and 0.0 to 1.0 atm CO_2. Am J Sci 278:179–216

Plummer LN, Parkhurst DL, Wigley TML (1979) Critical review of calcite dissolution and precipitation. In: Jenne EA (ed) Chemical modelling in aqueous systems, Am Chem Soc Sym Ser 93:537–573

Powell RL (1975) Theories of the development of karst topography. In: Melhorn WN, Flemal CF (eds) Theories of landform development, Allen and Unwin, London Boston Sidney pp 217–242

Powell RL (1977) Joint patterns and solution channel evolution in Indiana. In: Tolson IS, Doyle FL (ed) Karst hydrology, IAH, 12, UAH press, Huntsville, Alabama pp 255–269

Price NJ (1966) Fault and joint development in brittle and semi-brittle rock. Pergamon, Oxford New York

Quinlan JF, Ewers RO, Ray JA, Powell RL, Krothe NC (1983) Groundwater hydrology and geomorphology of the Mammoth Cave Region, Kentucky, and of the Mitchell Plain, Indiana. In: Shaver RH, Sunderman JA (eds) Field trips in midwestern geology. Bloomington, Ind Geol Soc of America and Indiana Geological Survey pp 1–85

Quinn JA, Otto NC (1971) Carbon dioxide exchange of the air–sea interface: flux augmentation by chemical reaction. J Geophys Res 76:1539–1549

Rauch HW, White WB (1977) Dissolution kinetics of carbonate rocks 1. Effects of lithology on dissolution rate. Water Resourc Res 13:381–394

Reddy MM (1977) Crystallization of calcium carbonate in the presence of trace concentrations of phosphorus containing anions. J Crystal Growth 41:287–295

Reddy MM (1983). Characterization of calcite dissolution and precipitation using an improved experimental technique. Sci Geol Mem 71:109–107

Reddy MM, Nancollas GH (1971) The crystallization of calcium carbonate I:166–171

Reddy MM, Plummer LN, Busenberg E (1981) Crystal growth of calcite from bicarbonate solutions at constant p and 25°C: a test of calcite dissolution model. Geochim Cosmochim Acta 45:1281–1289

Rhoades R, Sinacori NM (1941) Patterns of groundwater flow and solution. J Geol 49:785–794

Rickard DT, Sjöberg EL (1983) Mixed kinetic control of calcite dissolution rates. Am J Sci 283:815–830

Riddiford AC (1966) The rotating disc system. In: Delahay P (ed) Advances in electrochemistry and electrochemical engineering 4:47–116, Interscience New York London

Röglic J (1965) The depth of fissure circulation of water and the evolution of subterranean cavities in the Dinaric karst. In: Panos V, Stelcl O (eds) Problems of speleological research. Academia Prague, Czechoslovakia

Schulz HD (1986) Schichtung von Kalksinter der römischen Wasserleitung nach Köln. Eine Hilfe zur Datierung. In: Grewe K Atlas der römischen Wasserleitung nach Köln, Rheinisches Landesmuseum, Bonn pp 263–268

Schulz HD, Bauman J (1985) Modellvorstellung zum Vorgang der Kalklösung in natürlichen Sanden. Z Dtsch Geol Ges 136:407–416

Shuster ET, White WB (1971) Seasonal fluctuations in the chemistry of limestone springs: a possible mean for characterizing carbonate aquifers. J Hydrol 14:93–128

Sjöberg EL (1976) A fundamental equation for calcite dissolution kinetics. Geochim Cosmochim Acta 40:441–447

Sjöberg EL (1978) Kinetics and mechanism of calcite dissolution in aqueous solutions at low temperatures. Stockholm Cont Geol 32; 1:1–92

Sjöberg EL (1983) Mixed kinetics control of calcite dissolution. Sci Geol Mem 71:119–126

Sjöberg EL, Rickard DT (1983) The influence of experimental design on the rate of calcite dissolution. Geochim Cosmochim Acta 47:2281–2286

Sjöberg EL, Rickard DT (1984a) Temperature dependence of calcite dissolution kinetics between 1 and 62°C at pH 2.7 to 8.4 in aqueous solutions. Geochim Cosmochim Acta 48:485–493

Sjöberg EL, Rickard DT (1984b) Calcite dissolution kinetics: surface speciation and the origin of the variable pH dependence. Chem Geol 42:119–136

Sjöberg EL, Rickard DT (1985) The effect of added dissolved calcium on calcite dissolution kinetics in aqueous solution at 25°C. Chem Geol 49:405–413

Skelland AHP (1974) Diffusional mass transport. Wiley, New York

Smith DI, Atkinson TC (1976) Process, landforms and climate in limestone regions. In: Derbyshire E (ed) Geomorphology and climate. Wiley, New York pp 367–409

Smith DI, Atkinson TC (1977) Underground flow in cavernous limestone with special reference to the Malham area. Field Stud 4:597–616

Smith DI, Atkinson TC, Drew DP (1976) The hydrology of limestone terrains. In: Ford TD, Cullingford CHD (eds) The science of speleology, Academic Press, London New York pp 179–212

Spöcker RG (1950) Ein Beitrag zur Frage der Kalkaflösung unter natürlichen Bedingungen (Eine zerlaugte Brunnenpackung). Mitt Dtsch Ges Karstforsch 4:4–8

Stephenson D (1984) Pipeflow analysis. Elsevier, Amsterdam

Stringfield VT, Rapp JR, Anders RG (1979) Effects of karst and geologic structure on the circulation of water and permeability in carbonate aquifers. J Hydrol 43:313–332

Stumm W, Morgan JJ (1981) Aquatic chemistry. Wiley, New York

Sturrock PLK, Benjamin L, Loewenthal RE, Marais GvR (1976) Calcium carbonate precipitation kinetics part 1 pure system kinetics. Water SA:101–109

Suarez DL (1983) Calcite supersaturation and precipitation kinetics in the lower Colorado River, All American-Canal and East Highland Canal. Water Resour Res 19:653–661

Sweeting MM (1972) Karst landforms. Macmillan, London

Swinnerton AC (1932) Origin of limestone caverns. Geol Soc Am Bull 43:662–693

Terjesen SC, Erga O, Thorsen G, Ve A (1969) Phase boundary processes as rate determining steps in reactions between solids and liquids. The inhibitory effect of metal ions on the formation of calcium bicarbonate by the reaction of calcite with aqueous carbon dioxide. Chem Eng Sci 14:277–289

Thrailkill J (1968) Chemical and hydrological factors in the excavation of limestone caves. Geol Soc Am Bull 79:19–46

Tien CL (1959) On the eddy diffusities for momentum and heat. Appl Sci Res Sect A 8:345–348

Troester WT, White WB (1986) Geochemical investigations of three tropical karst drainage basins in Puerto Rico. Ground Water 24:475–482

Trudgill ST (1985) Limestone geomorphology. (Geomorphology texts; 8). Longman, Essex New York

Trudgill ST, High CJ, Hana FK (1981) Improvements to the micro erosion meter. Br Cave Res Group Tech Bull 29:3–17

Tsang YW (1984) The effect of tortuosity on fluid flow through a single fracture. Water Resour Res 20:1209–1215

Tsang YW, Witherspoon PA (1981) Hydrolomechanical behaviour of deformable rock fracture subject to normal stress. J Geophys Res 86:9287–9298

Usdowski E (1982) Reactions and equilibria in the systems $CO–H_2O$ and $CaCO_3–CO_2–H_2O$. A review. Neues Jahrb Mineral Abh 144:148–171

Usdowski E, Hoefs J, Menschel G (1979) Relationship between ^{13}C and ^{18}O fractionation and changes in major element composition in a recent calcite depositing spring. A model of geochemical variations with inorganic $CaCO_3$ precipitation. Earth Planet Sci Lett 42:267–276

Waltham AC (1971) Controlling factors in the development of caves. Trans Cave Res Group Great Britain 13:73–80

Waltham AC (1981) Origin and development of limestone caves. Prog Phys Geogr 5:242–256

Warwick GT (1962) The origin of limestone caves. In: Cullingford CHD (ed) British caving: an introduction to speleology, 2nd edn. Routledge and Paul, London pp 55–82

Wermund EG, Cepeda JC (1977) Regional relation of fracture zones to the Edwards limestone aquifer, Texas. In: Tolson JS, Doyle FL (eds) Karst hydrolology, UAH Press Huntsville Alabama pp 239–253

Weyl P (1958) The solution kinetics of calcite. J Geol 66:163–176

White WB (1976) Cave minerals and speleothems. In: Ford TD, Cullingford TDH (eds) the science of speleology. Academic Press London New York pp 267–327

White WB (1977a) Conceptual models for carbonate aquifers: revisited. In: Dilamarter RR, Csallany SC (eds) Hydrolologic problems in karst regions. Western Kentucky Univ, Bowling Green pp 176–187

White WB (1977) Role of solution kinetics in the development of karst aquifers. In: Tolson JS, Doyle FL (eds) Karst hydrology. UAH Press, Huntsville, Alabama pp 503–517

White WB (1979) Water balance, mass balance, and time scales for cave system development. Natl Speleol Soc Bull 41:115

White WB (1984) Rate processes: chemical kinetics and karst landform development. In: La Fleur RG (ed). Groundwater as a geomorphic agent. Allen and Unwin, London Boston Sydney pp 227–248

White WB, Longyear J (1962) Some limitations on speleogenetic speculation imposed by the hydraulics of groundwater flow in limestone. Nittany Grotto Newl 10:155–167

Wiechers HNS, Sturrock P, Marais GVR (1975) Calcium Carbonate crystallization kinetics. Water Res 9:835–845

Wigley TML (1975a) Speleogenesis: a fundamental approach. Proc 6th Int Congr Speleol. Olomonc, 1973, 3:317–324

Wigley TML (1975b) Carbon 14 dating of groundwater from closed and open systems. Water Resourc Res 11:324–328

Wigley TML, Plummer LN (1976) Mixing of carbonate waters. Geochim Cosmochim Acta 40:989–995

Williams PW (1983) The role of the subcutaneous zone in karst hydrology. J Hydrol 61:45–67

Wissbrun KF, French DM, Patterson A (1954) The truce ionisation constant of carbonic acid in aqueous solution from 5 to 45°C. J Phys Chem 58:693–695

Yevjenich V (ed) (1981) Karst water research needs. Water Resource Publ, Littleton, Col USA

Subject Index

activation energy 121, 122
activities 14
activity coefficients 15
advective transport 47
allogenic recharge 206
anastomoses 192, 207
aquifer 5
    conduit 7
    confined 94
    diffused 5, 201
    diffuse flow 185, 186
    fracture 6, 203
    karst 7, 92, 200
    limestone 202
    mixed 203
authigenic recharge 206, 216

back reaction 59, 64
base level 3, 4, 186
batch experiments 108
bedding plane 8, 140, 191, 200, 210, 219,
    224, 227
    partings 189, 191, 199, 210, 241
beddings 80, 180
beds 188, 189
bed segments 195
bed thickness 189
Bernoulli equation 84
Boltzmann statistics 64
borehole 204, 205, 215
boundary layer
    diffusion 104, 106, 107, 109, 122, 154,
    170–172
    hydrodynamic 104, 106, 108, 171
    laminar 106
    viscous 87, 88
breakthrough 240
    times 239, 241, 242, 247
bypass tubes 213

calcite
    equilibrium curve 29, 32
    saturation index 248, 268

saturation state 16, 113
saturation value 27
solubility product 16, 27
supersaturated solutions 167
supersaturation 29
undersaturation 29
calcite deposition
    in caves 258
    rates 156, 157, 175, 271
    rates in closed systems 168, 169
    rates in open systems 175
    rates in turbulent flow 157, 168, 174
    in Roman aqueducts 272
    on stalagmites 267
    in surface streams 268
calcite dissolution
    back reactions of 119, 120, 126
    close to equilibrium 136
    elementary reactions of 119
    experimental verification of theoretical
    calcite dissolution rates 158, 168, 175
    far from equilibrium 139, 143, 160,
    176
    forward reactions of 59, 119, 120, 122,
    125, 126
    fourth-order kinetics 237
    free-drift experiment 111, 117
    influence of CO_2 conversion 110, 143,
    150, 164
    influence of lithology on 178
    influence of stirring rates 108, 109, 110
    inhibition mechanism 136
    kinetic constant 155, 255
    kinetics 7, 112, 130, 139, 140, 144, 229
    kinetics close to equilibrium 139, 160, 231
    linear rate law 231
    measurement of dissolution rates 110
    mechanism 118
    numerical values of kinetic constants 155,
    165, 167
    in porous calcite 181
    in porous media 180
    PWP equation 151

calcite dissolution
    PWP model   116
    rates   59, 103, 110, 113, 126, 144, 150,
    163, 254
    rates in closed systems   161, 163, 166
    rates in open systems   174
    rates in the presence of foreign ions   177,
    178
    rates under turbulent flow   154, 163,
    164, 166, 174
    reaction zone in   151, 153, 164, 173
    regimes of   112, 150
    second-order rate equation   231
    temperature dependence of the rates
    151
    time constant   161, 169, 170, 178
calcite dissolution models
    Davies and Jones   131, 135
    Nancollas and Reddy   135
    Plummer-Wigley-Parkhurst (PWP)   132,
    135
    Reddy and Nancollas   133
calcite equilibrium
    acid effect on   33
    base effect on   33
    common-ion effect   33
    concentrations   28, 32, 250
    curve   29, 32
    ionic strength effect on   14, 20, 33
    ion-pair effect on   33
    mean time T for approach to   158
calcite-precipitating stream   139, 268, 270
calcite precipitation   9, 21, 30, 103, 130,
    140, 143, 160, 248, 256
    free-drift seeding technique   133
    kinetics   130, 135, 144
    kinetics in the closed system   161
    rates   132, 144, 156, 165
    rates in turbulent flow   156, 170
canyon   140, 191, 192, 201, 207
carbonate equilibria   13–16
    temperature dependence of mass action
    constants   17–20
carbonate rocks   186
carbon dioxide see CO_2
carbonic acid   15
cave conduit
    development of   253, 254
    solutional   210
cave development   5, 209
    control by beds   195, 199
    control by joints   195, 196
    deep phreatic theory   210
    experimental models   243

four-state model   210, 241
high dip model   225
influence of faults on   190
influence of fissure frequency on   211
initial state of   207
intermediate state   212, 240
joint orientation in   198
low dip model   223
mature state   185, 236
network linking model   223, 241
patterns of   196, 223
restricted input model   227, 237
vadose theory   209
water table theory   210
cave passages   5, 81
    dip-oriented   211, 226, 227, 277
    orientation of cave passages in relation
    to the orientation of fractures and
    joints   176–178
    strike-oriented   210, 226, 227
cave streams   201
cave system   241
    development of large   189
    vadose   216
caves   3, 5, 198, 206
    bathyphreatic   211
    deep phreatic   210, 211
    dendritic
    denudation rates in   252, 253
    developed on bed partings   198
    drawdown vadose   217
    minimum   7
    network   206
    phreatic conduit   254
    tectonic control of   191
    vadose   209, 216, 217
    water table   210, 212–214
changes in the tube geometry   91
chemical control of reactions   65, 108
chemical equilibrium   13, 59, 65
    mean lifetime to obtain   66, 170
chemical kinetics   59
chemically enhanced diffusion   77
circulation of water   5
closed depression   203, 246
closed system   21, 29, 141, 163
    initial CO_2 pressure in   21, 28
CO_2 1, 2, 15
CO_2 contents in soil   257
CO_2 conversion   70, 103, 143, 151, 154,
    162
    half-lifetime for   74, 75
    kinetics of   70, 73

CO_2 conversion
  temperature dependence of rate
    constants 72, 151
CO_2 -outgassing 151, 167, 269
CO_2 pressure 1
  in the cave atmosphere 256
  in the soil 250
  in various environments 2
CO_3^{2-} 15, 16
Colebrook-White formula 90
concentration profiles 152, 164, 174
conduit drainage system 5
conduit flow system 81
convective flow 48
corrosion in caves 252
cross joints 189
cross sections of cave passages 192, 193

Darcy's law 93, 200
Darcy-Weisbach equation 90
denudation equation 250, 251
denudation rates 140, 248, 249
  on bare rock 251, 252
  in cave conduits 253
depth of karstification 214, 216
detailed balance 63
development of cave patterns 223
development of caves, see cave development
development of distribution of hydraulic
    heads 222
development of early channels 218
development of large caves 189
diffuse flow 80, 203
diffuse flow system 81, 204
diffuse recharge 206
diffuse storage 204
diffusion
  in confined conduits 50
  controlled mechanism 65
  decay time in 50
  eddy 54, 58
  laws of 46 - 48
  through a layer 52
  molecular 54, 56
  from point source 48
  with prescribed flux 51, 146
  into a semi-infinite body 49
  in turbulent flow 54
diffusion coefficient 47, 56
  due to longitudonal dispersion 56
  due to transverse dispersion 56
  eddy 58, 154, 171
  joint 57
  molecular 57
diffusion length 76, 151

diffusional flux from a sphere 114
dip tubes 207, 210, 241
dissociation 15
dissolution (see calcite dissolution)
dissolution of limestone 51, 59, 62, 179,
    206
distribution of joint orientations 196
distribution of structural elements 194
doline 246
  formation of 247
drainage system 203
  conduit 204
  diffuse flow 204
  subsurface 1
  surface 204

earliest initial flow 218
earliest karst channels 218, 236
earliest recognizable flow path 193, 241
early conduits 191
eddies 82, 104
electric analogue experiments 220
electric analogues 219
elementary reactions 59
empirical rate laws 79
entrenchment of canyons 192, 212
entrenchment of valley 192, 236
equation of continuity 94
equilibrium shape of stalagmites 260, 262
equipotential lines 96, 97
evapotranspiration 248

faults 187, 189, 190, 194
Fick's first equation 47
Fick's second equation 48, 76
  basic solutions of 48
fissures 6
  apertures 236
  frequency of 211, 215
  penetrable by water 6, 185, 210
  primary 2
flat-lying rocks 216
flood pulse 80, 201
floodwater zone 201
flow 6
  convective 48
  domains 97, 223
  fracture 203
  initial pathways of 191
  laminar 6, 53, 80, 82, 85
  laminar diffused 200
  lines 97, 105, 200, 203
  nets 96
  open channel 92
  plug 53

flow
      in porous media 93
      rate through fracture 239
      resistance 237
      systems 80
      turbulent 53, 58, 80, 86, 154, 160
      turbulent conduit 200
      velocity 80, 83
flux 46, 47, 53, 171
foreign ions 30, 35, 176
fracture 187, 199
      enlargement of 237, 238, 247
      flow 203, 238
      formed by compression 107
      formed by extension 107, 198
      natural 218
      penetrable by water 214, 241
      pressure fields in 218
      short 242
      traces 198

Gauss distribution 45, 49
Gauss error function 49
geological setting of the limestone 206
geomorphological theories 240
ground-air 251
      CO_2 250
groundwater 200
      basin 224, 225
      drainage 217
      flow velocities 81
      hardness of 250
      storage of 200
      transmission of 200

half-lifetime of reactions 66, 68
half tube 192
HCO_3^- 15, 16
H_2CO_3 15, 16
head 83
      loss 85, 91
Henry's law 13
hydraulic conductivities 94
hydraulic gradient 6, 85, 86, 91, 94,
      207, 239
      field of 218
hydraulic head 2, 85, 89, 221
      isolines of 220–222
hydraulic potential 93, 94
      distribution of 210, 220, 221
hydraulic radius 91, 92
hydrodynamic dispersion 54, 55, 58
hydrostatic pressure 84

idealized conduit stream 201

infiltration zone 216
inhibitor 137, 138
initial CO_2 pressure in the closed system
      21
initial flow routes 189
initial input-output geometry 228
initial phase of karstification 6, 240, 241
initial state of karstification 1, 160, 186,
      214, 234, 236
initial tubes 243
initiation of karstification 242
input by swallets 201
input-output configuration 2, 3, 99
      in a bedding plane 219
      initial 228
interception segments 193
ionic strength 14, 20

joint 8, 80, 167, 180, 187, 188, 200,
      219
      apertures 189
      chimneys 210
      cross 188, 189, 196
      distance 189
      frequencies 189, 197
      master 188, 189, 196
      orientation 196–198
      sets of 187

karst 1, 4, 201
      aquifer 7, 92, 200
      channel 201, 247
      deep- 4,
      denudation 248
      development 236
      geosyncline- 4
      holo- 4
      initial state of 234, 240
      initiation 210
      intermediate state of 141
      mature state of 1, 185, 236
      mero- 4
      permeability 204, 205
      platform- 4
      porosity 204
      processes on the surface 245
      shallow- 4
      springs 5, 92, 200, 201, 209
      systems 5, 185
      waters 16
karstification 2, 5
      coefficient of 215
      depth of 216
      initial state of 1, 234
      initiation of 210, 240

minimum time for 241
process 80

laminar flow (see flow)
Laplace equation 94
law of Bernoulli 83
limestone 1, 200
    dissolved 80
    primary porosity of karstifiable 185,
    186
    secondary porosity 185, 186
    standard tablets 252, 254
    steeply dipping 225
limestone aquifer 2, 202
limestone rocks 187

mass action constants 17, 20
mass action equation 14
mass action law 61
mass flux 49, 51
mass flux from a rotating disc 106
mass transfer coefficient 78
mass transport 43, 51
    diffusional 43
    diffusive and advective 53
    equation 142
master conduit 207
micro-erosion meter 252, 254
Mischungskorrosion 7
mixed kinetics 75, 129
mixing corrosion 242, 243
mixing length 54
mixing of saturated waters 29, 242

Nernst model 106, 107, 171

open system 21, 140, 150
output of water 2, 220−222

Peclet number 56, 58, 142, 180
penetration distance 140
penetration length 229, 230, 232, 233,
    236, 237, 239
    in turbulent flow 234
percolation input 201
permeability 186, 205
    primary 185
    secondary 186, 218, 243
phreatic loops 212, 213
phreatic tubes 217
phreatic zone 5, 141, 201, 207, 213
pH-stat method 111, 117
Plummer-Wigley-Parkhurst Equation 120,
    140, 144, 254
point inputs 200, 207

pore size 186
porosity
    primary 204
    secondary 186
    total effective secondary 204
porous medium 22, 54, 55, 93
    calcite-containing 182
    two-dimensional 140, 142, 219
pre-cave state 210
precipitation (see calcite precipitation)
pressure 83
pressure field 220
    distribution in fractures 218
primary tubes 244, 245
principle of detailed balance 63, 123

random walk 43
reaction 59, 67
    activation energy of 64, 65
    back 59, 64
    chemical control of 65
    chemically controlled 108
    diffusion controlled 65
    forward 59, 64
    heterogeneous surface 104, 155
    homogeneous 78
    irreversible 65, 66
    mixed kinetics in 76, 151
    overall 61
    principle of detailed balance 63, 123
    rate constant of 60
    rate-determining 62
    rate-limiting 67, 110
    reversible 65, 69
    surface 108
    temperature dependence of rate
    constants 64
    transport control 108
recession curves 203
Reynolds number 82, 171
rocks
    flat lying 216
    permeable 200
    steeply dipping 216
rotating disc 105, 175

saturated $CaCO_3$ solutions 7, 26
saturation length 181
saturation state 16
saturation value of $[Ca^{2+}]$ 27
Schmidt number 106, 171
seeded growth technique 130
shaft 200, 246, 248
    widening of 248

Sherwood number   109, 171, 172
    correlation to Schmidt- and Reynold
    number   172
sinkhole   3, 198, 247
smooth and rough pipe walls   90
solutional conduits   210
solutional widening   185
speleothems   256
springs   201, 209
    conduit   80
    diffuse flow   80
    response to flood pulses   201, 204
stalagmite   257, 267
    diameter of   266, 267
    growth   260
    growth rates of   156, 259, 264, 267
    maximal growth rates of   267, 268
    morphology of   257
    radius   260, 261
steeply dipping rocks   216
strata
    flat lying   211, 216
    homoclinally dipping   211
    horizontal   4
    steeply dipping   212, 216
streamline   54, 82, 95, 96
strike-oriented elements   227
strike-oriented tubes   207, 211
structural segments   191
    bed   193, 195
    bed-joint   102, 193

    fault   194, 195
    fault-joint   194, 195
    joint   193, 195
subsurface drainage system   1
surface denudation   7

transport equation   47, 143
transport mechanism   43
transverse dispersion   56
tubes
    dip   207, 210, 241
    oriented along strike   207
    primary   244
    secondary   244
turbulent core   52
turbulent flow   86
    in rough tubes   90, 91, 235
    in smooth tubes   90, 91, 235

vadose entrenchment   211
vadose zone   140, 201
viscosity   82, 85
viscous sublayer   87, 88
volumetric flow rate   86

water divide   97, 203
water drainage systems   5
water input   186
water output   2
water table   3, 201, 209
widening of a fracture   237

CPSIA information can be obtained at www.ICGtesting.com
Printed in the USA
LVOW10s1025270714

396238LV00006B/182/P